Structural Equation Modelling with Partial Least Squares Using Stata and R

Structural Equation Modelling with Partial Least Squares Using Stata and R

Mehmet Mehmetoglu
Department of Psychology,
Norwegian University of Science and Technology

Sergio Venturini
Department of Management,
Università degli Studi di Torino

CRC Press is an imprint of the
Taylor & Francis Group, an **informa** business
A CHAPMAN & HALL BOOK

First edition published 2021
by CRC Press
6000 Broken Sound Parkway NW, Suite 300, Boca Raton, FL 33487-2742

and by CRC Press
2 Park Square, Milton Park, Abingdon, Oxon, OX14 4RN

CRC Press is an imprint of Taylor & Francis Group, LLC

ISBN: 9781482227819 (hbk)
ISBN: 9780429170362 (ebk)

Typeset in CMR10
by KnowledgeWorks Global Ltd.

To Rannvei Sæther [M]

A Deborah,
il mio tesoro più prezioso,
per quello che fai,
per quello che sei [S]

Contents

Preface

Structural equation modelling (SEM) has in the past three decades probably been one of the most often used statistical techniques in the social science research. The main reason for this is that SEM is a linear modelling technique encompassing all the traditional statistical techniques (e.g., linear regression, ANOVA) as well as allowing for estimating complex models including latent variables. In this respect, SEM should indeed be viewed rather as an overarching statistical framework or even a toolbox than as a single statistical technique. The popularity of SEM has of course been facilitated by a continuously growing number of both commercial and open-source software and packages.

Although the traditional SEM approach, which usually goes under the rubric of covariance-based SEM (CB-SEM), has received the initial and most attention from researchers, partial least squares SEM (PLS-SEM) has nevertheless also started drawing the attention of academics. However, PLS-SEM has met some serious criticisms from applied statisticians/econometricians. As a result, even a hostile attitude has grown towards PLS-SEM in some social science fields and journals rejecting the validity of results from PLS-SEM. The main reason for this criticism is the fact that PLS-SEM should not be used as an alternative to CB-SEM. Some scholars in the PLS-SEM camp has taken this criticism seriously and accordingly invented new algorithms (i.e., consistent PLS-SEM) making it possible to use PLS-SEM instead of CB-SEM.

In other words, we have got two versions of PLS-SEM: a traditional one, which we refer to as PLS-SEM, and a modern one which we refer to PLSc-SEM. The natural to ask question is "When do we use these two versions of PLS-SEM?". Our answer is that you would use PLSc-SEM instead of CB-SEM whenever the latter fails to converge (we provide detailed motivations in Chapter 1). This suggestion is also corroborated by recent simulations studies. Further, we suggest that one uses traditional PLS-SEM when the purpose of the study is to explain as much variance as possible of the endogenous (dependent) variable. The reason is that PLS-SEM algorithm seeks parameter estimates that maximize explained variance.

The antagonism between PLS-SEM and CB-SEM does have consequences for the terminology used in each of these two frameworks. For instance, a construct will typically be referred to as a latent variable in CB-SEM and as a composite in PLS-SEM. This distinction may sometimes also lead to some confusion and even unnecessary strong disagreements among researchers. We have instead adopted a pragmatic approach not distinguishing between the two terminologies. This choice is further strengthened by the fact that PLSc-SEM with its algorithm is indeed more comparable with CB-SEM.

This reasoning is also the rationale behind our way of assessing and reporting a PLS-SEM study as it resembles the criteria used for CB-SEM studies. Our intention has consistently been not to dwell on the technical or conceptual differences between PLS-SEM and CB-SEM. Instead, we take a practical approach to explaining and demonstrating how PLS-SEM can be performed using some powerful user-written packages that can be accessed in Stata and R.

Software and Material

As it is conveyed in the title, we use and explain Stata and R software for partial least squares structural equation modelling in this book. We use Stata in the main text whereas R equivalent codes are provided in the appendices at the end of each chapter. Both software are given equal priority in that all Stata solutions and estimations are replicated using R codes in detail as well. As you already may know, Stata and R are both code-based software. There are thankfully many volunteer academics around the world who continuously develop and make available packages in these software for solving a variety of statistical tasks including PLS-SEM. In fact, there are several alternative packages both in Stata and R for PLS-SEM analysis.

One of these packages is developed and maintained by the authors of this book. We have named our package `plssem`, which can readily be installed for free by those who already have purchased and have access to Stata's version 15 or above. There are two ways of installing `plssem` in Stata. The first and suggested option (as it provides the most updated version) is to install it from GitHub. To do that, you need to run the following code directly in Stata:

```
net install github, from("https://haghish.github.io/github/")
```

This command installs a package called `github`, which enables installing Stata packages from GitHub. Then, to install our `plssem` package you need to execute the following code:

```
github install sergioventurini/plssem
```

As an alternative to GitHub installation, you can get `plssem` by typing the following code in Stata:

```
ssc install plssem
```

When it comes to packages in R, we have chosen two packages that complement each other. These are `cSEM` and `plspm`. The former is developed by Manuel E. Rademaker and Florian Schuberth whereas the latter is developed by Gaston

Sanchez. Here too, you can install these two packages in similar ways to those above. The first option is to install each of these packages directly from GitHub by typing:

```
install.packages("devtools")
```

The previous code installs a package called `devtools`, which enables installing `R` packages from GitHub. The following two lines install the packages `cSEM` and `plspm` from the corresponding GitHub repositories into `R`:

```
devtools::install_github("M-E-Rademaker/cSEM")
devtools::install_github("gastonstat/plspm")
```

Alternatively, you could install the currently stable versions of these packages from the Comprehensive R Archive Network (CRAN) using the following compact code in `R`:

```
install.packages(c("cSEM", "plspm"))
```

In addition to these packages, we have also programmed some functions for `R` users so that we could replicate all the Stata estimations in `R` as well. As we develop our `plssem` package in Stata, we may continue to develop some more functions for `R` users. Both existing and possible future functions for `R` users can be followed and installed from the book's GitHub repository available at the following link:

`https://github.com/sergioventurini/SEMwPLS`

This repository will also contain all the necessary material related to the book including:

- Datasets
- `R` functions and additional Stata packages (if needed)
- `R` and Stata codes
- Figures
- Errata
- Frequently asked questions regarding installation of `plssem` and `cSEM` and `plspm`
- Auxiliary chapters
- References to PLS-SEM applications
- Notices regarding new developments in PLS-SEM and software

- Suggestions for the next edition of the book
- Other supporting material

Despite our efforts with the above measures to make the reader as independent of us as possible, if you still would like to reach us with your questions/comments, we provide here our email addresses:

mehmetm@ntnu.no and sergio.venturini@unito.it.

Overview of Chapters

In Chapter 1, we define partial least squares structural equation modelling (PLS-SEM) and compare it with covariance-based structural equation modelling (CB-SEM). Further, we provide guidelines as to when to use PLS-SEM instead of CB-SEM. In Chapter 2, we treat in detail several multivariate topics (bootstrapping, principal component analysis, etc.) that are of direct relevance to understanding and interpreting PLS-SEM estimations in subsequent chapters. Readers with previous knowledge of multivariate statistics can, if preferred, readily jump to the next chapter. In Chapter 3, we explain thoroughly the PLS-SEM algorithm using an example application based on a real dataset. Subsequently, the `plssem` Stata package (`R`'s `cSEM` and `plspm` in chapter appendices) is presented with all its options. In so doing, several salient modelling issues (e.g., bootstrap, missing data) and types of models (e.g., higher order, mediation) are treated in detail as well. This chapter does also present both the algorithm and package for the newly emerged technique called consistent PLS-SEM. The foregoing chapters in tandem lay down a solid basis for assessing PLS-SEM models coming next.

In Chapter 4, we elucidate all the concepts (e.g., convergent validity) and criteria (e.g., average variance extracted) to be used in the assessment of measurement and structural parts of PLS-SEM models. Instead of providing a series of different models to show different aspects and cases of PLS-SEM models, we specify and estimate a complex model encompassing different aspects (formative, single item, higher order, reflective, etc.) of PLS-SEM models. In assessing this complex model, readers will readily pick up and adopt the adequate criteria necessary for assessing their own PLS-SEM models.

In Chapter 5, we explain what mediation analysis is and present the commonly known Barron-Kenny approach and its alternative approach to testing mediational hypotheses using PLS-SEM. In so doing, several examples are provided. Chapter 6 goes through another important topic, namely moderation/interaction analysis. Here, we explain thoroughly three different procedures used to test interactions effects with relevant examples using PLS-SEM. These are product-indicator approach, two-stage approach and multi-sample approach (also called multi-group analysis). As it is a prerequisite prior to sample/group comparisons, we also show how to examine measurement model invariance.

In Chapter 7, we present two specific approaches to detecting latent classes in the data based on a PLS-SEM model. These are response-based unit segmentation (REBUS) and finite mixture (FIMIX) partial least squares approaches. Chapter 8 provides a standard template for writing up a PLS-SEM based research. To do so, an actual published work of one of the authors is used as a framework to show how to structure and present each section of a typical PLS-SEM publication.

As we already remarked, in all chapters Stata codes are provided in the main text whereas corresponding `R` codes are provided at the end of each chapter. We have also tried to include as little mathematical details as possible in the main text. Most of these details are often included in some technical appendices. Every chapter is nevertheless written in such a way that readers (not interested in mathematical details) will still be able to understand and use PLS-SEM with the chosen software without any difficulty. Finally, to make the book self-contained we also added an appendix which presents more basic material about correlation and linear regression.

Acknowledgement

We would first like to thank the anonymous reviewers for their constructive comments and suggestions. We further thank Florian Schuberth for his useful comments on Chapter 1. We also feel that special thanks must go to Gaston Sanchez for the development of the `plspm` package, and to Manuel E. Rademaker and Floarian Schuberth for the `cSEM` package, both of which we have built this book around together with our own `plssem` package. We also want to thank our editor Rob Calver for encouraging us to write the book. Finally, we would like to heartedly thank our families (Rannvei and Deniz of Mehmet) and (Deborah, Irene and Carlo of Sergio) for letting us have the time to write the book.

Trondheim (Norway), *Mehmet Mehmetoglu*
Cremona (Italy), *Sergio Venturini*

December 2020

Authors

Mehmet Mehmetoglu is a professor of research methods in the Department of Psychology at the Norwegian University of Science and Technology (NTNU). His research interests include consumer psychology, evolutionary psychology and statistical methods. Mehmetoglu has co/publications in about 30 different refereed international journals such as Journal of Statistical Software, Personality and Individual Differences, and Evolutionary Psychology.

Sergio Venturini is an associate professor of Statistics in the Management Department at the Università degli Studi di Torino (Italy). His research interests include Bayesian data analysis methods, meta-analysis and statistical computing. He coauthored many publications that have been published in different refereed international journals such as Annals of Applied Statistics, Bayesian Analysis and Journal of Statistical Software.

List of Figures

List of Tables

List of Algorithms

Abbreviations

2SLS Two-stage least squares

AIC Akaike's information criterion

ANOVA Analysis of variance

ANCOVA Analysis of covariance

AVE Average variance extracted

BIC Bayesian information criterion (or Schwarz's information criterion)

CA Cluster analysis

CAIC Consistent Akaike's information criterion

CFA Confirmatory factor analysis

COM Communality

CB-SEM Covariance-based structural equation modelling

CM Closeness measure

EFA Exploratory factor analysis

EM Expectation-maximization algorithm

EN Entropy criterion

ESS Explained sum of squares

FIMIX-PLS Finite mixture partial least squares

FMM Finite mixture model

FPLS-LCD Fuzzy PLS path modelling for latent class detection

FOLV First-order latent variable

GA Genetic algorithm

GoF Goodness-of-fit

GLS Generalized least squares

GQI Group quality index

GSCA Generalized structured component analysis

HOLV Higher order latent variable

LCA Latent class analysis

LISREL Linear structural relations

LRT Likelihood ratio test

LV Latent (unobserved) variable

MDL Minimum description length

MIMIC Multiple indicators and multiple causes

ML Maximum likelihood

MLR Multiple linear regression

MGA Multi-group analysis

MV Manifest (observed) variable

MVREG Multivariate regression

NEC Normed (or normalized) entropy criterion

OLS Ordinary least squares

PA Path analysis

Pathmox Path modelling segmentation trees

PCA Principal component analysis

PCR Principal component regression

PLS Partial least squares

PLSR Partial least squares regression

PLS-GAS Genetic algorithm segmentation for PLS-SEM

PLS-PM Partial least squares path modelling (synonym for PLS-SEM)

PLS-POS Prediction-oriented segmentation in PLS-SEM

PLS-R Partial least squares regression

PLS-TPM Partial least squares typological path modelling

PLS-SEM Partial least squares structural equation modelling

PLSc-SEM Consistent partial least squares structural equation modelling

REBUS-PLS Response-based procedure for detecting unit segments in PLS-SEM

RSS Residual sum of squares

SEM Structural equation modelling

SLR Simple linear regression

SOLV Second-order latent variable

SVD Singular value decomposition

SUR Seemingly unrelated regression

MLR Multiple linear regression

TSS Total sum of squares

ULS Unweighted least squares

VIF Variance inflation factor

WLS Weighted least squares

Greek Alphabet

Upper case letter	*Lower case letter*	*Greek letter name*	*English equivalent*
A	α	Alpha	a
B	β	Beta	b
Γ	γ	Gamma	g
Δ	δ	Delta	d
E	ε	Epsilon	e
Z	ζ	Zeta	z
E	η	Eta	h
Θ	θ	Theta	th
I	ι	Iota	i
K	κ	Kappa	k
Λ	λ	Lambda	l
M	μ	Mu	m
N	ν	Nu	n
X	ξ	Xi	x
O	o	Omicron	o
Π	π	Pi	p
R	ρ	Rho	r
Σ	σ	Sigma	s
T	τ	Tau	t
Υ	υ	Upsilon	u
Φ	ϕ	Phi	ph
X	χ	Chi	ch
Ψ	ψ	Psi	ps
Ω	ω	Omega	o

Part I

Preliminaries and Basic Methods

1

Framing Structural Equation Modelling

In this chapter we will describe what structural equation modelling (SEM) is by first relating it to traditional single-equation techniques such as regression analysis. We then present the two main approaches to doing SEM analysis, namely covariance-based SEM and partial least squares SEM. Although we assume that the reader has already made her/his decision to use PLS-SEM, we still in a compact manner provide suggestions (mainly based on existing simulation studies) as to when to use PLS-SEM. In so doing, we also present consistent PLS-SEM as a valid method to match CB-SEM. For the reading of this chapter, we also assume that the reader has a sound background in linear regression analysis[1].

1.1 What Is Structural Equation Modelling?

Let us start this section by first defining what a *structural equation* is. Intuitively put, a structural equation refers to a statistical association whereby one variable (X) influences another one (Y). When there is only one dependent variable predicted by one (or several) independent variables in our statistical model[2], then the structural equation is univariate and it can be expressed as follows[3]:

$$Y_i = \beta_0 + \beta_1 \cdot X_i + \varepsilon_i. \tag{1.1}$$

Based on your previous statistics knowledge, you will easily recognize equation (1.1). It represents a simple regression model including two variables (X and Y). The significance testing of β_1 will provide the empirical evidence as to whether X (e.g., education level) influences Y (e.g., hourly wage) or not. Incidentally, if X is a dichotomous variable (e.g., gender) or a polytomous variable (e.g., types of occupation), equation (1.1) will then correspond to an independent t-test and analysis of variance (ANOVA) respectively. Extending our regression model in equation (1.1)

[1] Appendix A provides a brief review of the linear regression model, while Section 2.2 summarizes the basics of principal component analysis.

[2] A model represents the hypothesized relationships between a set of variables.

[3] In this book we will frequently use letters taken from the Greek alphabet. For your convenience, we provide it on page xxxvii.

TABLE 1.1: Examples of univariate and multivariate statistical techniques.

Univariate	Multivariate
Multiple regression	Multivariate regression
Logistic regression	Seemingly unrelated regression
Multinomial logistic regression	Path analysis (also called simultaneous equation models)
Discriminant (function) analysis	Structural equation modelling
Survival analysis	
Poisson regression	
Log-linear analysis	
Independent *t*-test	
Analysis of variance	
Multilevel regression	

with one more independent variable will turn it into a multiple regression (or ANCOVA[4] or two-way ANOVA[5] for that matter).

Increasing the number of independent variables does turn our model into a *multivariable* one. However, it corresponds still to a *univariate* model in that we still have just one dependent variable to predict. Examples of such models including only one dependent variable are listed in Table 1.1. Even when we view Table 1.1 at a glance, we will notice that the *univariate* modelling techniques (e.g., linear regression, logistic regression) are clearly the ones that have been used most commonly in quantitative research regardless of social or natural sciences.

Nonetheless, the need to be able to estimate models including more than one dependent variable has also led to the invention of some useful *multivariate* modelling techniques in quantitative research (see Table 1.1). One of these techniques is multivariate regression (MVREG), which corresponds to a multiple regression facilitating more than one dependent variable. The advantage of this technique is that it allows for cross-equation comparisons. Two limitations of MVREG are that all the dependent variables must be predicted by the same set of predictors, and that the correlations among the residuals are not taken into consideration. Seemingly unrelated regression (SUR) does readily surmount these limitations. These features make SUR a simultaneous[6] multivariate modelling technique.

Although SUR (and MVREG for that matter) has helped quantitative researchers solve some demanding research questions, its limitation is that it allows for estimating only direct effects. Path analysis (PA) however allows to estimate both direct and indirect effects, making it a technique for estimating complicated models. The fact that PA estimates indirect effects allows researchers test mediational hypotheses. As such, PA cannot only include multiple dependent variables, but it can also allow a variable to be both a dependent and an independent variable in the same model.

[4]ANCOVA is basically an ANOVA with an additional continuous independent variable in the model.
[5]Two-way ANOVA is an ANOVA with an additional categorical independent variable in the model.
[6]Simultaneous means that a series of equations are estimated at the same time.

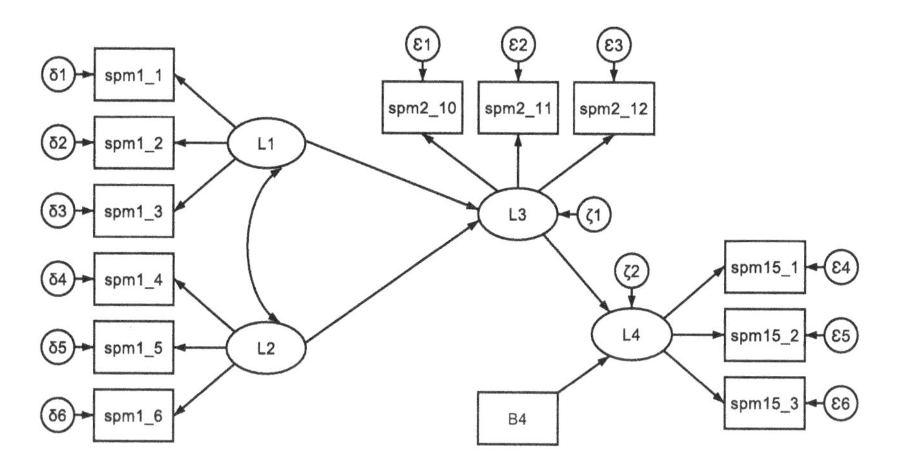

FIGURE 1.1: Graphical representation of a structural equation model. In particular, this diagram represents an example of covariance-based SEM. However, without the indicator error variances, it can be used for partial least squares SEM as well (see the next section).

One limitation of PA is that it works only with observed variables (i.e., single-item questions).

Structural equation modelling (SEM) overcomes the above-mentioned limitations effectively. Consequently, we can define SEM as a simultaneous multivariate technique that can be used to estimate complex[7] models including observed and latent variables. We illustrate all these features of SEM through a hypothetical example in Figure 1.1. In this model, it is not necessary, but if needed, you could also correlate the error terms (i.e., ζ_1 and ζ_2) to serve the same purposes as SUR does.

SEM being a simultaneous multivariate technique makes all the statistical techniques listed in Table 1.1 redundant. That is, SEM can easily be used for the same purposes that the traditional univariate (e.g., regression) and multivariate (e.g., seemingly unrelated regression) techniques are used for. This specific feature lifts SEM from being a routine statistical technique up to an overarching framework for statistical modelling. Yet, this is not the novel contribution of SEM. What makes SEM a special statistical technique is the fact that it can handle *latent variables.*

A latent variable is a hypothetical or an unobservable concept[8] (e.g., happiness) that we measure using a set of observable variables (e.g., satisfaction with work, family, goals achieved). The main reason why we want to use latent variables is that many

[7]Complex means that we have for instance a form of a path model (e.g., mediation analysis).

[8]We note that some authors in the social science literature typically distinguish between the concept (i.e., the theoretical entity) one wants to investigate, and a corresponding latent variable (i.e., the statistical entity), which is used to model the concept itself (Henseler, 2017; Benitez et al., 2020). Even if we agree that this distinction is useful from a theoretical point of view, in this book we adopt a more pragmatic perspective and so we will not pursue it any longer.

concepts like happiness are multifaceted and cannot just be represented by a single indicator (item or question) as they encompass more than one aspect. Researchers have traditionally used different types of factor analytic procedures to discover one or more number of latent variables among a set of indicators. Whereas the traditional approaches to factor analysis are of exploratory nature, the SEM technique addresses the factor analysis in a confirmatory fashion.

Exploratory factor analysis (EFA) assumes that all the indicators correlate with all the latent variables, while confirmatory factory analysis (CFA) lets the researcher decide which indicators to correlate with which latent variables. Earlier we mentioned that SEM can replace all the traditional regression techniques, and now we can add that SEM can also replace the traditional factor analytic techniques. It follows from the above that SEM is a technique that can be used either for doing regression/path analysis or doing CFA, or doing a combination of these two analyses in one operation (see Figure 1.1).

1.2 Two Approaches to Estimating SEM Models

Before the invention of SEM, researchers would first perform a factor analysis to ascertain which items represent which latent variables. They would then compute an index/sum score for these latent variables (e.g., extraversion, neuroticism). Finally, they would use these index/sum scores as independent/dependent variables in a regression analysis in line with their hypotheses (e.g., extraversion related to dining out)[9]. Due to the demonstrated drawbacks[10] of this rather crude statistical approach (Hsiao et al., 2018; Steinmetz, 2013), quantitative researchers would directly use structural equation modelling technique.

1.2.1 Covariance-based SEM

The most common structural equation modelling approach is the one based on covariance matrices estimating accordingly model parameters by only using common variance (Hair et al., 2018b). This *confirmatory* approach is widely known as **covariance-based structural equation modelling**[11] (CB-SEM) and it has been developed by Jöreskog (1969). The fact that CB-SEM uses common variance means that *measurement error* (i.e., unreliability) of indicators is taken into account during the model estimation (Mehmetoglu and Jakobsen, 2016). This specific feature makes CB-SEM model estimates less biased compared to techniques (e.g., index/sum score) assuming no measurement error at all (Harlow, 2014).

[9]In Section 2.5 we provide a summary of the most popular approaches for computing sum scores.

[10]In addition to the fact that estimates are subject to bias, we cannot easily perform path analysis including mediational analysis.

[11]The fact that the standard estimation method of SEM is maximum likelihood (ML), this type of SEM goes also under the name of ML-SEM or ML-CBSEM.

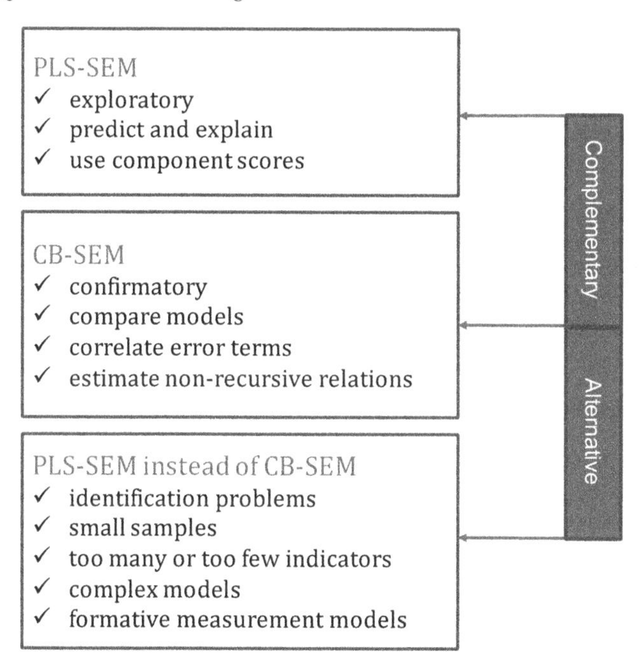

FIGURE 1.2: When to use PLS-SEM and CB-SEM.

The main purpose (and advantage) of CB-SEM is to statistically *test theories* (hypothesized models) in their entirety. One way to assess the adherence of the model's estimates to the data is through goodness-of-fit measures. These fit measures are obtained by computing the discrepancy between the estimated model-implied variance–covariance matrix ($\hat{\mathbf{\Sigma}}$) and the sample variance–covariance matrix ($\boldsymbol{S}$). The smaller the difference between $\hat{\mathbf{\Sigma}}$ and $\boldsymbol{S}$, the better the theoretical model fits the data (Mehmetoglu and Jakobsen, 2016). Therefore, the CB-SEM algorithm seeks to find the set of parameter estimates (loadings and coefficients) that minimize the difference between $\hat{\mathbf{\Sigma}}$ and $\boldsymbol{S}$ (Chin, 2010).

The fact that CB-SEM makes use of omnibus fit measures makes it an appropriate technique to statistically *compare alternative models*. These measures will then find out which of the competing models fits the data best. Moreover, as far as the model set up is concerned, CB-SEM allows for *correlating error terms* as well as *specifying non-recursive structural relations* (Grace, 2006). The above-mentioned features (in italic) of CB-SEM are the reasons for the increasing popularity and application of CB-SEM techniques in social science publications.

Despite the invaluable novel features of CB-SEM, it does often suffer from non-convergent and improper solutions (Bagozzi and Yi, 1994; Rindskopf, 1984). Solutions are non-convergent when an estimation method's algorithm is unable to arrive at values that meet predefined termination criteria, whereas solutions are improper when the values for one or more parameter estimates are not feasible (Anderson and Gerbing, 1988; Gerbing and Anderson, 1987). Nonconvergent/improper solutions may be caused by small sample size[12] (Anderson and Gerbing, 1988), complex models (many structural relations) (Chin, 2010), too few indicators[13] (Hoyle, 2011), too many indicators (Deng et al., 2018) or formative measurement models[14] (Sarstedt et al., 2016).

A further limitation of CB-SEM is that it produces latent variable predictions, so called factor scores, that are not unique. This issue has to do with *factor indeterminacy* (Bollen, 1989; Ringdon, 2012), which refers to the fact that in factor-based models an infinite number of different sets of values can be obtained for the factor scores which will fit the model equally well (for a detailed technical explanation of this problem we suggest to see Mulaik, 2010, in particular Chapter 13).

1.2.2 Partial least squares SEM

The second approach gaining popularity is that of partial least squares structural equation modelling (PLS-SEM) developed by Wold (1975) [15]. PLS-SEM is referred to as **variance-based structural equation modelling** as it uses the total variance to estimate model parameters (Hair et al., 2018b). The fact that PLS-SEM uses total variance means that measurement error of indicators is ignored in model estimation, which is the reason why PLS-SEM produces biased parameters when mimicking CB-SEM (Aguirre-Urreta and Rönkkö, 2018). This is however an expected outcome since PLS-SEM is more orientated towards optimizing predictions (i.e., explained variance) than statistical accuracy of the estimates (Esposito Vinzi et al., 2010)[16].

Unlike its confirmatory counterpart CB-SEM, PLS-SEM is mainly an *exploratory* technique (Hair et al., 2012). PLS-SEM should accordingly be used when the phenomenon in question is relatively new or changing and theoretical model or measures are not well formed (Chin and Newsted, 1999; Reinartz et al., 2009). In fact, PLS-SEM should further be used instead of CB-SEM when the researcher uses it in an exploratory manner (revising the model based on modification indices) to get a better model fit (see Chin, 1998a and Chin, 2010, pp. 658–659). It follows from the above that PLS-SEM and CB-SEM are complementary (see Figure 1.2) rather than competitive techniques as stated also by Wold himself (Barroso et al., 2010, p. 432).

Due to the nature of its algorithm, PLS-SEM is able to avoid non-convergent and improper solutions that often occur in CB-SEM (Sirohi et al., 1998). Thus, PLS-SEM is suggested to be used when working with small samples[17] including less than 250 observations (Chin and Newsted, 1999; Reinartz et al., 2009). PLS-SEM is

[12]There is a general consensus that a small sample includes less than 250 observations.

[13]As widely suggested in CB-SEM, a latent variable must have at least three indicators.

[14]In a formative measurement model, indicators are causing the latent variable.

[15]For a historical overview of the development of PLS-SEM we suggest to read the appendix of Sanchez (2013) as well as the corresponding expanded book-length version of Sanchez (2020).

[16]Note that CB-SEM also provides an "R-squared" index as part of the output. For example, Stata calls it the "coefficient of determination" and reports it among the model's goodness-of-fit measures. However, in the case of CB-SEM the index has a different aim and can't be interpreted as the amount of explained variance.

[17]Researchers should still follow some formal procedures to decide the exact required sample size (Hair et al., 2017; Kock and Hadaya, 2018).

further recommended to be used to estimate complex models including large number of indicators (Haenlein and Kaplan, 2004), large number of latent variables, as well as large number of structural relations among latent variables (Chin, 2010). Relatedly, PLS-SEM can also handle few indicators (less than three) per latent variable. Moreover, PLS-SEM should be the clear choice to estimate formative models as the MIMIC approach in CB-SEM imposes constraints that often contradict the theoretical assumptions (Hair et al., 2018b; Sarstedt et al., 2016). Finally, PLS-SEM produces scores that can be readily used for predictive purposes in subsequent analyses if needed (Chin, 2010). More specifically, a critical difference between CB-SEM and PLS-SEM regards how the two methods conceive the notion of latent variables. CB-SEM considers constructs (i.e., the representation of a concept within a given statistical model) as **common factors**, which are assumed to explain the association between the corresponding indicators. Differently, in PLS-SEM constructs are represented as **composites**, that is as weighted sums of the indicators. As such, CB-SEM is also referred to as a **factor-based** while PLS-SEM as a **composite-based** SEM method.

1.2.3 Consistent partial least squares SEM

The fact that the PLS-SEM algorithm does not take into account measurement error leading to inconsistent and biased estimates, makes it an inferior alternative to CB-SEM from a statistical point of view. Recognizing this limitation, Dijkstra and Henseler (2015b) developed a new version of PLS-SEM based on disattenuated correlation matrix of the composites. They named this new method as **consistent PLS** (PLSc-SEM)[18]. Moreover, the simulations run by Dijkstra and Henseler (2015b) showed that PLSc-SEM has advantages when using non-normally distributed data as well as non-linear models (Dijkstra and Schermelleh-Engel, 2014).

Aguirre-Urreta and Rönkkö (2018) provide simulation-based evidence supporting the use of PLSc-SEM with common factor models. In the same study, they stress the importance of employing bootstrap confidence intervals in conjunction with PLSc-SEM. These findings contribute to making PLSc-SEM a much-needed flexible alternative to CB-SEM. PLSc-SEM can then readily be used instead of CB-SEM whenever CB-SEM suffers from the known issues of non-convergence and improper solutions (see Figure 1.2). In so doing, the researcher mimics CB-SEM successfully as well as benefiting from the previously mentioned advantages (small sample, formative measures, etc.) of the traditional PLS-SEM. PLSc-SEM is nonetheless closer to CB-SEM than the traditional PLS-SEM.

[18]Another PLS technique, referred to as PLSF, that accounts for measurement error has recently been developed by Kock (2019). The technique estimates factors which are subsequently used in the estimation of parameters.

1.3 What Analyses Can PLS-SEM Do?

PLS-SEM, as opposed to CB-SEM, found its way into the toolbox of researchers rather late. The main reason for this is that there was no easily available software for PLS-SEM until about 10 years ago. Due to the increasing number of both commercial software (SmartPLS, WarpPLS, XLSTAT, ADANCO, etc.) and open-source packages (`cSEM`, `plspm`, `semPLS` and `matrixpls` in `R`, and `plssem` in Stata[19]) in the past decade, PLS-SEM has finally gained the deserved growing attention from academic scholars. The technique appears to have been used in disciplines/fields such as marketing, psychology, management, political science, information systems, medicine/health, tourism and hospitality, and education. This trend is confirmed by ample applications of PLS-SEM in research published in international journals.

PLS-SEM is a composite-based modelling technique. However, as we already stated previously, it can readily be used with observed variables. It can thus be used instead of many of the traditional techniques shown in Table 1.1. What analyses PLS-SEM can do depends, to a larger degree, on the analytic options offered by different software. Most of the software for PLS-SEM will however have the options to perform the following analyses:

- Regression analysis (RA)
  It can estimate linear regression models including one or more than one dependent variable and continuous or/and categorical independent variables.

- Analysis of variance (ANOVA)
  It can compare means of two or more than two groups – also by controlling for categorical or/and continuous variables as well as testing interactions effects.

- Path analysis (PA)
  It can estimate models including at least one chain-relationship in which X facilitates M, which then influences Y – by allowing for decomposing direct and indirect effects.

- Latent structural analysis (LSA)
  It can estimate full SEM models including one or more than one latent dependent variable, and one or more than one latent independent variable.

- Latent mediation analysis (MED)
  It can estimate mediation models including indirect relationships (effects) between latent variables.

- Latent moderation analysis (MOD)
  It can estimate moderation models including a latent variable influencing the effect of a latent independent variable on a latent dependent variable.

[19]The book that you hold in your hands is based on the `plssem` package for Stata and the `cSEM`/`plspm` packages for `R`. Similar to `R` and all its packages, the `plssem` Stata package is open-source and thus freely available.

- Multi-group SEM analysis (MG-SEM)
  It can estimate latent variable means of two or more than two groups – also by controlling for categorical or/and continuous variables as well as testing interactions effects.

- Higher order latent analysis (HLA)
  It can estimate higher order latent variable models including latent variables expressed by first-order observed indicators and higher order (second, third, etc.) latent variables that subsumes these (see for instance Wetzels et al., 2009).

- Latent class analysis (LCA)
  It can detect latent classes based on the size, sign, and significance of regression coefficients (relationships) in an estimated latent variable model (see for instance Mehmetoglu, 2011).

As you already might have figured it out from the above, LSA, MED, MOD, MG-SEM and LCA can all be done with observed variables or combination of observed and latent variables.

1.4 The Language of PLS-SEM

In Figure 1.3 we provide what is commonly known in the literature as LISREL[20] notation which can be used to graphically portray and mathematically specify all types of SEM including PLS-SEM[21]. The depicted model shows the relationships between two *exogenous* (i.e., independent) latent variables[22] and two *endogenous* (i.e., dependent) latent variables as well as their relationships with their respective *manifest variables*, also called *indicators*[23]. Figure 1.3 demonstrates in fact a typical example of a latent structural analysis (LSA).

According to this conventional representation, latent variables are depicted as ovals while manifest variables are shown by rectangles. One-way arrows ($\longrightarrow$) represent *direct effects* whereas two-way arrows ($\longleftrightarrow$) represent covariance/correlations. However, in PLS-SEM covariances between exogenous variables are by default taken into account in the estimation. This is also the reason why we, in real applications, rarely see these covariances depicted specifically in the study models. Error and disturbance terms are exhibited by small circles. Error terms representing indicator measurement error are not shown in diagrams that represent PLS-SEM models.

[20]LISREL (LInear Structural RELationships) is a CB-SEM software developed by Jöreskog and Sörbom (1989).

[21]Since most of the literature is using the LISREL notation, it is useful to get accustomed to it.

[22]Alternative terms are factor, construct, hypothetical, and unobservable.

[23]Other alternative terms are observed, and measured variables.

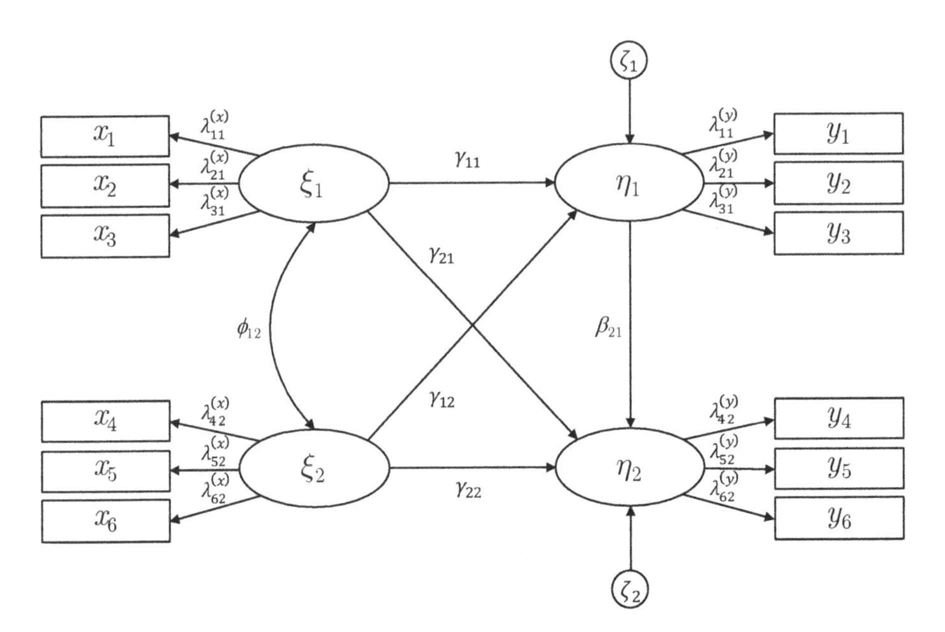

FIGURE 1.3: PLS-SEM model with LISREL notation. In particular, ξ (ksi) indicates an exogenous (independent) latent variable, η (eta) indicates an endogenous (dependent) latent variable, x represents an indicator of an exogenous variable, y represents an indicator of an endogenous variable, ϕ (phi) indicates the correlation between two exogenous variables, γ (gamma) indicates the coefficient between an exogenous and an endogenous variable, β (beta) indicates the coefficient between two endogenous variables, ζ (zeta) represents the unexplained variance in an endogenous variable, λ (lambda) indicates the coefficient (or loading) between indicators and latent variables.

1.5 Summary

In this chapter, we have reasoned that PLS-SEM should be used for exploratory (i.e., contribute to theory-building) and CB-SEM for confirmatory (test theory) purposes. When you use CB-SEM and slide into the exploratory mode (i.e., revising models based on modification indices) or/and encounter non-convergent or improper solutions, you should consider using PLS-SEM instead. In that case, you should rather use consistent PLS-SEM since it is developed as a direct alternative to CB-SEM. By doing so, you will not only get the model converged but also get as close as possible to the results that you would have gotten if you had CB-SEM converged. Finally, we have also shown how PLS-SEM can relate to traditional statistical techniques as well as its potential to estimate more advanced models.

2

Multivariate Statistics Prerequisites

In this chapter we present some preparatory material that is not usually covered in basic statistics courses on which we will build the rest of the book. In particular, we assume that the reader is already familiar with the basic theory of linear regression, so that this chapter is dedicated to introduce concepts from multivariate statistics[1]. After reviewing the bootstrap method, we introduce principal component analysis, which provides a simple and intuitive technique for reducing the dimensionality of a dataset by eliminating the redundancy in the data caused by the correlations among the variables. Next, we illustrate the most popular statistical approaches for identifying clusters of observations. Then, we present path analysis, also known as simultaneous equation modelling in the economics and econometrics literature, which is used to describe the interdependencies among a set of manifest variables (path analysis does not involve unobserved latent variables). Finally, in the last section we briefly address the common practice of using sum scores in a subsequent analysis such as linear regression or path analysis. In the chapter we show a number of examples using both real and simulated data. While these examples will be illustrated using Stata, in the concluding appendix we will also show how to perform them in R.

2.1 Bootstrapping

What you have learnt in basic statistics courses is that the data you analyse are a small portion of those available. In other terms, in practice you only have access to a (random) *sample* taken from a *population* of interest. So, the focus of your analysis typically involves estimating the value of unknown features of the population (e.g., the average of a certain quantity, the correlation index between two quantities or the corresponding linear regression coefficients) by "approximating" them using the sample data. For this reason, you also need to assess the reliability of the sample estimates. This process is usually referred to as *statistical inference*. The classical approach in statistics is to model this problem using probabilities, and the easiest way to do that is by assuming that the population data are distributed according to a normal distribution. The big advantage of the normality assumption is that

[1]Readers that have no background in linear regression, as well as those that simply need to review it, can find a brief introduction in Appendix A.

computations are straightforward producing closed-form expressions for standard errors, confidence intervals and test statistics in many useful cases. However, as you probably already know, reality is not that simple and the assumption that the data in the population are distributed according to a normal distribution is often inappropriate. Therefore, in these cases we are not allowed to perform inference assuming the data are normally distributed. If you find yourselves in these situations, you typically have two options. The first consists in transforming the variables in your analysis with the hope that the distribution of the transformed variables get close to a normal. Even if this option is quite popular when dealing with non-normally distributed data, it suffers from several drawbacks:

- It is not clear a priori which one is the best transformation to use, that is which transformation allows us to get the closest to the normal distribution[2].
- Often, even transforming the quantities of interest doesn't guarantee to get close enough to the normal situation.
- Sometimes, transformed variables make the interpretation of the results more involved and in some cases even useless.

The second option you have in case the normality assumption is unsuitable consists in using a more general approach, called **bootstrap**. The bootstrap is a technique for approximating the sampling distribution of an estimator. More practically, the bootstrap is typically used to compute:

- standard errors for estimators,
- confidence intervals for unknown population parameters,
- p-values for test statistics under a null hypothesis.

The bootstrap, which belongs to the broader class of *resampling methods* together with the jackknife, cross-validation and permutation tests[3], is particularly important to us, because PLS-SEM uses it routinely. For this reason, we provide here an introduction to this important topic and we show how it can be used to perform inference in the linear regression context. In particular, we discuss the most common form of bootstrap called the *non-parametric bootstrap*, in which no distributional assumption is posed, while we skip the other version, the so called *parametric bootstrap*, in which it is assumed that the data follow a known probability distribution (e.g., the Gamma distribution), whose parameters are estimated using the bootstrap principles. In the appendix at the end of this chapter we provide more technical details about the non-parametric bootstrap, but these can be safely skipped now without compromising the understanding of the rest of the book.

[2]Stata provides some commands to find the optimal transformation to normality, namely `ladder`, `gladder` and `qladder`. Moreover, it also includes the command `boxcox` which directly looks for the best transformation and then performs linear regression. The latter command implements the so called *Box-Cox transformation* (see for example Fox, 2016, Section 12.5.1).

[3]For a general presentation of resampling methods you can refer to Good (2006). We provide an introduction to permutation tests in a technical appendix on page 274.

TABLE 2.1: Numerical example of the bootstrap method for approximating the sampling distribution of the sample mean. The original sample of size $n = 5$ is provided in the second column, while to save space only few of the $B = 200$ bootstrap samples are given in the next columns. The last row reports the sample mean corresponding to each set of data.

	Original sample	Bootstrap samples (b)						
		1	2	3	4	5	...	$B = 200$
	7	1	1	-5	-2	1		7
	4	-5	1	4	7	-2		-5
	-5	-2	7	1	-5	-5	...	4
	1	7	-2	-2	7	1		-2
	-2	4	-5	1	1	1		1
Sample mean	1	1	0.2	-0.2	1.6	-0.8	...	1

The basic idea of the bootstrap is simple but extremely powerful at the same time: since the sample we collected provides an approximation of the population from which it was drawn, we can consider it a proxy of the population itself. Then, we can mimic the standard (i.e., frequentist) inferential reasoning by repeatedly drawing random samples from the original sample. These simulated samples are usually called **bootstrap samples**, or resamples. For each bootstrap sample we can then compute the statistic of interest, such as the sample mean. The distribution of the statistic values obtained from the different resamples is called the **bootstrap distribution** of the statistic, and it provides an approximation to its true sampling distribution.

To be more specific, the bootstrap samples are all the same size as the original sample, that is n, and they must be taken *with replacement*, which simply means that after drawing an observation from the original sample, we put it back and proceed by drawing the next observation. The fact that we draw with replacement implies that a specific observation may be included more than once in a given bootstrap sample.

Let's consider now a simple example using the following fictitious data, $(y_1 = 7, y_2 = 4, y_3 = -5, y_4 = 1, y_5 = -2)$, which are also reported in Table 2.1 under the column "Original sample". With the aim of estimating the unknown mean of the hypothetical population from which the data have been drawn, we use the bootstrap to get an approximation of the unknown sampling distribution of the sample mean. We achieve this by sampling with replacement these data 200 times (we explain why 200 in the next paragraphs). To save space, the rest of the table reports only some of the 200 bootstrap samples, while the last row provides the corresponding means. Figure 2.1 shows the histogram of the 200 bootstrap means, that is the bootstrap distribution of the sample mean for these data.

Note that in the example above, because of the very small sample size, we might have enumerated all the possible $5^5 = 3125$ bootstrap samples. However, in general this is not feasible because the number of bootstrap samples, n^n, grows very quickly

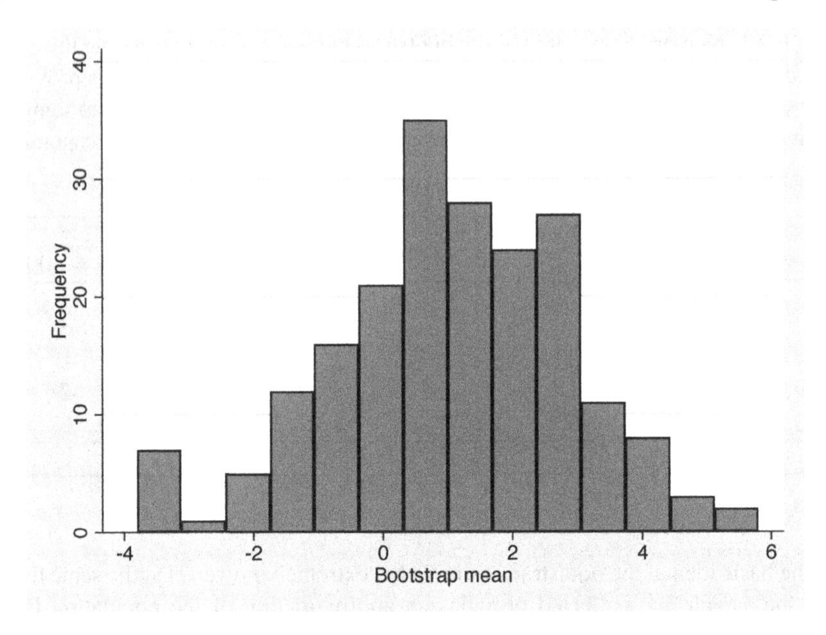

FIGURE 2.1: Bootstrap distribution of the sample mean for the data shown in Table 2.1.

with the sample size[4]. So, in practice we can only rely on simulation, that is on drawing randomly a finite number B, called the **number of bootstrap replications**, of the possible resamples[5]. As you can imagine, B is a critical number for the success of the bootstrap. Even if it is true that larger values of B provide more accurate estimates of the statistics of interest, there are diminishing returns from increasing it beyond a certain point, and it has been shown that $B = 200$ suffices in most cases (see Efron and Hastie, 2016, Chapter 10). However, we must say that there is no general consensus in the literature on the value to use for B with other authors that recommend drawing at least 10000 resamples (Cihara and Hesterberg, 2019, Section 5.9).

Once the bootstrap distribution has been obtained, we can proceed to solve our inferential problem by computing the necessary quantities. In particular:

- We can estimate the standard error of the statistic of interest through the standard deviation of the bootstrap distribution.

- We can approximate the $100(1-\alpha)\%$ confidence interval for the statistic of interest by computing the percentiles of order $100\alpha/2$ and $100(1-\alpha/2)$ of the bootstrap distribution. This approach is known as the **percentile method**, but

[4]Try for example calculating this number when n is as small as 20.

[5]This means that the histogram shown in Figure 2.1 represents an approximation of the "theoretical" bootstrap distribution which comprises n^n values, which in turn approximates the sampling distribution of the statistic.

other more accurate methods have also been developed (see the technical appendix at the end of the chapter for more details).

- We may compute the p-value in a test, but this operation is more complicated because a p-value requires that the null distribution must be true and thus resampling should be performed assuming that this holds. Since we won't use bootstrap for finding p-values, we do not discuss this topic and refer you to the literature (see for example Boos and Stefanski, 2013, Section 11.6).

Stata implements the bootstrap through the `bootstrap` prefix. After the latter has been executed, the `estat bootstrap` postestimation command displays a table of confidence intervals for each statistic[6].

The bootstrap principles we illustrated so far can be directly applied to the linear regression model. More specifically, if $z_i = (y_i, x_{1i}, x_{2i}, \ldots, x_{pi})$ represents the vector of response and predictor values for the ith observation, with $i = 1, \ldots, n$, we just need to resample with replacement the z_is and compute the regression estimates for each bootstrap sample. Then, the methods we described above to compute standard errors and confidence intervals can be applied to the bootstrapped regression coefficients[7]. Clearly, the bootstrap idea is so general that it can also be used with more complicated regression models.

In Stata you can use the general `bootstrap` prefix with a linear regression problem, but if you only need to use the bootstrap for estimating the coefficients' standard errors, we suggest to exploit the more convenient `vce(bootstrap)` option of the `regress` command. We show a practical application of these commands in Section A.2.7.

2.2 Principal Component Analysis

Two variables are strongly correlated when they tend to move together in a systematic way. In this case, we can get accurate predictions of one of the variables using the information coming from the other. Put it differently, when two variables are strongly correlated, the information they convey is partially "overlapped". A typical situation where this may occur and produce undesirable effects is in linear regression when some of the predictors are strongly linearly associated (for a review see Section A.2.6). A common solution to get rid of the redundant information in a set of variables is to preprocess them using a *dimensionality reduction* technique. One

[6]For a detailed presentation of the bootstrap in Stata, we suggest to look at Cameron and Trivedi (2009), Chapter 13.

[7]This approach to bootstrapping in the regression setting is called *random pairs* bootstrap and it is specifically suited when the predictors are assumed to be *random*, as it is reasonable in most cases in the social sciences. If predictors are instead assumed to be *fixed*, as it is common in an experimental setting, then *residual-based* bootstrap is usually suggested, in which we resample the residuals from the original sample and use them to form the bootstrap responses. For more details about this distinction we suggest to see Fox (2016), Section 21.3.

of the most popular method for reducing the dimensionality of a set of data is **principal component analysis** (PCA)[8]. The origin of PCA goes back to the beginning of the 20th century with the work of Pearson and its first application in psychology by Hotelling. PCA has recently seen a resurgence of interest because of the rapid growth of the machine learning field, where it is now included among the feature engineering tasks. Nonetheless, PCA is a tool that is routinely used in many fields such as marketing, finance, biology and psychology[9].

Starting from a set of correlated variables, PCA combines them producing new quantities, called *principal components* (or simply components), that have the following properties:

- they are ordered in terms of their "importance", with the first component summarizing most of the information from the original variables, the second component providing the next most informative piece of information, and so on,
- the components are uncorrelated to each other.

The main idea of PCA is thus that of retaining a number of uncorrelated components that allow to account for a given portion of the total original information. In this sense PCA reduces the complexity of the dataset removing the redundant information shared by the original variables.

More technically, PCA linearly combines the original variables to produce a new set of orthogonal (i.e., uncorrelated) axis that are orientated towards the directions of maximal variability in the data. The direction with the largest variance is called the first principal component, the orthogonal direction with the second largest variance is called the second principal component, and so on. Figure 2.2 provides a graphical representation of PCA for a pair of positively correlated variables X_1 and X_2. The original data are shown in panel 2.2a, while panel 2.2b shows the corresponding principal components, that is the orthogonal directions with the greatest variance. Note that before computing the new axes, the data have been centred thus moving the origin of the system in correspondence of the averages of the original variables. In PCA it is always assumed that the variables are mean centred and all software perform this preliminary operation before finding the new axes.

In panel 2.2c we highlighted one point with a big cross to show that each observation can be projected onto the new coordinate system to get a new set of coordinates called the *component scores*. The scores are typically added as new columns in the dataset, which can then be used in a subsequent analysis (e.g., as the predictors of a linear regression model). Panel 2.2d shows the same data but after projecting them in the new space spanned by the principal components.

[8]Other widespread dimensionality reduction techniques are multidimensional scaling, Sammon mapping, self-organizing maps, projection pursuit, principal curves and kernel principal components. For an introduction to these topics we suggest to see Hastie et al. (2008), Chapter 14.

[9]In psychology factor models are more often used than PCA. Factor models are statistical models whose main purpose is to describe the covariance relationships among the observed variables in terms of a set of unobservable quantities called factors. For more details on factor models you can look at Mehmetoglu and Jakobsen (2016, Chapter 11) or at Brown (2015) for a longer treatment.

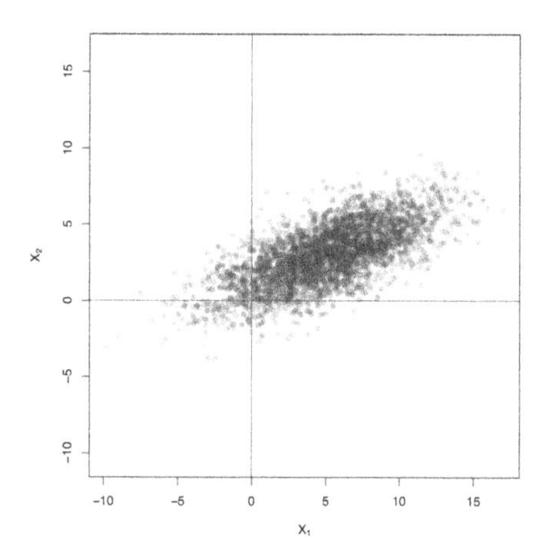

(a) Original data simulated from a bivariate normal distribution with means equal to 5 and 3, standard deviations equal to 4 and 2 and linear correlation index equal to 0.7.

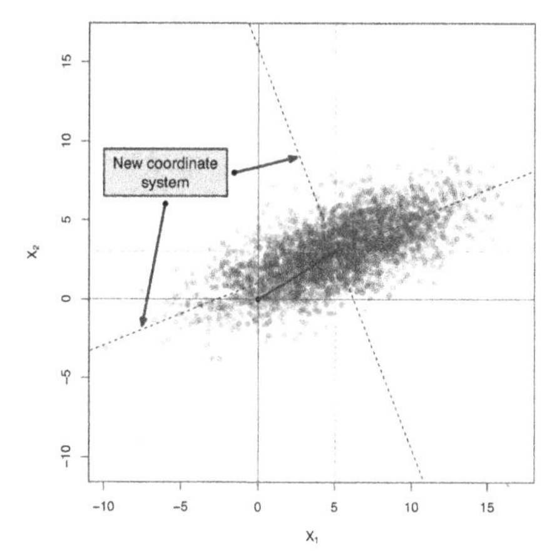

(b) Original data together with new coordinate system.

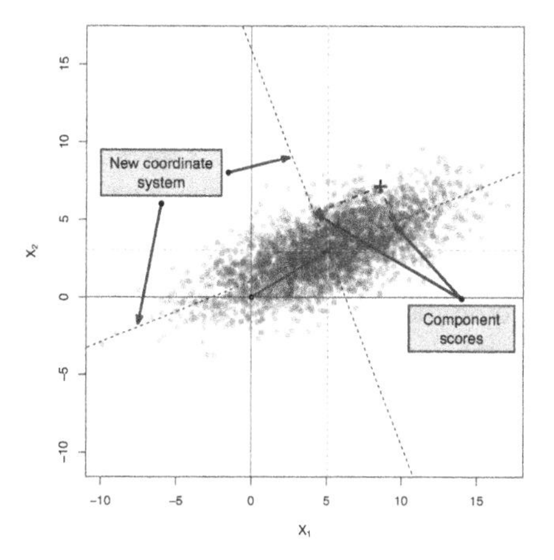

(c) Original data together with new coordinate system and component scores for one observation.

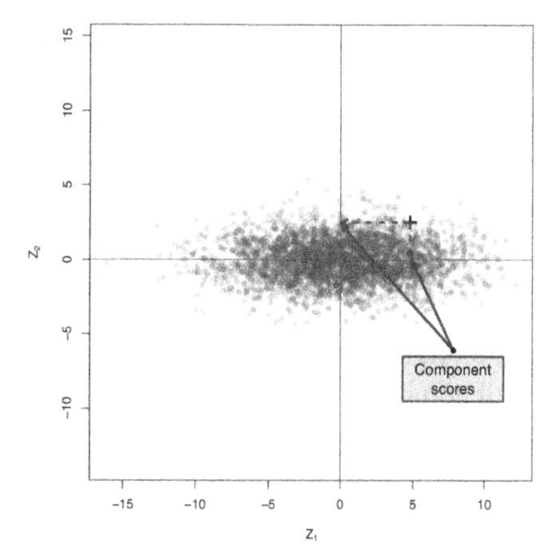

(d) Projections of original data in the new coordinate system and corresponding component scores for one observation.

FIGURE 2.2: Geometric interpretation of PCA for some fictitious data on two positively correlated variables X_1 and X_2. The corresponding principal components are denoted as Z_1 and Z_2.

It can be shown (see the technical appendix at the end of this chapter for more details) that the principal components correspond to the so called **eigenvectors** of the original covariance matrix, while the component variances correspond to the **eigenvalues** of the same matrix. Eigenvectors provide the coefficients for the linear combinations of the original variables that produce the components. As we said, principal components are ordered by their variance, that is by their eigenvalues. Moreover, the sum of the eigenvalues is equal to the sum of the variances of the original variables. A simple way to assess the relative importance of the components is to compute the proportion of the total variance accounted for by each component, which can be simply computed as the ratios of the eigenvalues to their sum (i.e., to the total variance). For the fictitious data in Figure 2.2, the eigenvalues are equal to 17.285 and 1.864 respectively, so the proportions of the total variance explained by each component are equal to $17.285/(17.285+1.864) = 0.903$ and $1.864/(17.285+1.864) = 0.097$. So, the first component summarizes around 90% of the original information. This is clearly due to the strong linear association between the two variables.

The previous comment brings in the issue related to the variable scales. More specifically, if one variable has a greater scale compared to the others, it will dominate in the calculations producing a first principal component that substantially corresponds to that variable only. In this case, the other components will inevitably show much smaller variances. A similar situation may occur when the units of measure of the original variables are too diverse. In these cases it is a good suggestion to standardize the variables before performing PCA. The covariances of variables that have been standardized correspond to their linear correlations (see Section A.1). Therefore, performing PCA on the covariance matrix of a set of standardized variables is equivalent to do that on the corresponding correlation matrix. So, unless there is any specific reason to act differently, we recommend to perform PCA on the correlation matrix instead of the covariance matrix.

Since PCA is typically used to reduce the dimensionality of a set of data, the hope in general is that the first few components account for most of the total information in the original variables. In practice, this principle is put into action by selecting a small number of components to keep in the analysis, thus discarding the other ones. The choice of the number of components to retain is tricky and no general answer is available unless strong assumptions are advanced[10]. However, the following guidelines are frequently used by practitioners:

- Retain a number of components that allow to account for a large percentage of the total variation of the original variables. Typical values used here are in between 70 and 90%.

- Retain the components whose eigenvalue is larger than the average of the eigenvalues. When PCA is performed on the correlation matrix, this average is equal to 1.

[10] A typical assumption used to facilitate the choice of the number of components in PCA is multivariate normality of the original variables. This is an assumption that rarely holds in practice, but if you are inclined to accept it, a number of tests are available. You can find more details on these tests in Mardia et al. (1979), Section 8.4.

- Use the **scree plot**[11], which is a diagram reporting the eigenvalues on the vertical and the component number on the horizontal axis. According to this criterion, it is suggested to retain a number of components found as one fewer than that at which the plot shows an "elbow" (i.e., a sudden change of slope). A more specific suggestion recommends looking for the point beyond which the curve becomes nearly flat.

We remind that these criteria are exploratory and often lead to contrasting conclusions.

In Stata we can perform PCA with the `pca` command. Let's consider an example using the data available in the `ch2_Rateprof.dta` file[12]. These data include the summaries of the ratings of 366 instructors (207 male, 159 female) at a large campus in US collected by the `http://www.ratemyprofessors.com` website over the period 1999–2011. Instructors included had 10 or more ratings on the site. Students provided ratings on a scale from 1 to 5 on quality, helpfulness, clarity, easiness of instructor's courses, and rater-interest in the subject matter covered in the instructor's courses. These data originate from the study Bleske-Rechek and Fritsch (2011) whose aim was to test the assumption that students' ratings are unreliable. Here, we use the data merely to illustrate how to perform and interpret PCA with Stata.

We start by inspecting the scatter plot matrix for the five ratings, which is provided in Figure 2.3. This diagram clearly shows that there is a strong positive association between the items `quality`, `helpfulness` and `clarity`, while the correlations of the other ratings are weaker. So, we expect that the former items will be combined by PCA in a single component.

We now perform PCA using Stata's `pca` command. The command's default is to compute PCA on the correlation matrix, but the `covariance` option allows to run it on the covariance matrix. In our example all the variables are measured on the same scale and have approximately the same variability, so it is practically indifferent to use one or the other. However, we follow our own suggestion and compute PCA on the correlation matrix (the corresponding results are shown in Figure 2.4):

```
1 use ch2_Rateprof, clear

2 pca quality-raterinterest
3 screeplot
```

The first table in the output reports the eigenvalues together with their differences, the proportion of explained variances and the corresponding cumulative proportions, while the second table contains the eigenvectors. As you see, the first component accounts for around 72% of the total variance, with an eigenvalue much larger than 1. The second component accounts for 16% of the total variability, which together with the first one allows to reach a cumulative proportion of 88%. The third component

[11]The name comes from its similarity to a cliff with rocky debris at its bottom.

[12]All data files and code shown in this book are available at `https://github.com/sergioventurini/SEMwPLS`.

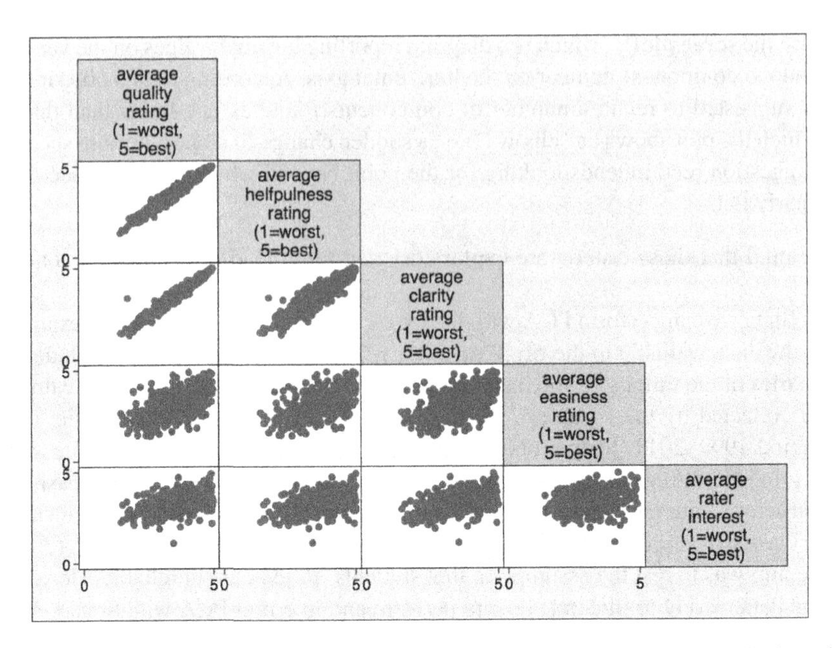

FIGURE 2.3: Scatter plot matrix for some variables included in the `ch2_Rateprof.dta` data file.

```
Principal components/correlation                  Number of obs    =        366
                                                  Number of comp.  =          5
                                                  Trace            =          5
    Rotation: (unrotated = principal)             Rho              =     1.0000

    --------------------------------------------------------------------------
       Component |   Eigenvalue   Difference         Proportion   Cumulative
    -------------+------------------------------------------------------------
           Comp1 |      3.58011      2.77825             0.7160       0.7160
           Comp2 |      .801854      .264397             0.1604       0.8764
           Comp3 |      .537456       .45877             0.1075       0.9839
           Comp4 |     .0786866      .076788             0.0157       0.9996
           Comp5 |    .00189859            .             0.0004       1.0000
    --------------------------------------------------------------------------

Principal components (eigenvectors)

    ------------------------------------------------------------------------------
        Variable |    Comp1     Comp2     Comp3     Comp4     Comp5 | Unexplained
    -------------+--------------------------------------------------+-------------
         quality |   0.5176   -0.0384   -0.2666   -0.0362   -0.8113 |           0
     helpfulness |   0.5090   -0.0436   -0.2451   -0.6977    0.4384 |           0
         clarity |   0.5053   -0.0241   -0.2893    0.7148    0.3867 |           0
        easiness |   0.3537   -0.5582    0.7498    0.0322    0.0043 |           0
    raterinter~t |   0.3042    0.8273    0.4722    0.0042   -0.0004 |           0
    ------------------------------------------------------------------------------
```

FIGURE 2.4: Principal component analysis from the correlation matrix of the ratings `quality`, `helpfulness`, `clarity`, `easiness`, and `raterinterest` in the `Rateprof` example.

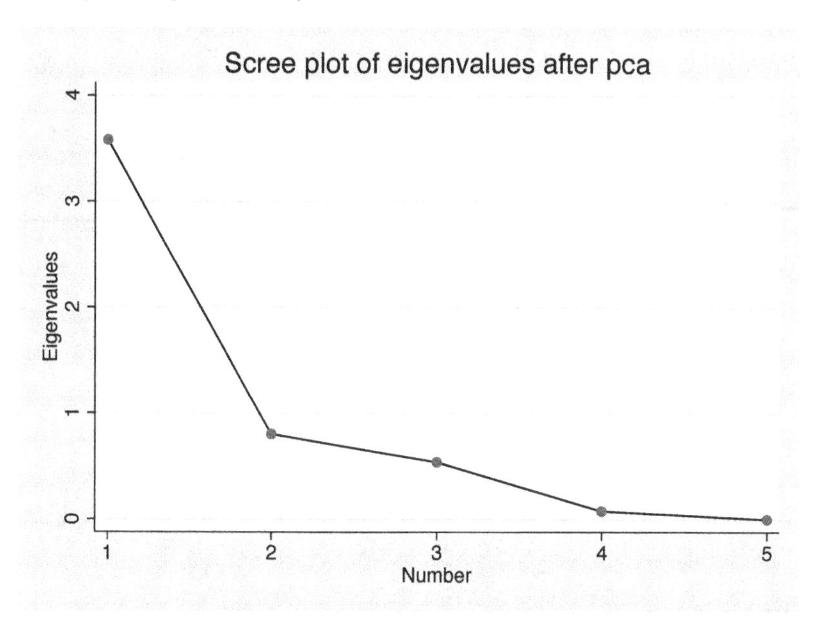

FIGURE 2.5: Scree plot for the PCA of the ratings `quality`, `helpfulness`, `clarity`, `easiness` and `raterinterest` in the `Rateprof` example.

explains approximately 11% of the total variance, contributing to reach a cumulative proportion of about 98%. Finally, the remaining two components contribute by explaining a very small portion, that is around 1.6%. According to the guidelines provided above, we should retain either 1 or 2 components (the scree plot for this example, obtained with the `screeplot` postestimation command, is shown in Figure 2.5).

Then, we execute the `pca` command again but now adding the option `components(2)` (the results are shown in Figure 2.6):

```
1  pca quality-raterinterest, components(2)
```

The first table is identical to the previous one, while the second one only contains the first two columns, that correspond to the requested components to retain. While these columns report the same numbers as in the previous output, the column labelled `Unexplained` now shows that not all of the variance of the original variables has been accounted for. We see from this column that the amount of unexplained variance for the `easiness` rating is 30%, a fairly large value. Therefore, it may be a good suggestion to override the guidelines above and redo the analysis using 3 components (Figure 2.7):

```
Principal components/correlation                 Number of obs    =        366
                                                 Number of comp.  =          2
                                                 Trace            =          5
    Rotation: (unrotated = principal)            Rho              =     0.8764

    --------------------------------------------------------------------------
       Component |   Eigenvalue   Difference         Proportion   Cumulative
    -------------+------------------------------------------------------------
           Comp1 |      3.58011      2.77825             0.7160       0.7160
           Comp2 |      .801854      .264397             0.1604       0.8764
           Comp3 |      .537456       .45877             0.1075       0.9839
           Comp4 |     .0786866      .076788             0.0157       0.9996
           Comp5 |    .00189859            .             0.0004       1.0000
    --------------------------------------------------------------------------

Principal components (eigenvectors)

    ------------------------------------------------
        Variable |    Comp1     Comp2 | Unexplained
    -------------+--------------------+-------------
         quality |   0.5176   -0.0384 |      .03954
     helpfulness |   0.5090   -0.0436 |      .07094
         clarity |   0.5053   -0.0241 |      .08545
        easiness |   0.3537   -0.5582 |       .3022
    raterinter~t |   0.3042    0.8273 |       .1199
    ------------------------------------------------
```

FIGURE 2.6: Principal component analysis from the correlation matrix of the ratings `quality`, `helpfulness`, `clarity`, `easiness`, and `raterinterest` in the `Rateprof` example retaining only two components.

```
1 pca quality-raterinterest, components(3)
```

To complete the example, we now compute the so called **loadings**, which provide a very handy tool to interpret the components. Loadings are computed by appropriately rescaling the eigenvectors[13] and they can be interpreted as the correlation of each component with the original variables. In Stata we can get the loadings through the `estat loadings` postestimation command as follows (see Figure 2.8):

```
1 estat loadings, cnorm(eigen)
```

The new output shows that the first component is mainly correlated with the ratings `quality`, `helpfulness` and `clarity`. Therefore, we can interpret the first component as an index of the overall goodness of the professors as instructors. The second and third components instead load mainly on the `raterinterest` and `easiness` ratings respectively, and therefore we can interpret them as single-item components.

Finally, we can save the component scores in the dataset with the `predict` command:

[13] See the technical appendix at the end of this chapter for more details.

```
Principal components/correlation                  Number of obs    =        366
                                                  Number of comp.  =          3
                                                  Trace            =          5
    Rotation: (unrotated = principal)             Rho              =     0.9839

    --------------------------------------------------------------------------
       Component |   Eigenvalue   Difference         Proportion   Cumulative
    -------------+------------------------------------------------------------
           Comp1 |      3.58011      2.77825             0.7160       0.7160
           Comp2 |      .801854      .264397             0.1604       0.8764
           Comp3 |      .537456       .45877             0.1075       0.9839
           Comp4 |     .0786866      .076788             0.0157       0.9996
           Comp5 |    .00189859            .             0.0004       1.0000
    --------------------------------------------------------------------------

Principal components (eigenvectors)

    ----------------------------------------------------------
        Variable |    Comp1     Comp2     Comp3 | Unexplained
    -------------+------------------------------+-------------
         quality |   0.5176   -0.0384   -0.2666 |     .001353
     helpfulness |   0.5090   -0.0436   -0.2451 |      .03867
         clarity |   0.5053   -0.0241   -0.2893 |      .04048
        easiness |   0.3537   -0.5582    0.7498 |   .00008182
    raterinter~t |   0.3042    0.8273    0.4722 |   1.372e-06
    ----------------------------------------------------------
```

FIGURE 2.7: Principal component analysis from the correlation matrix of the ratings `quality`, `helpfulness`, `clarity`, `easiness`, and `raterinterest` in the `Rateprof` example retaining only three components.

```
Principal component loadings (unrotated)
    component normalization: sum of squares(column) = eigenvalue

    --------------------------------------------
                 |    Comp1     Comp2     Comp3
    -------------+------------------------------
         quality |    .9794   -.03434    -.1954
     helpfulness |    .9631   -.03903    -.1796
         clarity |    .9561   -.02161    -.2121
        easiness |    .6692    -.4999     .5497
    raterinter~t |    .5756     .7408     .3462
    --------------------------------------------
```

FIGURE 2.8: Principal component analysis from the correlation matrix of the ratings `quality`, `helpfulness`, `clarity`, `easiness`, and `raterinterest` in the `Rateprof` example retaining only three components.

```
predict pc*, scores
```

Before moving to the next topic, we provide a last remark on using PCA to deal with multicollinearity in linear regression analysis through a procedure called *principal component regression* (PCR). The common practice in this context is to use as predictors only the components that contribute to explain a sufficiently large portion of the total variance of the X variables. The fact that not all the components are retained is often justified to avoid reducing too much the number of residual degrees of freedom. This is critical especially in fields where the number of predictors in the model can be very large, e.g., in genetics or chemometrics. However, the aim of regression analysis is to predict the response variable and it is not necessarily the case that the most important components for the predictors are also those with the highest predictive power. An alternative to PCR is *partial least squares* (PLS) *regression*[14], which explicitly aims to find the unobserved components in such a way as to capture most of the variance in the X variables, while also maximizing the correlation between the X and the Y variables. For more details on PLS regression see for example Hastie et al. (2008, Chapter 3).

2.3 Segmentation Methods

Segmentation is the process of dividing the sample cases into meaningful and homogeneous subgroups based on various attributes and characteristics. The most popular application of segmentation methods is in marketing, where companies frequently need to identify such groups to better understand their customers and build differentiated strategies (Wind, 1978; Wedel and Kamakura, 2000; Kotler and Keller, 2015; Dolnicar et al., 2018). In the following sections we provide an overview of the segmentation methods that will be used in the rest of the book. In particular, we first describe **distance-based** methods, that is methods that use a notion of similarity or distance between the cases. Then, we move to **model-based** methods, which instead are based on a probabilistic assumption on the nature of the segments.

2.3.1 Cluster analysis

Cluster analysis (CA) is one of the most widely applied statistical techniques in many fields such as marketing, genetics and biology. The aim of CA is to find a classification in which observations are sorted into a usually small number of homogeneous and exclusive (i.e., non-overlapping)) groups[15]. Despite its popularity,

[14]While sharing a common origin with the main topic of this book, PLS regression implements a completely different algorithm and has a different aim.

[15]There exist also many overlapping clustering algorithms in which an observation may belong to multiple groups, but they are beyond the scope of this review (see Everitt et al., 2011, in particular Section 8.4).

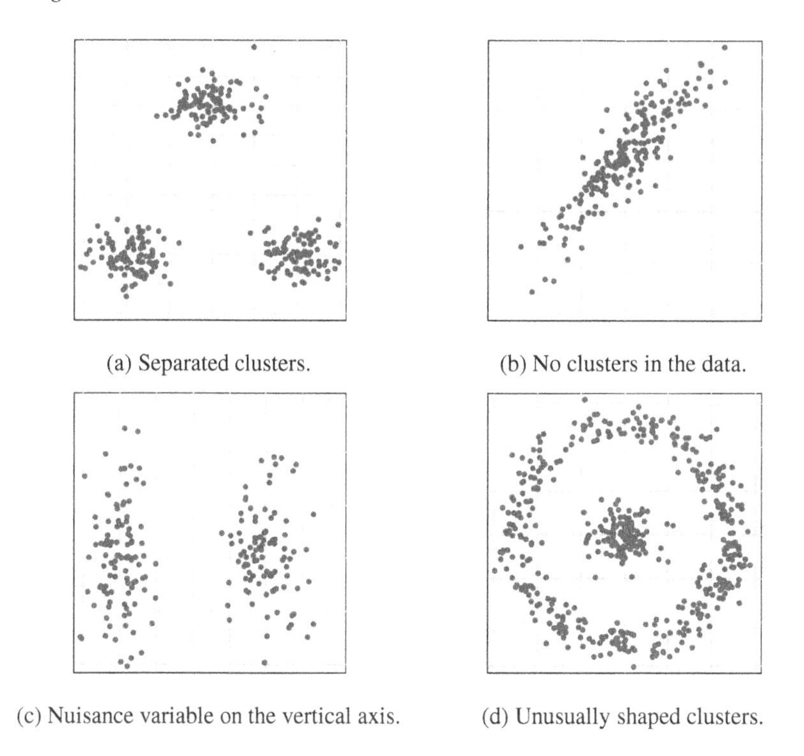

(a) Separated clusters. (b) No clusters in the data.

(c) Nuisance variable on the vertical axis. (d) Unusually shaped clusters.

FIGURE 2.9: Examples of data structures.

the application of CA can be very subtle in some cases. This may be due to many reasons. First, when we run a CA, we generally take for granted that the data are naturally composed of homogeneous groups, but often they are not. An example of this situation is shown in Figure 2.9b. Since clustering algorithms always produce an output, we are spontaneously driven to the conclusion that the groups actually exist. So, it is the responsibility of the researcher to critically validate the results. The second reason why CA may provide misleading results is that we don't know a priori the variables according to which the groups are defined. For example, in a customer segmentation we don't know a priori whether the purchasing habits of the subjects are different because of their incomes or their education or the perception they have of the product features. The standard practice is to use as many information are available for discovering the clusters. However, the inclusion of irrelevant (also called *nuisance*) variables can radically change the interpoint distances and thus undermine the clustering process. A simple example of such a situation is provided in Figure 2.9c, where the variable reported on the vertical axis is nuisance since it is irrelevant for identifying the clusters. Moreover, the grouping structure can be complicated with oddly shaped clusters, such as those reported in Figure 2.9d. Unfortunately, classical clustering algorithms are not able to identify these structure types. Finally, each clustering algorithm depends on some tuning parameters whose settings may modify the solution sometimes in a critical way. Again, the researcher should assess the extent of these differences by comparing the solutions corresponding to different choices of the parameters.

Classical CA algorithms are usually classified in **hierarchical** and **partitional** algorithms. The main difference between these approaches is that in the former case the number of clusters must not be chosen in advance. Hierarchical algorithms start from a solution with n groups each containing a single observation and produce a hierarchy (i.e., a tree) of solutions where at each iteration the two closest groups are joined together, until all observations are merged in a single cluster. Partitional algorithms seek to partition the n cases into a predefined number of groups, usually denoted K. In this case, the algorithm returns a single solution (no hierarchy), but clearly one can run them for different values of K.

2.3.1.1 Hierarchical clustering algorithms

Hierarchical clustering techniques are usually subdivided into **agglomerative** (or bottom-up) methods, which proceed by a series of successive fusions of the n observations into groups, and **divisive** (or top-down) methods, which work the other way round splitting one of the existing cluster into two new clusters. In both cases, it is up to the researcher deciding which solution represents the "natural" grouping of the data. A long list of indexes and tests to assist the researcher in this decision have been developed in the literature. Since divisive methods are less popular in practice, in the following we review only the agglomerative ones.

Hierarchical CA algorithms typically return a tree-based visualization of the solution hierarchy known as **dendrogram**, which shows the groups merged at each stage and their reciprocal distances. Central to the illustration of agglomerative algorithms are the notions of **dissimilarity** and **linkage**.

The dissimilarity between two observations is a measure of how much they differ in terms of the variables used in the analysis. Sometimes dissimilarities are also referred to as *distances*, even if there is no perfect match between the two notions[16]. All agglomerative algorithms use a symmetric matrix of dissimilarities $\boldsymbol{D}$ with non-negative entries (i.e., the dissimilarity between two cases can only be zero or a positive number) and zero diagonal elements (i.e., the dissimilarity between an observation and itself is zero) as input. A large number of dissimilarity measures have been introduced in the literature depending on the type of the variables (categorical or numerical). It is not our aim here to provide a detailed examination of all these measures, for which we suggest to see Everitt et al. (2011, Chapter 3)[17]. The only dissimilarity measure that we review is the **Euclidean distance**, because it is the most

[16]From a purely mathematical point of view, a measure of the diversity between observations is a distance only if it satisfies a series of technical conditions. However, some dissimilarity measures used in CA do not fully comply with these requirements, and thus cannot be called distances in the strict sense. In particular, some dissimilarity measures do not satisfy the so called *triangle inequality*, which requires that the distance between x and z must be at most as large as the sum of the distances between x and y and y and z, or in more technical terms, $d(x,z) \leq d(x,y) + d(y,z)$.

[17]For a list of the dissimilarity measures available in Stata, type `help measure_option`.

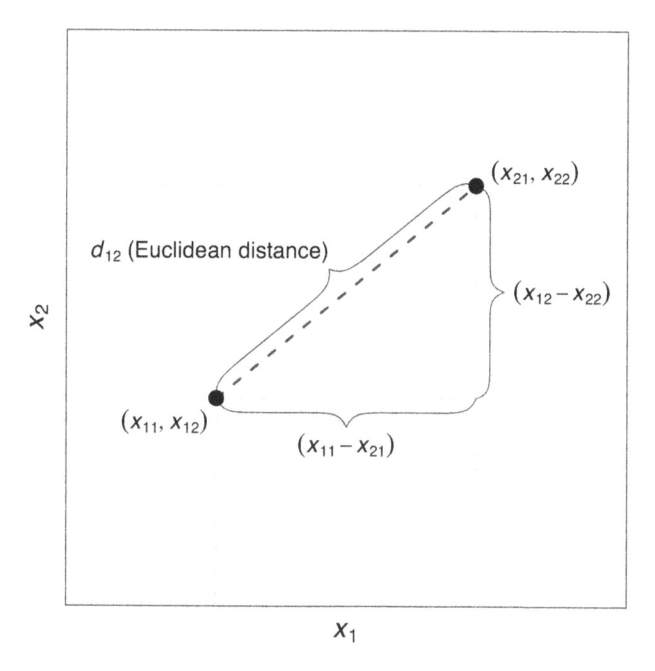

FIGURE 2.10: A graphical representation of the Euclidean distance with $p = 2$ variables.

popular. Given p variables $x_1, \ldots, x_p$ measured on each observation, the Euclidean distance between observations i and j, with $i, j = 1, \ldots, n$, is defined as

$$d_{ij} = \left[\sum_{k=1}^{p} \left(x_{ik} - x_{jk} \right)^2 \right]^{1/2}, \tag{2.1}$$

where x_{ik} and x_{jk} are the kth variable value for observations i and j respectively. This dissimilarity measure has the appealing property that it can be interpreted as the physical distance between the two p-dimensional points $(x_{i1}, \ldots, x_{ip})$ and $(x_{j1}, \ldots, x_{jp})$ in the space defined by the variables $x_1, \ldots, x_p$ (see Figure 2.10 for the case of $p = 2$ variables).

Once the software has computed the dissimilarities between each pair of observations, it forms n groups each composed by a single observation. Then, at each step the distances between all pairs of groups are computed[18]. Linkage is the term used in agglomerative CA to refer to the specific way the distances between groups are computed. There are different type of linkage notions which produce potentially different clustering solutions. The most common linkage measures are:

[18]We call your attention on the fact that these are the distances between the *groups* and not between the individual *observations*. We remind that the distances between observations are the dissimilarities.

- **single linkage**, which computes the distance d_{AB} between two groups A and B as the *smallest* distance from each observation in group A to each one in group B. In technical terms, single linkage calculates

$$d_{AB}^{\text{single}} = \min_{i \in A, j \in B} (d_{ij}).$$

- **complete linkage**, which computes the distance between two groups A and B as the *largest* distance from each observation in group A to each one in group B, that is

$$d_{AB}^{\text{complete}} = \max_{i \in A, j \in B} (d_{ij}).$$

- **average linkage**, which computes the distance between two groups A and B as the *average* of the distances from each observation in group A to each one in group B, that is

$$d_{AB}^{\text{average}} = \frac{1}{n_A n_B} \sum_{i \in A, j \in B} d_{ij},$$

  where n_A and n_B are the number of observations in each group.

- **Ward's method**, which is different from the previous approaches because it doesn't focus directly on distances between groups, but rather it considers the heterogeneity within the groups. In particular, at each iteration Ward's method joins those two groups that contribute to the smallest increase in the total (i.e., across groups) within-cluster error sum of squares, defined as

$$E = \sum_{g=1}^{K} E_g,$$

  where

$$E_g = \sum_{i=1}^{n_g} \sum_{k=1}^{p} \left(x_{ik}^g - \bar{x}_k^g\right)^2$$

  represents the within-cluster error sum of squares for cluster g, where x_{ik}^g is the value on the kth variable for the ith observation in cluster g and $\bar{x}_k^g$ is the corresponding mean.

A graphical comparison of the single, complete and average linkage methods is provided in Figure 2.11.

Each agglomerative algorithm has pros and cons, in particular (for a more detailed discussion see Everitt et al., 2011, especially Table 4.1):

- **single linkage** produces solutions that often present a phenomenon known as *chaining*, which refers to the tendency to incorporate intermediate points between clusters into an existing cluster rather than starting a new one. As a result, single linkage solutions often contain long "straggly" clusters.

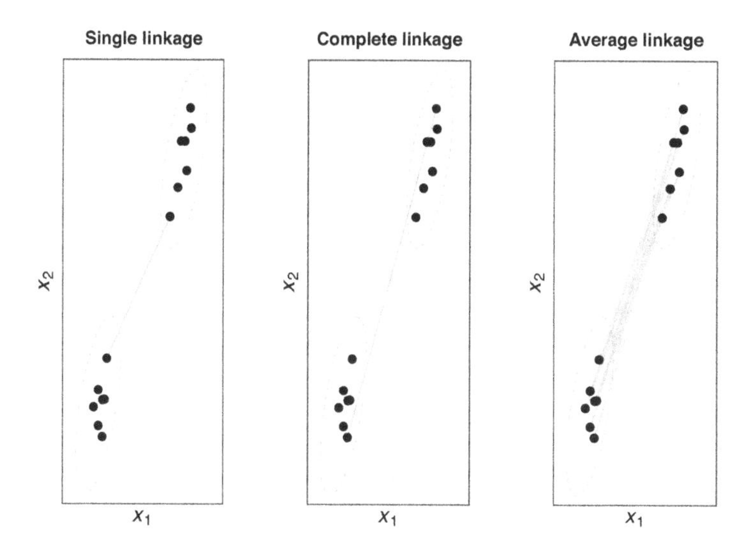

FIGURE 2.11: A comparison of the single, complete and average linkages for agglomerative clustering.

- **complete linkage** tends to create cohesive groups with similar diameters (i.e., largest distance of the objects in a cluster).
- **average linkage** tends to merge groups with low internal variance. This method is a compromise between single and complete linkage and it is more robust to outliers.
- **Ward's method** finds spherical groups with similar size, but it is rather sensitive to outliers.

As we already mentioned above, the full hierarchy of solutions produced by agglomerative algorithms is typically represented graphically with a dendrogram, whose nodes correspond to clusters and the stem lengths represent the distances at which clusters are joined. Figure 2.12 shows the dendrograms produced by the linkage methods described above for some fictitious data, whose dissimilarity matrix is given by

$$\boldsymbol{D} = \begin{array}{c} \\ 1 \\ 2 \\ 3 \\ 4 \\ 5 \end{array} \begin{array}{c} \begin{array}{ccccc} 1 & 2 & 3 & 4 & 5 \end{array} \\ \left[\begin{array}{ccccc} 0 & & & & \\ 9 & 0 & & & \\ 1 & 7 & 0 & & \\ 6 & 5 & 9 & 0 & \\ 10 & 11 & 2 & 8 & 0 \end{array} \right] \end{array}.$$

At the bottom of the dendrogram we find the first step of the procedure, where

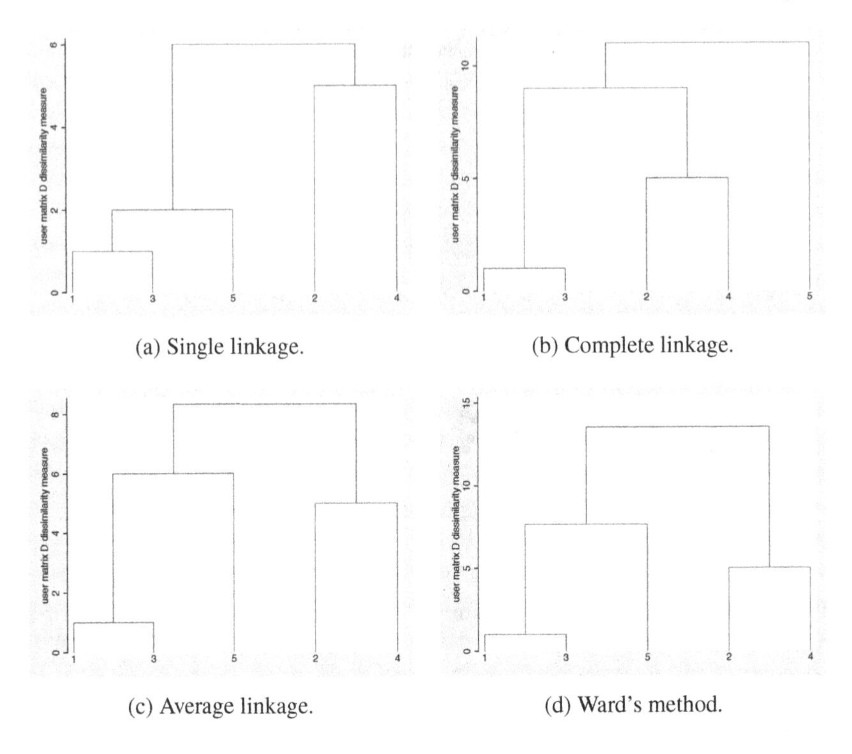

(a) Single linkage. (b) Complete linkage.

(c) Average linkage. (d) Ward's method.

FIGURE 2.12: Dendrograms produced by different agglomerative algorithms for some fictitious data.

each observation has been placed in a separate cluster. Then, the distances between each pair of groups is computed and the two closest groups are joined to form a new larger cluster. This process continues until the last iteration where all the objects are merged together in a single cluster. As we see from the dendrograms in Figure 2.12, different algorithms produce different solutions. In this simple example average linkage and Ward's method provide the same hierarchy, while single and complete linkage do return different answers.

Since agglomerative hierarchical algorithms provide the entire hierarchy of clustering solutions, the researcher must then choose a criterion for selecting the number of clusters. Informally, this is achieved by "cutting" a dendrogram at a particular height. Large changes in fusion levels (i.e., big jumps in the diagram) are usually taken to indicate the best cut. Formal methods known as **stopping rules** have been introduced in the literature (a classic reference is Gordon, 1999, in particular Section 3.5), but they rarely lead to an unambiguous conclusion. Apart from the now dated study by Milligan and Cooper (1985), which compared a list of 30 methods, there have been in general only a limited investigation into the properties of these rules (for a more recent survey for clustering of binary data see Dimitriadou et al., 2002). Tibshirani et al. (2001) introduced a new procedure for selecting the optimal number of clusters based on the so called **gap statistic**, whose main advantage is to allow the evaluation of the quality of the single cluster solution. This represents an important information for the researcher to confirm or not the existence of distinct groups in the data.

The study by Milligan and Cooper (1985) identified the stopping rules developed by Caliński and Harabasz (1974) and Duda et al. (2001) among the best ones to use when clustering numerical data. In both cases, large values of the stopping rule indexes indicate a distinct cluster structure, so we identify as optimal the solution corresponding to the largest values of these indexes. The Caliński-Harabasz and Duda-Hart rules are available in Stata through the `cluster stop` command, for which we provide more details in the technical appendix at the end of this chapter.

Once the researcher has identified and validated the stability of the clustering solution, a CA typically terminates with the **profiling** of the groups. This procedure allows to interpret the clusters in light of the observed variables (e.g., sociodemographics) and it typically consists of simple graphical and numerical summaries for each cluster.

Hierarchical cluster analysis: An example for simulated data using Stata

As an illustrative example of the agglomerative clustering algorithms, we consider the simulated data available in the file `ch2_SimData.dta` and shown in Figure 2.13, from which we see a group structure with three segments[19].

Stata allows to perform a CA through the `cluster` suite of commands. More specifically, the agglomerative CA algorithms described above are implemented in Stata with the subcommands `singlelinkage`, `completelinkage`, `averagelinkage`, `wardslinkage` respectively[20]. After the algorithm has run, we can visualize its dendrogram with the command `cluster dendrogram`.

The code below computes the solutions and the corresponding dendrograms for the simulated data using the four algorithms illustrated above. Since the variables involved in the analysis are all numerical, we adopt the Euclidean distance as the dissimilarity measure, which represents the default in Stata[21]. The resulting dendrograms are reported in Figure 2.14. Note that, we performed the analysis on the standardized variables to avoid weighting them differently. This operation is not strictly necessary, but it is a common practice in CA.

```
1  use ch2_SimData, clear

2  /* the following two lines are useful if you already ran the
3     code; they should be executed before running it again */
4  // capture drop *_std
```

[19]These data have been simulated from three four-dimensional multivariate normal distributions with different mean vectors and variance matrices, whose values can be found in the Stata code available on the book's GitHub repository.

[20]Other agglomerative algorithms are also available in Stata (type `help cluster_linkage`).

[21]You can choose a different dissimilarity measure with the option `measure()`.

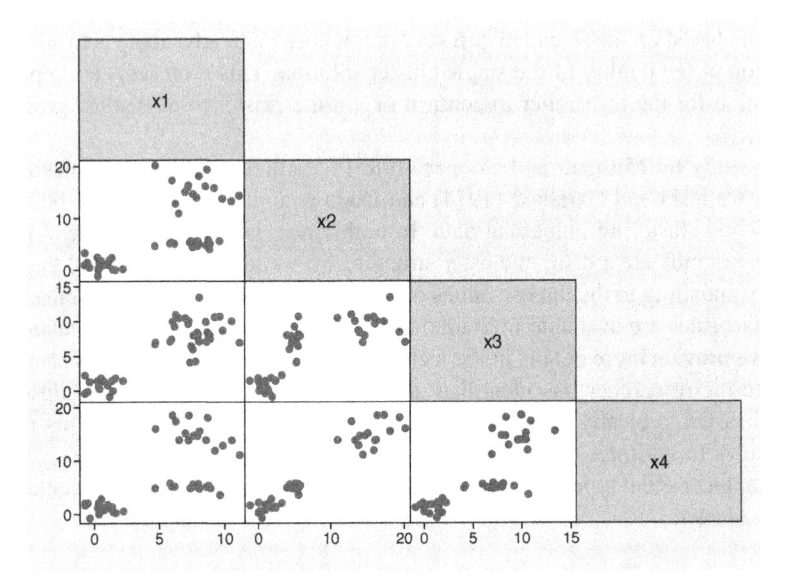

FIGURE 2.13: Simulated data used to illustrate different clustering algorithms.

```
5  // cluster drop _all

6  foreach var of varlist x* {
7    egen `var'_std = std(`var')
8  }

9  cluster singlelinkage *_std, name(single_linkage)
10 cluster dendrogram

11 cluster completelinkage *_std, name(complete_linkage)
12 cluster dendrogram

13 cluster averagelinkage *_std, name(average_linkage)
14 cluster dendrogram

15 cluster wardslinkage *_std, name(ward_linkage)
16 cluster dendrogram
```

As the previous code shows, the `cluster` command allows to specify a `name()` option under which we store the clustering solution in the Stata's memory. This is useful because if we need to use the results later, we can recall them by their names thus avoiding to rerun the whole algorithm.

Next, for each algorithm we can compare the solutions corresponding to different group numbers with the command `cluster stop`. The Caliński-Harabasz and Duda-Hart stopping rules are available with the options `rule(calinski)` (the default) and `rule(duda)` respectively. The code below applies these rules to our

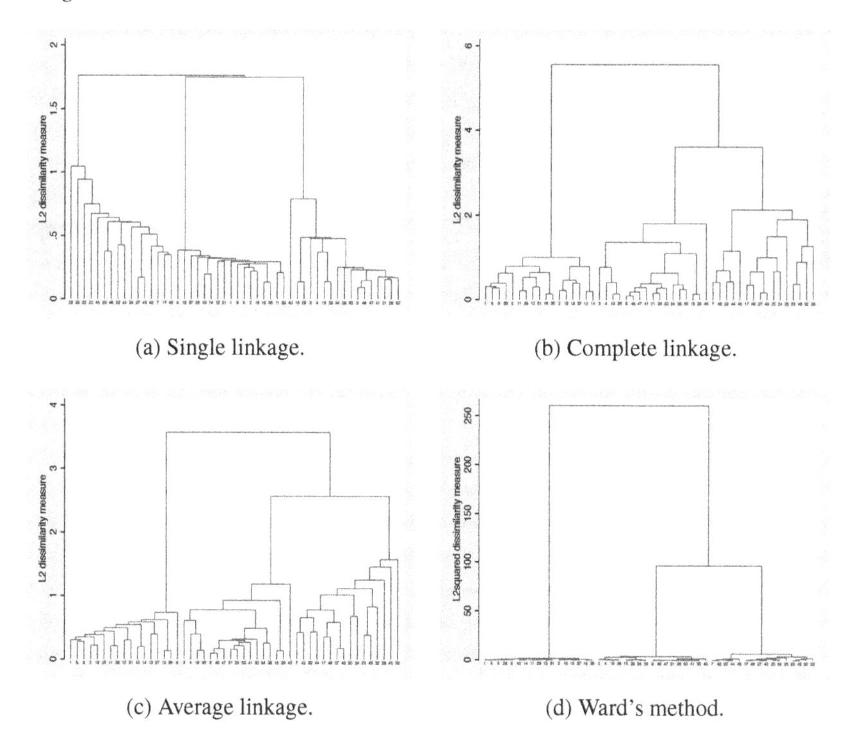

(a) Single linkage. (b) Complete linkage.

(c) Average linkage. (d) Ward's method.

FIGURE 2.14: Dendrograms produced by different agglomerative algorithms for the simulated data shown in Figure 2.13.

example using a number of groups from 2 to 8 (see Figures 2.15 and 2.16 for the results):

```
cluster stop single_linkage, rule(calinski) groups(2/8)
cluster stop complete_linkage, rule(calinski) groups(2/8)
cluster stop average_linkage, rule(calinski) groups(2/8)
cluster stop ward_linkage, rule(calinski) groups(2/8)

cluster stop single_linkage, rule(duda) groups(2/8)
cluster stop complete_linkage, rule(duda) groups(2/8)
cluster stop average_linkage, rule(duda) groups(2/8)
cluster stop ward_linkage, rule(duda) groups(2/8)
```

For all algorithms, the Caliński-Harabasz rule provides the same answer, that is three groups, which also corresponds to the true number of groups (i.e., the number of groups we have generated the data from). The results of the Duda-Hart rule are less easy to be interpreted. A conventional approach for deciding the number of groups based on the Stata's Duda-Hart table is to find one of the largest `Je(2)/Je(1)` values that corresponds to a low `pseudo-T-squared` value with much larger

Number of clusters	Calinski/ Harabasz pseudo-F
2	72.89
3	238.51
4	176.21
5	140.78
6	120.85
7	106.48
8	94.44

(a) Single linkage.

Number of clusters	Calinski/ Harabasz pseudo-F
2	95.02
3	238.51
4	193.97
5	164.79
6	143.49
7	132.17
8	130.03

(b) Complete linkage.

Number of clusters	Calinski/ Harabasz pseudo-F
2	95.02
3	238.51
4	176.21
5	146.25
6	124.68
7	112.67
8	113.90

(c) Average linkage.

Number of clusters	Calinski/ Harabasz pseudo-F
2	95.02
3	238.51
4	193.97
5	165.15
6	152.91
7	143.01
8	139.02

(d) Ward's method.

FIGURE 2.15: Stata results for the Caliński-Harabasz stopping rule applied to the simulated data shown in Figure 2.13.

`pseudo-T-squared` values next to it. According to this approach, again three is the answer obtained for all algorithms.

Even if both stopping rules arrive at the same conclusion, we are not sure that the four algorithms provide the same partitioning of the data. To check this, we now use the `cluster generate` command that append to the dataset a new column containing the cluster memberships (i.e., the number of the cluster to which each observation has been assigned). Then, we compare the classifications using cross tabulations:

```
1 cluster generate single3 = groups(3), name(single_linkage)
2 cluster generate complete3 = groups(3), ///
3   name(complete_linkage)
4 cluster generate average3 = groups(3), name(average_linkage)
5 cluster generate ward3 = groups(3), name(ward_linkage)

6 table single3 complete3
7 table single3 average3
8 table single3 ward3
```

Number of clusters	Duda/Hart Je(2)/Je(1)	pseudo T-squared
2	0.1092	260.94
3	0.8146	3.19
4	0.8571	2.17
5	0.7512	4.97
6	0.8673	1.84
7	0.8944	1.30
8	0.7246	3.80

(a) Single linkage.

Number of clusters	Duda/Hart Je(2)/Je(1)	pseudo T-squared
2	0.2375	99.53
3	0.6840	6.47
4	0.6834	4.17
5	0.7512	4.97
6	0.5090	4.82
7	0.6036	9.19
8	0.3535	3.66

(b) Complete linkage.

Number of clusters	Duda/Hart Je(2)/Je(1)	pseudo T-squared
2	0.2375	99.53
3	0.8146	3.19
4	0.7964	3.32
5	0.8369	2.34
6	0.7512	4.97
7	0.6543	5.81
8	0.6222	4.86

(c) Average linkage.

Number of clusters	Duda/Hart Je(2)/Je(1)	pseudo T-squared
2	0.2375	99.53
3	0.6840	6.47
4	0.6145	9.41
5	0.6834	4.17
6	0.3450	11.39
7	0.5090	4.82
8	0.3535	3.66

(d) Ward's method.

FIGURE 2.16: Stata results for the Duda-Hart stopping rule applied to the simulated data shown in Figure 2.13.

```
9 table single3 truegroup
```

The cross tabulations (not reported here) show that actually the four algorithms provide the same data partition, which also corresponds to the true grouping structure (reported in the dataset as the variable labelled `truegroup`). Figure 2.17 shows the three clusters.

2.3.1.2 Partitional clustering algorithms

Partitional algorithms seek to partition the n observations into a predefined number of non-overlapping groups. Therefore, contrary to hierarchical algorithms, they do not return a tree of solutions, even when they are performed sequentially for a set of distinct cluster numbers. The most popular partitional algorithm is ***K*-means**, an iterative procedure that looks for the cluster centres that minimize the total within-cluster variance. More specifically, given an initial set of centres (usually chosen randomly), the algorithm alternates the following two steps until convergence:

Step 1 – centres update: the cluster centres are calculated using the most recent allocation,

Step 2 – reallocation: each observation is assigned to the cluster centre to which it is closest.

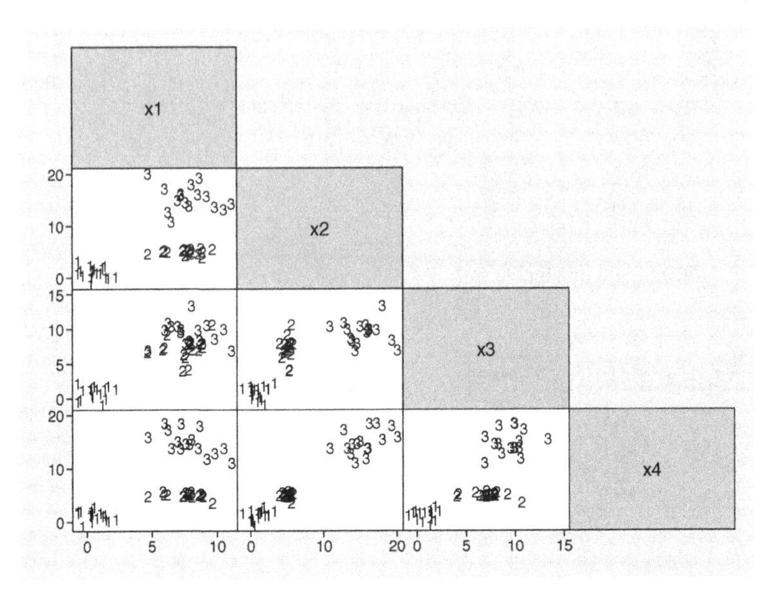

FIGURE 2.17: Simulated data illustrating the agglomerative clustering algorithms showing the best solution (i.e., three clusters, shown in the scatterplots as 1, 2 and 3).

A graphical illustration of the K-means algorithm for some fictitious data is provided in Figure 2.18.

Convergence criteria typically used are either cluster stabilization (i.e., all observations remain in the same group from the previous iteration), or the attainment of a given number of iterations. Unfortunately, the K-means algorithm does not guarantee to find the *global* optimal partition of the data (Everitt and Hothorn, 2011, Section 6.4). Moreover, different choices of the initial cluster centres may lead to different final partitions of the data corresponding to different *local* optimal divisions. To reduce this risk, it is usually suggested to restart the algorithm a certain number of times from different initial points to check that it converges to the same final solution.

As for hierarchical algorithms, K-means requires a criterion for choosing the best number of clusters. Most of the indexes that can be used with hierarchical clustering algorithms can also be used with K-means (a detailed account of these validity indexes can be found in Charrad et al., 2014). In particular, the Caliński-Harabasz stopping rule can also be used with the K-means algorithm[22].

[22]The Duda-Hart rule can't be used with partitional algorithms because it requires a hierarchical structure of the clustering solutions. Thus, the Duda-Hart index is *local* in the sense that it requires only the information regarding the cluster being split, while Caliński-Harabasz is *global* because it uses the information from each group.

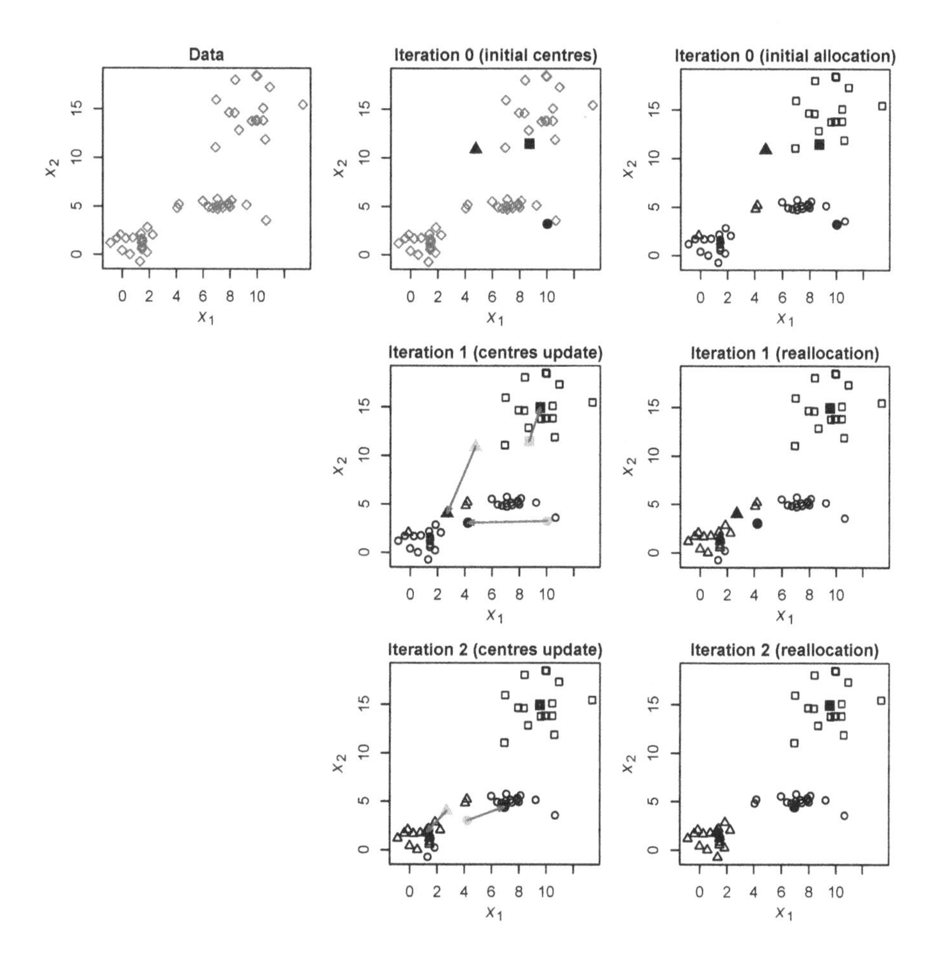

FIGURE 2.18: Graphical presentation of the K-means algorithm for some simulated data. The filled markers correspond to current cluster centres, while those grey shaded indicate the centres from the previous iteration. Grey arrows show the movement of the centres from iteration to iteration. After two iterations the algorithm converges to a stable solution.

Partitional cluster analysis: An example with simulated data using Stata

Using the simulated data shown in Figure 2.13, we can perform a K-means analysis in Stata with the command `cluster kmeans`[23]. The following code chunk performs K-means for a number of clusters from 2 to 8, saving the group memberships in new columns of the dataset, and then it compares the solutions by computing the Caliński-Harabasz index values (not reported here).

```
1  use ch2_SimData, clear

2  capture drop *_std
3  cluster drop _all

4  forvalues g = 2/8 {
5    cluster kmeans *_std, k(`g') start(krandom(301)) ///
6      name(km`g') generate(km`g')
7  }

8  forvalues g = 2/8 {
9    cluster stop km`g'
10 }
```

An examination of the Caliński-Harabasz indexes allows to conclude that the correct number of groups (three) is still recovered. Moreover, the partition found by K-means corresponds to that found by the hierarchical algorithms described above. Figure 2.19 shows the corresponding solution.

2.3.2 Finite mixture models and model-based clustering

Despite their simplicity, the CA methods described in the previous section are not based on a formal statistical model. An alternative approach that goes in this direction is represented by the **finite mixture models** (FMMs).

In a FMM it is assumed that the whole population consists of a number of subpopulations (corresponding to the clusters in the data), with each subpopulation that is assumed to follow a different probability distribution[24]. The distributions corresponding to the subpopulations are called the **mixture components**. Clustering approaches that use FMMs as the modelling framework are usually referred to as **model-based clustering** (see Banfield and Raftery, 1993; Fraley and Raftery, 2002; Dolnicar et al., 2018).

A simple example of a FMM is provided in Figure 2.20, where it is assumed

[23] Stata also includes the `cluster kmedians` command, that allows to perform the same analysis but computing the median of the observations assigned to each group instead of the mean. K-medians algorithm is more robust than K-means towards outliers.

[24] The probability model used for the subpopulations is usually the same (e.g., the normal distribution) but each subpopulation is assumed to have its own set of parameter values (for example, different means and variances).

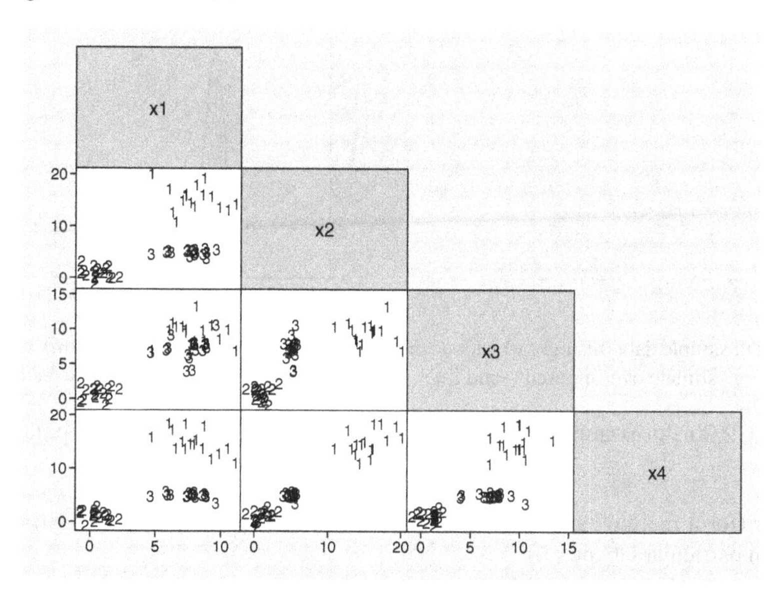

FIGURE 2.19: Simulated data illustrating the K-means partitional algorithm showing the best solution (i.e., three clusters, shown in the scatterplots as 1, 2 and 3).

that the data have been sampled from a population composed of two subpopulations following a normal distribution centred around the values 10 and 18 respectively.

Even if FMMs can use any probability model, mixtures of (multivariate) normal distributions are the most common choice used in practice for continuous quantities. This is because it can be shown that any probability distribution for continuous variables can be approximated to an arbitrary precision by a mixture of normals with enough components (see for example Rossi, 2014). Figure 2.21 shows some examples of the range of possible shapes that can be obtained with a mixture of two normal distributions in one and two dimensions.

A FMM is formally defined as a weighted average of the components distributions. More technically, a FMM for a p-dimensional vector of variables $\boldsymbol{X} = (X_1, \ldots, X_p)$ is defined as

$$f(\boldsymbol{x}) = \sum_{j=1}^{K} \pi_j f_j(\boldsymbol{x}; \boldsymbol{\theta}_j), \tag{2.2}$$

where $f_j(\boldsymbol{x}; \boldsymbol{\theta}_j)$ are the component distributions, each one depending on its own set of parameters $\boldsymbol{\theta}_j$, and $(\pi_1, \ldots, \pi_K)$ are the so called **mixture weights**, which must satisfy the requirements $0 \leq \pi_j \leq 1$ and $\sum_{j=1}^{K} \pi_j = 1$.

The output of an analysis based on a FMM is represented by the estimates of the mixture component parameters as well as the estimates of the mixture weights. Then, using the estimated parameters, the cluster membership are computed. Estimation of all these quantities can be performed using different methods with maximum likelihood, expectation-maximization (EM) and Bayesian inference being the most

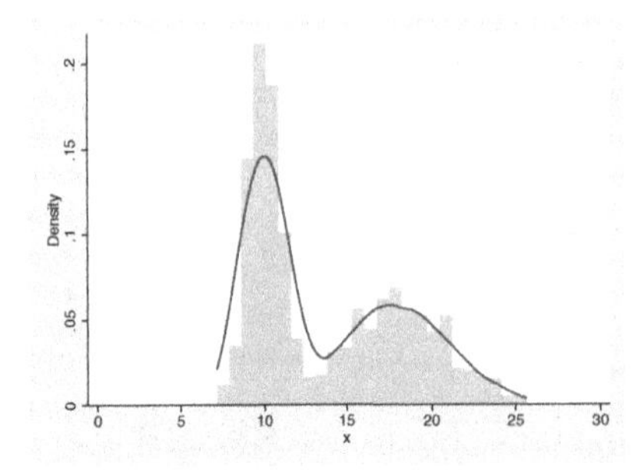

(a) Overall sample data (histogram) with a kernel density estimate overimposed (solid line).

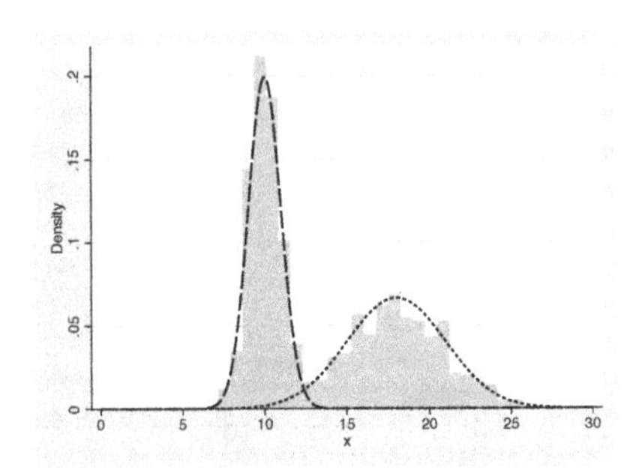

(b) Overall sample data (histogram) together with the generating populations (dashed lines).

FIGURE 2.20: An example of data originating from a mixture of two subpopulations.

popular (for a more detailed presentation of estimation of FMMs see Everitt et al., 2011, in particular Chapter 6).

Similar to K-means, FMMs also require the number of clusters K to be fixed in advance. Popular methods for selecting the best FMM are **likelihood ratio tests** (LRTs) and **information criteria**.

LRTs can be used for testing the null hypothesis that the FMM includes a given number of components, that is $H_0 : K = K_0$, against the alternative hypothesis that it involves a larger value, that is $H_1 : K = K_1$, with $K_1 > K_0$[25]. The main drawback of LRTs for FMMs is that, due to technical difficulties in the calculation of the likelihood, they tend to overestimate the actual number of groups K.

Information criteria are the most common approach for comparing different mixture models. The two criteria predominantly used in mixture modelling are the **Akaike's information criterion** (AIC) and the **Bayesian information criterion** (BIC). They both keep into account the lack of fit of the data and the model complexity. The model with the lowest information criteria values is preferred.

Here, we just scratched the surface of FMMs because they can be extended in many directions such as the possibility to let the component parameters as well as the mixture weights to depend upon a set of covariates. The technical appendix at the end of the chapter provides more details about FMMs, but for a complete presentation we suggest the monographs by McLachlan and Peel (2000) and Dolnicar et al. (2018).

Finite mixture models: An example with simulated data using Stata

Starting from the release 15, Stata allows to fit mixtures models via the prefix `fmm`. It is possible to include covariates for both the component distributions and the mixture weights[26]. To see the list of available models, type `help fmm_estimation`. For example, to fit a mixture of two normal distributions for a single continuous variable

[25]In practice the LRT approach is usually implemented by adding one component at a time, so that $K_1 = K_0 + 1$, and one keeps adding components until the null hypothesis is not rejected.

[26]Stata's `fmm` models the mixture weights using a multinomial logistic regression model.

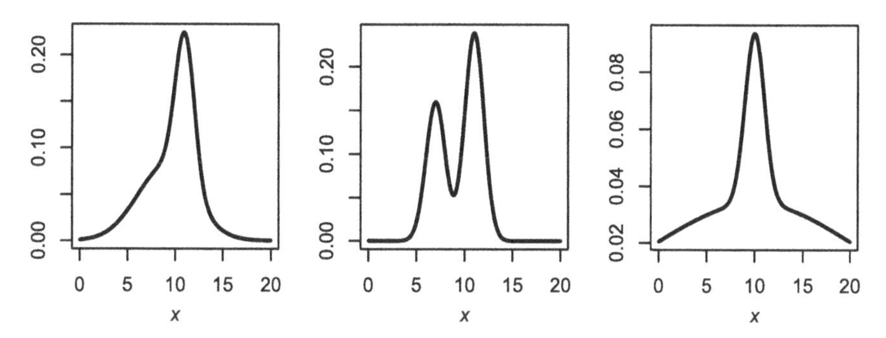

(a) Mixtures of two univariate normal distributions.

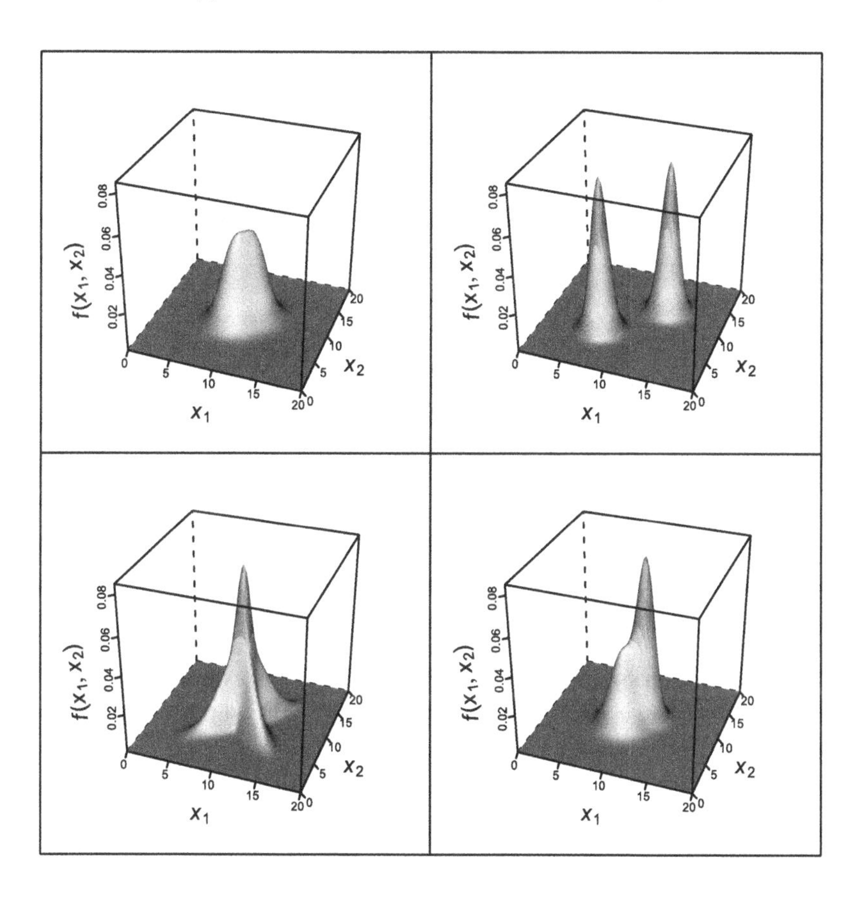

(b) Mixtures of two bivariate normal distributions.

FIGURE 2.21: Examples of finite mixtures of two normal distributions.

Model	Obs	ll(null)	ll(model)	df	AIC	BIC
fmm2	500	.	-4696.632	17	9427.264	9498.913
fmm3	500	.	-3469.94	26	6991.88	7101.46
fmm4	500	.	-3463.934	35	6997.869	7145.38
fmm5	500	.	-3452.817	44	6993.633	7179.076
fmm6	500	.	-3447.038	53	7000.075	7223.449
fmm7	500	.	-3433.418	62	6990.837	7252.142
fmm8	500	.	-3434.672	71	7011.344	7310.581

FIGURE 2.22: Results for the estimation of the FMMs using Stata's `fmm` for simulated data.

`y` we should use `fmm 2: regress y`. Similarly, if we are interested in fitting a mixture of three Bernoulli distributions for a single dichotomous (i.e., 0/1) variable, we need to run `fmm 3: logit y`. AIC and BIC values can be retrieved with the `estat ic` postestimation command.

Using the same data generating process as in Figure 2.13, but increasing the sample size to 500[27], the following code fits a mixture model for the variables `x1-x4` using normal components with K going from 2 to 8. Next, we compute the AIC and BIC values for each model and compare the results in a summary table, which is reported in Figure 2.22.

```
1 use ch2_SimData2, clear
2 forvalues K = 2/8 {
3   quietly fmm `K': regress (x1-x4)
4   estimates store fmm`K'
5 }
6 estimates stats fmm*
```

Both AIC and BIC indicate $K = 3$ as the optimal solution. Then, we ask for the estimates of both the component means[28] and the mixture weights using the `estat lcmean` and `estat lcprob` postestimation commands. Finally, we calculate the so called **class posterior probabilities**, which provide the estimated probabilities for each observation to belong to the different groups[29]. The outputs for the example above are shown in Figures 2.23 and 2.24.

[27] This is needed to let the algorithm have enough information to estimate the many parameters of the model. These data are available in the file `ch2_SimData2.dta`.

[28] To get also the variances, type `estimates restore fmm3`.

[29] Note that the last code line uses Mata, the advanced matrix language available in Stata (Gould, 2018) together with the `which` function that you can get by installing the `plssem` package described in the next chapters.

```
1  estimates restore fmm3

2  estat lcmean
3  estat lcprob

4  predict classpr*, classposteriorpr
5  generate fmm3 = .
6  mata: st_store(., "fmm3", .,
7    which(st_data(., "classpr*"), "max"))
```

```
. estat lcmean

Latent class marginal means                     Number of obs     =        500

------------------------------------------------------------------------------
             |            Delta-method
             |     Margin   Std. Err.      z    P>|z|     [95% Conf. Interval]
-------------+----------------------------------------------------------------
1            |
          x1 |   1.022167   .1091683     9.36   0.000     .8082009    1.236133
          x2 |   .9221559    .086716    10.63   0.000     .7521957    1.092116
          x3 |   .9245347   .0932485     9.91   0.000      .741771    1.107298
          x4 |   1.091596   .0862536    12.66   0.000     .9225416    1.260649
-------------+----------------------------------------------------------------
2            |
          x1 |   7.007023   .0938842    74.63   0.000     6.823014    7.191033
          x2 |   4.978933   .0290964   171.12   0.000     4.921906    5.035961
          x3 |   6.894333   .0916708    75.21   0.000     6.714661    7.074005
          x4 |   5.024879   .0336104   149.50   0.000     4.959004    5.090754
-------------+----------------------------------------------------------------
3            |
          x1 |   8.111524    .116786    69.46   0.000     7.882628     8.34042
          x2 |   14.93361   .1818318    82.13   0.000     14.57722    15.28999
          x3 |    10.1373   .1170033    86.64   0.000      9.90798    10.36662
          x4 |   15.01151   .1463245   102.59   0.000     14.72472     15.2983
------------------------------------------------------------------------------

. estat lcprob

Latent class marginal probabilities             Number of obs     =        500

--------------------------------------------------------------
             |            Delta-method
             |     Margin   Std. Err.     [95% Conf. Interval]
-------------+------------------------------------------------
       Class |
          1  |       .214   .0183414      .1802432    .2521344
          2  |       .498   .0223605      .4543014    .5417292
          3  |       .288   .0202512      .2499868    .3292559
--------------------------------------------------------------
```

FIGURE 2.23: Parameter estimates of the FMMs using Stata's `fmm`.

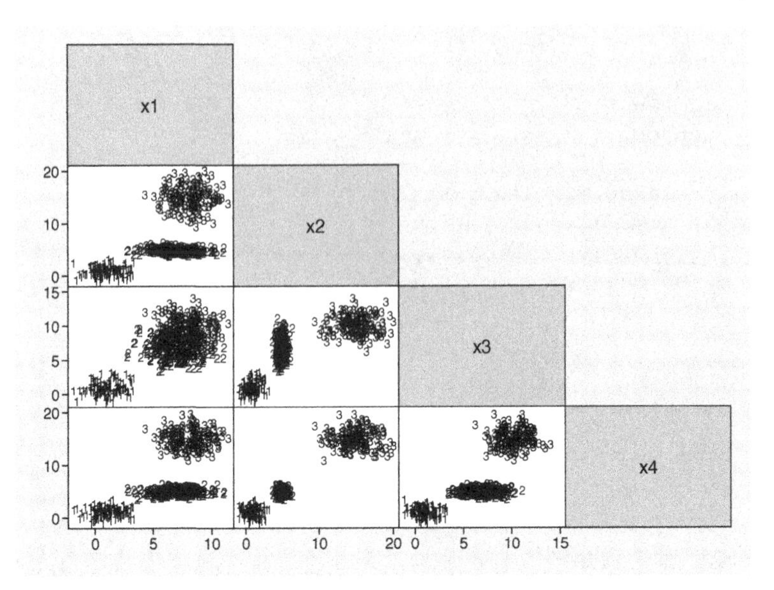

FIGURE 2.24: Simulated data illustrating the `fmm` command for fitting FMMs. The allocation for $K = 3$ groups are indicated in the scatterplots as 1, 2 and 3.

2.3.3 Latent class analysis

Finite mixture modelling can be seen as a form of latent variable analysis where the latent variable, represented by the cluster indicator, is categorical. Latent variable methods provide an important tool for the analysis of multivariate data and have become of primary importance in most social sciences. According to some authors, latent variable models can be unified under a general framework depending on the nature of the latent and manifest variables involved in the analysis. For example, when the latent and manifest variables are both numerical, the model is called the *linear factor model*, while if the latent variables are numerical but the manifest variables are categorical, then the model is called *latent trait analysis* (for a complete presentation of this unifying framework see Skrondal and Rabe-Hesketh, 2004 and Bartholomew et al., 2011). A specific approach falling under this general classification is **latent class analysis** (LCA). In LCA both manifest and latent variables are assumed to be categorical. More specifically, the response is modelled as a multivariate categorical variable, as it is the case for example in public opinion surveys or in consumer behaviour studies, while the latent variable is categorical nominal (i.e., unordered) representing the class membership. The additional requirement in LCA is that, conditionally upon the latent variable value, the manifest variables are assumed to be statistically independent. This is usually called **conditional** (or **local**) **independence assumption**[30]. Extensions of the basic model allow also for the possibility to

[30]Technically, the conditional independence assumption implies that the jth component distribution $f_j(\boldsymbol{x}_i;\boldsymbol{\theta}_j)$ in equation (2.2) can be written as a product over the p response variables.

include covariates for predicting the class membership. For a book-length treatment of LCA models we suggest to refer to the monographs by McCutcheon (1987) and Hagenaars and McCutcheon (2002).

Since LCA is a special type of mixture modelling where the response variables are assumed to be categorical and conditionally independent from each other given the class membership (and the class-specific covariates), all the tools we presented for FMMs (EM algorithm, information criteria, etc.) may be applied directly. Finally, if you have a software package that is able to fit (general enough) mixture models, you can also use it to estimate LCA models. Stata allows to perform LCA through `gsem`, the powerful command for generalized structural equation modelling, with the specification of the `lclass()` option.

2.4 Path Analysis

Classical linear models define a linear relationship between *one* dependent and one or more independent variables, which is thus specified through a single equation[31]. A potential failure of the linear regression assumptions is represented by the so called *endogeneity* issue, that is the situation in which some of the X variables are correlated with the error term[32]. Endogeneity may occur as a consequence of at least one of the following reasons:

- *Omitted variables*, which takes place when we cannot include in the model one or more relevant variables usually because of data unavailability or scarce knowledge of the problem. An example of omitted variable bias is when we try to explain the weekly wage of individuals with their characteristics, such as ability and years of education. Since ability is typically unobserved, we may choose to include only education. However, since ability and education are very likely to be correlated, this originates the bias.

- *Measurement error*, which occurs when we are not able to measure precisely the actual values of an independent variable[33], but we only collect inexact values. In other words, the independent variable measures contain a random error. This is a common situation in practice and a notable example is IQ score as a measure of ability. The bias in this case comes from the correlation between the observed (inexact) version of the independent variable (i.e., IQ) and the error term.

- *Simultaneity*, which happens when one or more independent variables are determined *jointly* with the dependent variable. In this case, the equation specifying the independent variable X_j also contains Y as a predictor, and this feedback effect is at the base of the bias. This situation arises very frequently, but not

[31] For a review of classical linear regression analysis see Section A.2.

[32] Endogeneity is the violation of the exogeneity assumption; see Section A.2 for more details.

[33] Measurement errors may also occur in the dependent variable, but this is not our focus here.

exclusively, in economics because most common economic quantities (labour, products, money, etc.) are the result of matching demand and supply in a given market.

If any of the explanatory variables is endogenous, then OLS estimates are biased and inconsistent[34]. In this section we elaborate on the simultaneity issue by generalizing single-equation linear models to the case of multiple equations, that is to the case where we have more than one dependent variable that may appear both on the left and right hand sides of the equations. These models are called **path analysis** (PA) **models** in the social sciences and **simultaneous equation models** in econometrics. For a detailed practical illustration of PA models, we suggest you to see Schumacker and Lomax (2016), Chapter 5, and Kline (2016), in particular Chapters 6 and 7, while for a technical treatment you can refer to the classic textbook by Bollen (1989), Chapter 4.

Path models postulate a set of relationships among a set of observed variables. In PA variables are classified as *endogenous* or *exogenous*. The former are determined within the model (i.e., they appear on the left hand side of the structural equations), while the latter are predetermined, which means that they are influenced by other variables not included in the model. On top of these two categories, we also have random errors which represent the unexplained part of the endogenous variables (assumed to be uncorrelated with the exogenous variables). A common way to represent graphically the structural relationships in a PA model is through a **path diagram**. In these diagrams observed variables are represented using squares or rectangles, endogenous variables have single-headed arrows pointing at them, while double-headed arrows are used to highlight associations (if any) between exogenous variables. A further distinction in PA is between *recursive* and *non-recursive* models. Recursive models are those in which the relationships between the endogenous variables are unidirectional, that is there are no feedback loops, and error terms are uncorrelated. In non-recursive models, instead, the relationships between the endogenous variables are bidirectional (i.e., the diagram contains feedback loops) or disturbances may be correlated. In general, non-recursive models are more difficult to analyse, because the presence of feedback loops between the endogenous variables implies a dynamic process that is operating underneath the system of equations. A simple example of a recursive path diagram for a hypothetical analysis with two endogenous variables, y_1 and y_2, and three exogenous variables, x_1, x_2 and x_3, is shown in Figure 2.25, while an example of a non-recursive model is provided in Figure 2.26.

The equations corresponding to the PA model in Figure 2.25 are

$$y_1 = \alpha_1 + \gamma_{11} x_1 + \zeta_1 \qquad (2.3)$$

$$y_2 = \alpha_2 + \beta_{21} y_1 + \gamma_{22} x_2 + \gamma_{23} x_3 + \zeta_2, \qquad (2.4)$$

where y_1 and y_2 are the observed endogenous variables, x_1, x_2 and x_3 are the observed exogenous variables, α_1 and α_2 are the structural intercepts, β_{21} is the coefficient relating the endogenous variables, γ_{11}, γ_{22} and γ_{23} are the coefficients relating

[34]We remind that an estimator is said to be consistent if its value gets closer and closer to the parameter value as the sample size increases indefinitely.

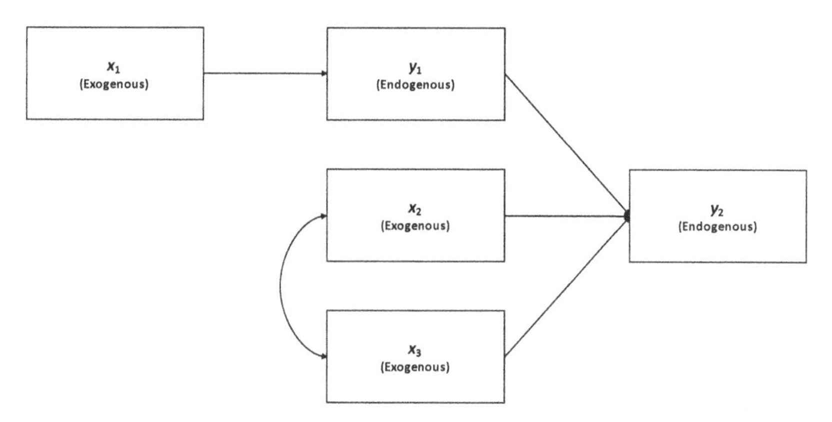

FIGURE 2.25: Path diagram for a hypothetical recursive path model with two endogenous variables, y_1 and y_2, and three exogenous variables, x_1, x_2 and x_3.

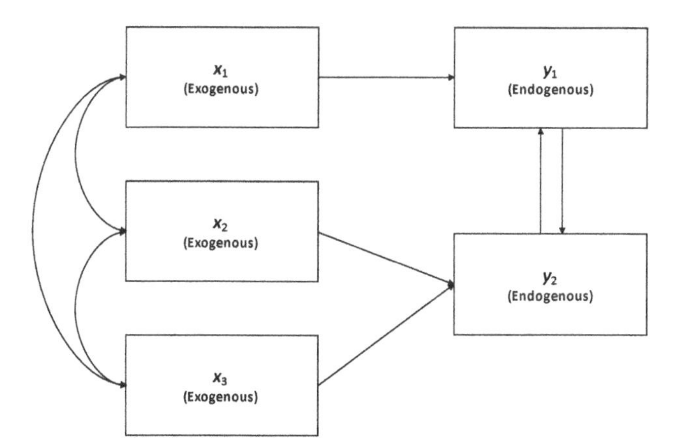

FIGURE 2.26: Path diagram for a hypothetical non-recursive path model with two endogenous variables, y_1 and y_2, and three exogenous variables, x_1, x_2 and x_3.

endogenous variables to the exogenous ones and ζ_1 and ζ_2 are the error terms, that are independent of the exogenous variables, but may or may not be correlated to each other. The equations above are said to be in *structural form* because they represent the structural relationships encoded by the model. Parameters in PA models can be broadly distinguished into three classes:

- *free* parameters, which must be estimated using the sample data; in the example above, these are α_1, γ_{11}, α_2, β_{21}, γ_{22} and γ_{23}.
- *fixed* parameters, that is parameters set to a predetermined value; the typical case is to set a parameter value to zero to indicate the absence of the relationship, but they could be fixed at any other value. In the example above, β_{11}, γ_{12}, γ_{13} and γ_{21} are fixed parameters because their values is set to zero (they do not appear in the equations).
- *constrained* parameters, when we require some of the coefficients to satisfy a given set of constraints. For example, we would have constrained parameters if we force coefficients γ_{22} and γ_{23} in the equations above to be the same[35].

An alternative approach to describe the same model is with the so called *reduced form*, with which we rearrange the equations so that the endogenous variables appear on the left hand sides only and the exogenous variables on the right hand sides. For the example above, the reduced form equations correspond to

$$\begin{aligned} y_1 &= \alpha_1 + \gamma_{11}x_1 + \zeta_1 \\ &= \pi_{01} + \pi_{11}x_1 + \zeta_1^* \\ y_2 &= (\alpha_2 + \beta_{21}\alpha_1) + (\beta_{21}\gamma_{11})x_1 + \gamma_{22}x_2 + \gamma_{23}x_3 + (\beta_{21}\zeta_1 + \zeta_2), \\ &= \pi_{02} + \pi_{21}x_1 + \pi_{22}x_2 + \pi_{23}x_3 + \zeta_2^*, \end{aligned}$$

where $\pi_{01} = \alpha_1$, $\pi_{11} = \gamma_{11}$, $\zeta_1^* = \zeta_1$, $\pi_{02} = \alpha_2 + \beta_{21}\alpha_1$, $\pi_{21} = \beta_{21}\gamma_{11}$, $\pi_{22} = \gamma_{22}$, $\pi_{23} = \gamma_{23}$ and $\zeta_2^* = \beta_{21}\zeta_1 + \zeta_2$. The coefficients π_{01} and π_{02} are the reduced form intercepts, π_{11}, π_{21}, π_{22} and π_{23} are reduced form slopes, while ζ_1^* and ζ_2^* are the reduced form error terms. As we see, the reduced form equations correspond to a system of linear regression models with correlated error terms[36].

One of the first issue to tackle in PA is checking whether the model is identified. In general, we say that a model is identified if it is possible to uniquely determine its parameters from the model structure and the sample data. For example, suppose that according to our model we know that $\mu = \theta + \tau$, where μ is the mean of an observed variable that we can estimate using the sample data. If there are no further information on θ and τ, there is no way to figure out uniquely their respective values. In other words, there are infinite many combinations of values for θ and τ that are all consistent with the model's structure, and we have no clue about which one to pick. In this case, we say that the model is non-identified. Identification is a highly

[35] In this brief presentation we skip the discussion of other parameters of the model, that is the variances and covariances between the exogenous variables and between the error terms.

[36] In econometrics these models are known as *seemingly unrelated regression* (SUR) models.

desirable property for any statistical model, and so it is important to check that it holds for PA models as well. However, since identification in PA is a very technical issue, we do not provide any further detail here and suggest you to see Bollen (1989), Chapter 4, for a comprehensive treatment.

After checking that the model is identified, we proceed to estimate its parameters. PA models in reduced form can be easily estimated using OLS, even if more efficient estimators are available. However, structural parameters are usually of interest for which OLS provide generally biased and inconsistent estimates[37]. Therefore, alternative approaches must be used to fit PA models in structural form. Standard estimation procedures in PA involve the minimization of the discrepancy between between sample variances and covariances and those implied by the model (see the technical appendix at the end of this chapter for more details). Note that this approach is similar to OLS in regression analysis, with the difference that regression coefficients are estimated by minimizing the sum of squared residuals, that is the differences between observed and predicted responses, while in PA the deviations that are minimized are those between observed and theoretical variances and covariances. The two most common discrepancy functions used in PA are[38]:

- *Maximum likelihood* (ML), the default method in PA, is based on the assumption that the endogenous and exogenous variables are jointly normally distributed. Even if ML estimates may be biased in small samples, they have appealing large sample properties, in particular they are consistent, asymptotically efficient and asymptotically normally distributed[39]. These results allow to compute approximate confidence intervals and tests based on the familiar normal-based theory. Finally, ML estimates are both *scale invariant* and *scale free*. Scale invariance means that the ML discrepancy function does not change if we modify the scale of one or more of the observed variables. Scale freeness is a similarly property but regarding the parameter estimates (instead of the discrepancy function value).

- *Weighted least squares* (WLS) focuses on the weighted squared differences between sample variances and covariances and those implied by the model. Different versions are available depending on the type of weights used. In particular, if the differences are not weighted, we get the so called *unweighted least squares* (ULS) estimation, while if the differences are weighted using the sample variances and covariances, we get the *generalized least squares* (GLS) approach.

[37] An exception to this statement is for recursive models with uncorrelated error terms across the different equations in the system.

[38] In econometrics, estimation methods for simultaneous equation systems are usually divided in single-equation (or limited information) methods and system (or full information) methods. In single-equation methods each equation is estimated individually, while in system methods all equations are estimated simultaneously. Examples of the former are indirect least squares and two-stage least squares (2SLS), while maximum likelihood is the most notable example of the latter. For more details about these estimation approaches see Greene (2018).

[39] We remind that an "asymptotic" property is one that holds when the sample size is assumed to grow up to infinity. In practice, infinity does not exist and so it is hard to say how large a sample should be to consider the property as attained. Usually, a number of observations of at least 50–100 is considered enough.

Even if ULS is a very simple and intuitive approach, it does not provide the most efficient estimates and moreover it is neither scale invariant nor scale free. On the other side, under the assumption of multivariate normality, the GLS method has the same asymptotic properties of ML and it is both scale invariant and scale free.

The discrepancy functions of both these approaches are complicated non-linear functions of the model's parameters, for which no closed-form solutions are available. Thus, they must be minimized using iterative numerical procedures such as the Newton–Raphson algorithm. Some care must be put in fixing the tuning parameters of the algorithm, because in some cases it shows difficulties in achieving converge.

Once the parameter estimates are available, we proceed to interpret them. The standard terminology used in PA (as well as in SEM) distinguishes between:

- *Direct effects*, which represent the effect that each endogenous and exogenous variables exert on the endogenous variables in the model without the mediation of any intermediate variable. For the model with equations (2.3) and (2.4), β_{21} represents the direct effect of y_1 on y_2, γ_{11} represents the direct effect of x_1 on y_1, while γ_{22} and γ_{23} provide the direct effects of x_2 and x_3 on y_2 respectively. Since these are merely coefficients of linear regression models, they can be interpreted in a similar way, that is as the increase in the average value of the endogenous variable on the left hand side for a unit increase in the value of the endogenous or exogenous variable on the right hand side, assuming all the rest is fixed.

- *Indirect effects*, which represent the effect of an independent variable (endogenous or exogenous) on an endogenous variable that is mediated by the effect of at least another variable. For our hypothetical example, there is no x_1 term in the second equation for y_2, so the direct effect of x_1 on y_2 is zero. However, x_1 has an indirect effect on y_2 through the mediation of y_1. Indirect effects are typically computed as products of direct effects. Indeed, in our example the indirect effect of x_1 on y_2 is given by $(\beta_{21} \times \gamma_{11})$[40].

- *Total effects*, that correspond to the sum of the direct and indirect effects.

The easiest way to fit PA models in Stata is using the powerful `sem` command, which provides a general tool for estimating structural equation models. It is not our intention to provide here a detailed presentation of `sem`, but we illustrate its basic features through a simple example we downloaded from the Stata portal of the UCLA Institute for Digital Research and Education[41]. These data, available in the `ch2_hsb2.dta` file accompanying this book, come from a national representative survey of high school seniors called High School and Beyond conducted by the National Center for Education Statistics (NCES)[42]. The data are a random sample of 200

[40] Total effects of exogenous variables correspond to the reduced-form coefficients (see the technical appendix for more details). Therefore, since in the example there is no direct effect of x_1 on y_2, the indirect effect is equal to the corresponding reduced-form coefficient.

[41] `https://stats.idre.ucla.edu/stata/`.

[42] `https://nces.ed.gov/surveys/hsb/`.

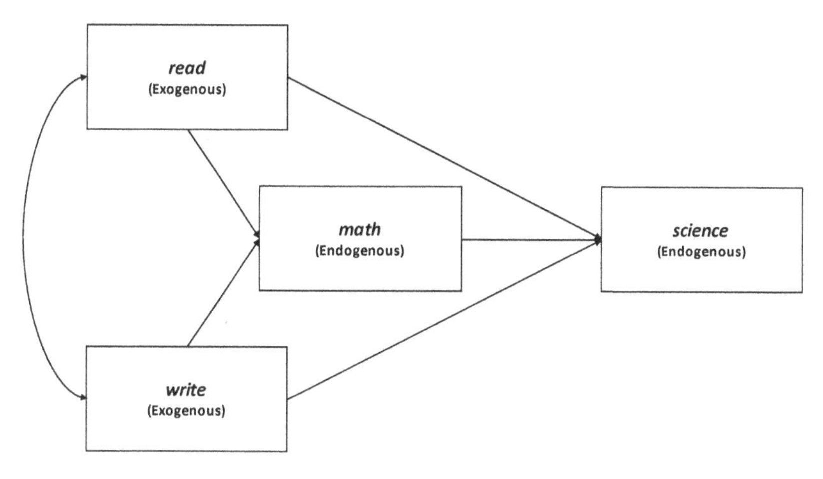

FIGURE 2.27: Path diagram for the path model described in the text that uses the HSB2 dataset.

students from the original dataset and contain information on student achievements in reading, writing, math, science and social sciences. The PA model we consider here is

$$\begin{aligned} \texttt{math} &= \alpha_1 + \gamma_{11}\,\texttt{read} + \gamma_{12}\,\texttt{write} + \zeta_1 \\ \texttt{science} &= \alpha_1 + \beta_{21}\,\texttt{math} + \gamma_{21}\,\texttt{read} + \gamma_{22}\,\texttt{write} + \zeta_2, \end{aligned}$$

whose path diagram is shown in Figure 2.27. This is a recursive model, so identification is not an issue. We estimate it using ML, the default in `sem`[43].

The syntax for `sem` requires to specify the individual equations by enclosing each within parentheses:

```
1 use ch2_hsb2, clear
2 sem (math <- read write) (science <- math read write)
```

Results, reported in Figure 2.28, show that reading and writing scores have a significant positive direct effect on both math and science achievement, with the latter being smaller than the former (we are allowed here to compare the coefficients because all scores are on the same scale). Similarly, the math score exerts a significant positive effect on science achievement. Equation-level goodness-of-fit statistics are available with the `estat eqgof` postestimation command (results not reported here), which provides an R-squared of approximately 0.52 and 0.5 for the

[43]Other estimation methods are available through the `method` option. Alternatively, the `reg3` command with option `sure` would provide the same results. `reg3` is a general interface available in Stata for fitting systems of equations. The command you should run here is `reg3 (math read write) (science math read write), sure`.

```
Endogenous variables

Observed:  math science

Exogenous variables

Observed:  read write

Fitting target model:

Iteration 0:   log likelihood = -2768.7045
Iteration 1:   log likelihood = -2768.7045

Structural equation model                       Number of obs     =        200
Estimation method  = ml
Log likelihood     = -2768.7045

------------------------------------------------------------------------------
             |                 OIM
             |      Coef.   Std. Err.      z    P>|z|     [95% Conf. Interval]
-------------+----------------------------------------------------------------
Structural   |
  math       |
        read |   .4169486   .0560586     7.44   0.000     .3070759    .5268214
       write |   .3411219   .0606382     5.63   0.000     .2222731    .4599706
       _cons |   12.86507   2.800378     4.59   0.000     7.376428    18.35371
  -----------+----------------------------------------------------------------
  science    |
        math |   .3190094   .0759047     4.20   0.000      .170239    .4677798
        read |   .3015317   .0679912     4.43   0.000     .1682715     .434792
       write |   .2065257   .0700532     2.95   0.003     .0692239    .3438274
       _cons |   8.407353   3.160709     2.66   0.008     2.212476    14.60223
-------------+----------------------------------------------------------------
  var(e.math)|   42.32758   4.232758                      34.79391    51.49245
var(e.science)|  48.77421   4.877421                      40.09314    59.33492
------------------------------------------------------------------------------
LR test of model vs. saturated: chi2(0)   =      0.00, Prob > chi2 =      .
```

FIGURE 2.28: Path analysis for the model shown in Figure 2.27 fitted using the `sem` command in Stata.

two equations respectively. Finally, we can get the direct, indirect and total effects with the `estat teffects` postestimation command, whose values are reported in Figure 2.29. For more details on how to use Stata's `sem` for fitting PA models, we suggest you to read Chapter 2 of Acock (2013).

2.5 Getting to Partial Least Squares Structural Equation Modelling

A common situation in social and behavioural sciences is to collect data by administering a questionnaire to individuals. Typically, items in the questionnaire are designed to measure a set of concepts that cannot be observed directly because of their complicated and intangible nature (the most notable example in psychology is intel-

```
Direct effects
------------------------------------------------------------------------------
             |                 OIM
             |      Coef.   Std. Err.      z    P>|z|     [95% Conf. Interval]
-------------+----------------------------------------------------------------
Structural   |
  math       |
        read |   .4169486   .0560586     7.44   0.000     .3070759    .5268214
       write |   .3411219   .0606382     5.63   0.000     .2222731    .4599706
  -----------+----------------------------------------------------------------
  science    |
        math |   .3190094   .0759047     4.20   0.000      .170239    .4677798
        read |   .3015317   .0679912     4.43   0.000     .1682715     .434792
       write |   .2065257   .0700532     2.95   0.003     .0692239    .3438274
------------------------------------------------------------------------------
```

```
Indirect effects
------------------------------------------------------------------------------
             |                 OIM
             |      Coef.   Std. Err.      z    P>|z|     [95% Conf. Interval]
-------------+----------------------------------------------------------------
Structural   |
  math       |
        read |          0  (no path)
       write |          0  (no path)
  -----------+----------------------------------------------------------------
  science    |
        math |          0  (no path)
        read |   .1330105   .0363514     3.66   0.000      .061763     .204258
       write |   .1088211   .0323207     3.37   0.001     .0454736    .1721686
------------------------------------------------------------------------------
```

```
Total effects
------------------------------------------------------------------------------
             |                 OIM
             |      Coef.   Std. Err.      z    P>|z|     [95% Conf. Interval]
-------------+----------------------------------------------------------------
Structural   |
  math       |
        read |   .4169486   .0560586     7.44   0.000     .3070759    .5268214
       write |   .3411219   .0606382     5.63   0.000     .2222731    .4599706
  -----------+----------------------------------------------------------------
  science    |
        math |   .3190094   .0759047     4.20   0.000      .170239    .4677798
        read |   .4345423   .0627773     6.92   0.000      .311501    .5575836
       write |   .3153468   .0679059     4.64   0.000     .1822536    .4484399
------------------------------------------------------------------------------
```

FIGURE 2.29: Estimated direct, indirect and total effects for the path model shown in Figure 2.27.

ligence). In principle, the easiest and most desirable case is when each concept is measured by a single item. However, often this is not possible in practice because the latent feature we are trying to quantify involves several dimensions. A popular approach consists in performing an exploratory factor analysis (EFA) on the items and choose an appropriate number of factors to represent the information underlying the original items. Then, the researcher computes so called *factor scores*, which represent the projection of the item values for each individual on the factor space identified from the EFA. These scores may be used as individual measures in applied settings or they can be included in subsequent analyses such as regression or path analysis. However, different approaches are available to compute the scores, each one with its own pros and cons. In this section we provide a brief summary of these approaches. More information can be found in Grice (2001); DiStefano et al. (2009) and Hair et al. (2018a).

Factor score computation methods are usually classified in two categories: *non-refined* (or *coarse*) and *refined*. The former are simple procedures that are easy to compute and interpret, while the latter are more complex but they also provide more accurate results. The most frequently used non-refined methods are:

- *Sum scores by factor*, also known as *summated scale*, involves summing or averaging items with high loadings[44] from the factor analysis. Sometimes items with a negative loading are included with a reversed sign. This process is known as *reverse scoring* and its aim is to prevent the values of the items to cancel out when they are summed. The average is usually preferred over the sum of the items because it preserves the scale of the items. Moreover, the average allows to get more consistent values across the sample in presence of missing values unless these are first imputed. Despite their simplicity, sum scores present a series of drawbacks. In particular, they assign the same weight[45] to all items and they do not account for the correlation among the items themselves. Finally, sum scores are not necessarily orthogonal (i.e., uncorrelated) and they should undergo an extensive assessment of their reliability and validity.

- *Sum scores using a cut-off value*, which include only items with loading values above a given threshold. This approach allows to consider an item's relationship to the factor, but it clearly wastes a lot of information and still suffers from the other drawbacks of the raw sum score. In addition, the cut-off value to use is subjective and no clear guidance is available in the literature.

- *Weighted sum scores*, which involves weighting the item values with the corresponding loading. One advantage of this method is that items with the highest loadings will have the largest impact on the factor score. However, due to the subjective choice of extraction model and rotation method, factor loadings may not be an accurate representation of the differences among factors.

[44] We remind that in factor analysis, as in PCA, loadings provide the correlation of the different items with each of the extracted factors.

[45] This concept is known as τ-*equivalence*.

Refined methods are more sophisticated and generate scores that are standardized (i.e., having mean zero e unit standard deviation). Different approaches are available in this case too, all of which compute the scores as linear combinations of the item values. Moreover, they respect the choice made in the EFA regarding the relationships between factors, that is they are uncorrelated when the solution uses orthogonal factors, or correlated when the solution is oblique. The most common refined method for computing scores after an EFA is the *regression* (or *Thompson*) *method*, which computes the factor scores using a linear regression model where the independent variables are the standardized item values and the dependent variables are the factor scores. It can be shown (see for example Johnson and Wichern, 2014, Section 9.5) that the regression coefficients are computed taking into account both the correlation between factors and items (through loadings), but also the item correlations.

Refined methods are generally preferable for computing scores both from the theoretical and practical point of views. Moreover, simulation studies have shown that non-refined methods perform poorly especially in presence of missing values (see for example Estabrook and Neale, 2013).

To conclude, refined methods provide a more reliable and theoretical grounded approach to generate factor scores. EFA represents the most popular choice for computing such scores, but other multivariate statistical techniques can be used. A more structured approach for computing scores is provided by so called **composite variables**, which are obtained as a weighted sum of indicators. As we already stated in Chapter 1, one of the most popular method that is based on the notion of composite variables is *partial least squares structural equation modelling*, that we introduce in the next chapter.

2.6 Summary

In this chapter we reviewed some building blocks that will be needed in the rest of the book. In particular, the bootstrap, PCA, cluster analysis and mixture models, together with correlation and linear regression analysis reviewed in Appendix A, are at the core of the methods we will present in the next chapters.

Appendix: R Commands

The aim of the R appendices is to show you how to perform the same analyses we discussed in the chapter using the R software. In this book we assume that you already have a basic familiarity with R. In particular, we take for granted that you know the main data structures available in R (i.e., vectors, factors, matrices, lists and data frames) and the basics of working with functions. In case you do not possess such skills, you can find tons of free material online if you simply type "learn R" in a search engine. One very handy introduction to R we feel to suggest is the manual by Grolemund (2014), a free online version of which is available at https://rstudio-education.github.io/hopr/. For a complete list of resources you can also look at https://education.rstudio.com/learn/.

The bootstrap

The basic implementation of non-parametric bootstrap is an easy exercise, since it involves the few steps summarized in Algorithm 2.1.

Algorithm 2.1 The basic bootstrap pseudo-code.

1: Given data for n observations, $\boldsymbol{x} = (x_1, x_2, \ldots, x_n)$. Specify the statistic $\widehat{\theta}$ to resample. Set the number of bootstrap replications B.

2: **for** $b \leftarrow 1, B$ **do**
3: Draw a random sample of size n, $\boldsymbol{x}^{*b} = (x_1^{*b}, x_2^{*b}, \ldots, x_n^{*b})$, with replacement from the original sample $\boldsymbol{x}$.
4: Using the sampled data $\boldsymbol{x}^{*b}$, compute the statistic's bootstrap replicate $\widehat{\theta}^{*b}$.
5: **end for**

As an example, consider the simple situation we discussed in Section 2.1, that is the estimation of the mean of a population using the sample data $(x_1 = 7, x_2 = 4, x_3 = -5, x_4 = 1, x_5 = -2)$. The R code for computing the bootstrap distribution (not reported here) of the sample mean using $B = 200$ replications is shown in the code below[46]:

```
1 set.seed(123)
2 x <- c(7, 4, -5, 1, -2)
3 B <- 200
4 mean_boot <- numeric(B)
5 for (b in 1:B) {
6   x_b <- x[sample(1:length(x), replace = TRUE)]
7   mean_boot[b] <- mean(x_b)
8 }
```

[46]The set.seed() function sets the random number generator seed. We set it every time to allow you reproducing the same results we report here. However, notice that in April 2019 R 3.6.0 changed the way it generates random numbers. Since we prepared most of the examples using the previous version of the random number generator, to recover the numbers shown here you need to run the command RNGversion("3.5.0") beforehand. If you're curious about the reason for the change, see the discussion at https://stat.ethz.ch/pipermail/r-devel/2018-September/076817.html.

```
9  hist(mean_boot, col = gray(0.5, 0.3))
```

Nonetheless, it is a well-known fact that loops in R are inefficient, so that a faster and more convenient approach to the bootstrap is often desirable. One of the most complete implementations of the bootstrap in R is provided by the boot package (Canty and Ripley, 2019), which also includes many other resampling plans (Davison and Hinkley, 1997). The main function in the package is boot() and it requires the following mandatory arguments:

- data, the data to resample (either a numeric vector or matrix, or a data frame),
- R, the number of bootstrap replications,
- statistic, the function implementing the calculations for the statistic of interest.

Optionally, you can also set other arguments, such as the type of resampling and whether the computation should rely on parallel computing, that are not described here (for more details see the boot documentation). A particularly tricky part is represented by the statistic argument, which must be a function with two arguments, the first corresponding to the data to use in the computation of the statistic, and the second being a vector of indices (or frequencies or weights) which define the bootstrap sample. The following code implements the same non-parametric bootstrap procedure described above but exploiting the boot() function[47]:

```
1  if (!require(boot, quietly = TRUE)) install.packages("boot")
2  library(boot)

3  set.seed(123)

4  mean_w <- function(x, i) mean(x[i])
5  mean_boot <- boot(data = x, statistic = mean_w, R = 200)

6  plot(mean_boot)
```

The corresponding output is as follows:

```
ORDINARY NONPARAMETRIC BOOTSTRAP


Call:
boot(data = x, statistic = mean_w, R = 200)


Bootstrap Statistics :
```

[47]The first line in the code checks if the boot package is available on your computer, and if it isn't, it installs the package from the internet.

```
    original  bias    std. error
t1*        1  -0.006    1.811511
```

In the code above, the `statistic` argument is set equal to the `mean_w()` function, which simply redefines the `mean()` function with the addition of a second argument, here called `i`, representing the indexes of the resampled data that will be used to compute the mean value in each replication.

Another advantage from using the `boot()` function is the possibility to use its output to perform further calculations. For example, we may compute the bootstrap confidence intervals by running the `boot.ci()` function[48]:

```
1 boot.ci(mean_boot, conf = 0.95,
2   type = c("norm", "basic", "perc", "bca"))
```

The corresponding output is given by:

```
BOOTSTRAP CONFIDENCE INTERVAL CALCULATIONS
Based on 200 bootstrap replicates

CALL :
boot.ci(boot.out = mean_boot, conf = 0.95,
    type = c("norm", "basic", "perc", "bca"))

Intervals :
Level      Normal              Basic
95%   (-2.5445,  4.5565 )   (-2.5838,  4.5838 )

Level     Percentile            BCa
95%   (-2.5838,  4.5838 )   (-3.5331,  3.4000 )
Calculations and Intervals on Original Scale
```

Principal component analysis

Different packages in `R` include interfaces for performing principal component analysis (PCA), but the most popular functions are `princomp()` and `prcomp()` available in the basic `R` installation. The former uses the spectral decomposition, while `prcomp()` is based on the so called singular value decomposition (SVD), a general type of matrix decomposition that applies also to rectangular (i.e., non-square) matrices[49]. Generally, the two functions provide the same results, but `prcomp()` should be preferred because the SVD guarantees a higher numerical accuracy. Both functions include the following methods:

- `print`, which prints the results in a nice format; for `princomp()` it returns only the standard deviations of the components (i.e., the square root of the

[48]For more details on the bootstrap confidence intervals, see the discussion in the technical appendix at the end of the chapter.

[49]For the more math-inclined readers, a good introduction to SVD is available at `https://en.wikipedia.org/wiki/Singular_value_decomposition`.

eigenvalues), while for `prcomp()` it also produces the matrix of the (unnormalized) eigenvectors.

- `summary`, which prints a table with the component standard deviations, the proportion of variance explained by each component and the corresponding cumulative proportions.

- `plot`, which produces the screeplot using either a bar or a line chart (the same plot can also be obtained with the `screeplot()` function).

- `predict`, which returns a numerical matrix containing the component scores.

- `biplot`, which produces the so called **biplot**, a plot that aims to represent both the observations and the variables on a single diagram by projecting the corresponding values onto a plane defined by two of the extracted components (for more details see for example Gower and Hand, 1996).

Additionally, `princomp()` has a `loadings` method, that produces an object of class `loadings` containing the (unnormalized) eigenvectors. A nice feature of the `loadings` objects is that they have their own `print` method that includes a `cut-off` argument, which allows to blank out the loadings below a given threshold. This simple trick permits to focus the attention on the patterns with larger loadings. Furthermore, to ease the components interpretation, some authors suggest to rotate them, but the correctness of this procedure is still debated (see Rencher and Christensen, 2012, Chapter 12). This can be achieved in `R` by using the `varimax()` or `promax()` functions directly on the output returned by `loadings()`. To perform the same analysis using the output of the `prcomp()` function, you first need to extract the eigenvectors from the output (see the next paragraph) and change its class to `loadings`.

The `princomp()` and `prcomp()` functions both outputs the results in the form of a list object whose elements can be directly accessed using the standard `R` list subsetting feature (i.e., either `$` or `[[]]`). However, the naming of these elements is not consistent in the two cases. In particular, the `princomp()` returns a list with the following elements:

- `sdev`, the standard deviations of the principal components,

- `loadings`, the matrix whose columns contain the eigenvectors,

- `center`, the vector of means that have been subtracted from the variables before performing PCA,

- `scale`, the vector of scaling measures applied to the variables before performing PCA,

- `n.obs`, the number of observations,

- `scores`, the matrix containing the component scores,

- `call`, the function call.

The list returned by `prcomp()` contains the following elements, instead:

- `sdev`, the standard deviations of the principal components,
- `rotation`, the matrix whose columns contain the eigenvectors,
- `center`, the vector of means that have been subtracted from the variables before performing PCA,
- `scale`, the vector of scaling measures applied to the variables before performing PCA,
- `x`, the matrix containing the component scores.

Finally, in `prcomp()` the variances are computed with the usual divisor $n-1$, while `princomp()` uses n.

Another package that we consider worth mentioning here is `FactoMineR`, which is completely dedicated to multivariate analysis and contains functions to perform PCA, exploratory factor analysis, simple and multiple correspondence analysis and clustering (for a full documentation see Husson et al., 2017). The same authors also developed a web-interface for the package called `Factoshiny`[50]. A further related package is `factoextra`[51], that provides `ggplot2`-based visualizations of the results produced by `FactoMineR`.

We now apply the `prcomp()` function to the `Rateprof` data presented in Section 2.2, which are available in the `alr4` package. The code below performs PCA on the correlation matrix for the variables `quality`, `helpfulness`, `clarity`, `easiness` and `raterInterest` by setting to `TRUE` both the `center` and `scale.` arguments and prints the results[52]:

```
1  if (!require(alr4, quietly = TRUE)) install.packages("alr4")
2  data(Rateprof, package = "alr4")
3  vars <- c("quality", "helpfulness", "clarity",
4    "easiness", "raterInterest")

5  prcomp_res <- prcomp(x = Rateprof[, vars], center = TRUE,
6    scale. = TRUE)
7  print(prcomp_res, digits = 5)
```

[50]`http://factominer.free.fr/graphs/factoshiny.html`.

[51]`https://rpkgs.datanovia.com/factoextra/index.html`.

[52]Note that for this example the results produced by `prcomp()` differ from those reported by Stata (see Figure 2.4) as for the signs of the eigenvectors, which are all inverted. Sign indeterminacy of the components is a well-known issue of PCA, which practically consists in the direction (sign) of principal components being arbitrary. Some software (e.g., Stata, but not `R`) return components signed so that the sum of each eigenvector's elements is strictly positive.

```
Standard deviations (1, .., p=5):
[1] 1.892117 0.895463 0.733114 0.280511 0.043573

Rotation (n x k) = (5 x 5):
                   PC1       PC2      PC3        PC4         PC5
quality       -0.51764 -0.038352  0.26656 -0.0361564  0.81130917
helpfulness   -0.50900 -0.043583  0.24505 -0.6976809 -0.43841916
clarity       -0.50529 -0.024135  0.28926  0.7147568 -0.38671397
easiness      -0.35369 -0.558242 -0.74981  0.0322385 -0.00427060
raterInterest -0.30422  0.827293 -0.47225  0.0041753  0.00035095
```

Next, the following code produces the screeplot, the scatterplot matrix of the component scores (both not reported here) and computes the loadings multiplying the matrix of eigenvectors by the diagonal matrix with the component standard deviations on the main diagonal (see equation (2.9) on page 84):

```
1  plot(prcomp_res, type = "lines")    # not shown
2  pairs(predict(prcomp_res))          # not shown
3  prcomp_res$rotation %*% diag(prcomp_res$sdev)
```

```
                    [,1]        [,2]       [,3]         [,4]          [,5]
quality       -0.9794289 -0.03434260  0.1954152 -0.010142277  0.0353510543
helpfulness   -0.9630870 -0.03902658  0.1796500 -0.195707371 -0.0191031732
clarity       -0.9560757 -0.02161203  0.2120591  0.200497360 -0.0168502307
easiness      -0.6692316 -0.49988458 -0.5496932  0.009043261 -0.0001860820
raterInterest -0.5756178  0.74080993 -0.3462129  0.001171206  0.0000152917
```

We conclude this section with a quick mention to a couple of packages, namely `pls` (Mevik et al., 2019; Wehrens, 2011) and `mixOmics` (Rohart et al., 2017), both of which provide fairly general interfaces to perform a wide range of multivariate regression analyses including principal component regression (PCR) and partial least squares regression (PLSR). In particular, `pls` provides the functions `pcr()` and `plsr()` that also implement model selection through cross-validation. The `mixOmics` package provides functions for the same purposes but with additional functionalities to deal with large sparse datasets.

Segmentation methods

The packages available in `R` to perform segmentation are so many that it is impossible to provide a comprehensive description here. Therefore, we present a selection of those we consider essential and invite you to visit the cluster CRAN task view at `https://cran.r-project.org/web/views/Cluster.html` for the complete list.

Cluster analysis

All the standard agglomerative hierarchical clustering algorithms are implemented in `R` through the `hclust()` function . `hclust()` requires in input the distance matrix, that can be computed through the `dist()` function, and the linkage measure, which is specified with the `method` argument. The `hclust()` function returns an object of class `hclust` whose `plot` method produces the corresponding dendrogram. After importing the data from the `ch2_SimData.dta` Stata file using the `read.dta()` function from the `foreign` package, the following code compares the four standard hierarchical clustering algorithms (i.e., single linkage, complete linkage, average linkage, and the Ward's method) on the same simulated data described in Section 2.3.1.1:

```
1  if (!require(foreign, quietly = TRUE)) install.packages("alr4")
2  library(foreign)
3  path_data <- ""                # write here the path for your data
4  simdata <- read.dta(file.path(path_data, "ch2_SimData.dta"))
5  simdata <- simdata[, 1:5]
6  pairs(simdata[, -1], pch = 19, col = gray(.5, .3))  # not shown

7  simdata_std <- scale(simdata[, -1])        # data standardization

8  simdata_dist <- dist(simdata_std, method = "euclidean")
9  single_linkage <- hclust(simdata_dist, method = "single")
10 complete_linkage <- hclust(simdata_dist, method = "complete")
11 average_linkage <- hclust(simdata_dist, method = "average")
12 ward_linkage <- hclust(simdata_dist, method = "ward.D2")
13 par(mfrow = c(2, 2))                                  # not shown
14 plot(single_linkage, main = "Single linkage", cex = .4)
15 plot(complete_linkage, main = "Complete linkage", cex = .4)
16 plot(average_linkage, main = "Average linkage", cex = .4)
17 plot(ward_linkage, main = "Ward method", cex = .4)
```

Two further functions that are useful after plotting a dendrogram are `rect.hclust()` and `cutree()`, which allow to "cut" the tree at a given height (with the argument `h`) or in correspondence of a given number of groups (with the argument `k`). More specifically, the former draws rectangles on the dendrogram around the selected groups, while the latter returns the corresponding group memberships. The code below generates the same comparison of the dendrograms as in the previous code but with the rectangles drawn for `k = 3` groups, together with the group memberships from the Ward's method:

```
1  par(mfrow = c(2, 2))                                  # not shown
2  plot(single_linkage, main = "Single linkage", cex = .4)
3  rect.hclust(single_linkage, k = 3)
4  plot(complete_linkage, main = "Complete linkage", cex = .4)
5  rect.hclust(complete_linkage, k = 3)
6  plot(average_linkage, main = "Average linkage", cex = .4)
7  rect.hclust(average_linkage, k = 3)
8  plot(ward_linkage, main = "Ward method", cex = .4)
9  rect.hclust(ward_linkage, k = 3)

10 ward_gm <- cutree(ward_linkage, k = 3)               # not reported
11 table(ward_gm, simdata[, 1])
```

A more flexible interface that implements many agglomerative hierarchical algorithms is `agnes()` (agglomerative nesting) from the `cluster` package (Maechler et al., 2019). Compared to `hclust()`, `agnes()` allows to specify directly the data in input, the distance measure to use, the clustering method and whether to standardize or not the data. The package also defines the corresponding `summary` and `plot` methods. The code reported below performs the analysis with the Ward's method on the same simulated data:

```
1  if (!require(cluster, quietly = TRUE)) install.packages("cluster")
2  library(cluster)
3  # agglomerative nesting (hierarchical clustering)
4  res_agnes <- agnes(x = simdata[, 2:5],     # data matrix
5                     stand = TRUE,           # standardize the data
6                     metric = "euclidean",   # distance measure
7                     method = "ward"         # linkage method
8  )
9  plot(res_agnes, main = "Ward method", which.plots = 2) # not shown
```

The `factoextra` package includes some functions to create nice `ggplot2`-based visualizations of the results. With regards to hierarchical clustering, the function `fviz_dend()` provides many arguments to format the final appearance of a dendrogram. The following code provides an example:

```
1  if (!require(factoextra, quietly = TRUE))
2    install.packages("factoextra")
3  library(factoextra)
4  fviz_dend(res_agnes,                          # not shown
5            k = 3,                       # cut in three groups
6            cex = 0.5,                           # label size
7            color_labels_by_k = TRUE, # color labels by groups
8            rect = TRUE          # add rectangles around groups
9  )
```

Another package that provides additional functionalities to represent dendrogram objects is `dendextend`. Finally, the `cluster` package also includes the `diana()` function that implements a divisive hierarchical algorithm. More details can be found in Kaufman and Rousseeuw (1990).

For what regards partitional clustering algorithms, the `kmeans()` function implements the standard *K*-means algorithm we discussed in Section 2.3.1.2. The `kmeans()` function requires in input the data matrix (x argument) and the number of clusters to look for (`centers` argument). Optionally, one may provide a vector of user-defined cluster centres (still with the `centers` argument), the maximum number of iterations (`iter.max`, which defaults to 10) and the number of algorithm restarts (`nstart`) to increase the chances of convergence towards the global optimal solution (it is suggested to set it at least to 10). The `kmeans()` function returns a `kmeans` object, a list that includes components such as `withinss`, a vector containing the within-cluster sum of squares for each cluster, and `tot.withinss`, the total within-cluster sum of squares . The latter can be useful to compare solutions with different values of *K*. Another useful component of a `kmeans` object is `cluster`, the vector of cluster memberships. The following code performs a K-means analysis for the same simulated data as above using a number of clusters from 1 (i.e., no cluster structure) to 10 and produces a graph (not reported here) that shows how the corresponding total within-cluster sum of squares decreases with K:

```
1  set.seed(101)                          # for reproducibility
2  wss <- numeric(10)                     # allocate memory for TWSS
3  for (k in 1:10) {                      # loop over different K
4    res_k <- kmeans(x = simdata[, 2:5],  # data matrix
5                    centers = k,         # number of clusters
6                    iter.max = 100,      # max num. of iterations
7                    nstart = 10          # number of restarts
8              )
9    wss[k] <- res_k$tot.withinss         # stores TWSS for given K
10 }
11 plot(wss, type = "b", lwd = 2, pch = 20, # not shown
12   xlab = "K", ylab = "Total within-cluster sum of squares")
```

The plot of the total within-cluster sum of squares shows that there is a clear "elbow" in correspondence of $K = 3$, which corresponds to the actual number of groups used to simulate the data and to the same groups identified by the hierarchical algorithms.

The `cluster` package includes two functions that perform partitional clustering, that is `pam()` (acronym for *p*artitioning *a*round *m*edoids) and `clara()` (which stands for *c*lustering *lar*ge *a*pplications), with the latter being a specialized version of the former that is particularly efficient for large datasets. These methods are fully described in Kaufman and Rousseeuw (1990).

Another package that is focused on partitional methods is `flexclust` (Leisch, 2006), which together with K-means also implements more flexible algorithms like hard competitive learning, neural gas and the possibility to perform "bagged clustering", which consists in running repeatedly the specified method on bootstrap samples from the original data and combining the resulting cluster centres using hierarchical clustering. The main function in `flexclust` is `kcca()`, which returns richer objects compared to `kmeans()` whose contents can be subsequently analysed using many accessory functions. A detailed description of the methods and features available in the `flexclust` package is provided by Dolnicar et al. (2018).

Finally, a lot of functions for identifying a suitable number of clusters and for validating a cluster solution are spread in different packages. To mention a few of these alternatives, the `NbClust()` function in the same package (Charrad et al., 2014) is particularly useful since it implements 30 different indexes for supporting the user to select the optimal number of clusters. Among these you can find the Caliński-Harabasz pseudo F and Duda-Hart pseudo t^2 indexes (see page 84), the gap statistic and the silhouette method. The following code computes the indexes with the `NbClust()` function for the simulated data we described above using the complete linkage method:

```
1  if (!require(NbClust, quietly = TRUE)) install.packages("NbClust")
2  library(NBClust)
3  res <- NbClust(data = simdata_std,            # data matrix
4                 distance = "euclidean",        # distance measure
5                 min.nc = 2,                    # minimum num. of clusters
6                 max.nc = 8,                    # maximum num. of clusters
7                 method = "complete",           # clustering algorithm
8                 index = "alllong")             # compute all indexes
```

The corresponding results (with the exception of the plot) are reported below:

```
*******************************************************************
* Among all indices:
* 2 proposed 2 as the best number of clusters
* 23 proposed 3 as the best number of clusters
* 1 proposed 5 as the best number of clusters
* 1 proposed 8 as the best number of clusters

                   ***** Conclusion *****

* According to the majority rule, the best number of clusters is  3
*******************************************************************
```

So, in this example 23 of the indexes computed by the `NbClust()` function agree in indicating that $K = 3$ is the best number of clusters for these data. To get the detailed results for each index, you can look inside the list object returned by the function.

We conclude this section with a brief mention to the `clValid` (Brock et al., 2008) and `fpc` (Hennig, 2020) packages, which provide tools for validating a cluster solution, that is for evaluating the goodness of the clustering results with respect to the compactness, connectedness, separation and stability of the cluster partitions. In particular, the `cqcluster.stats()` function from the `fpc` package provides a quite long list of validation statistics, while `clusterboot()` assesses the clusterwise stability of a clustering solution by resampling the data.

Finite mixture models

Finite mixtures (FMMs) are popular tools with applications in different fields. This fact is also confirmed by the existence of many `R` packages that implement a wide range of FMMs. Here, we briefly present the most acknowledged of these packages, that is `flexmix`, `mclust` and `mixtools`.

We start from `mclust`, the most popular package in `R` for finite mixture modelling (Scrucca et al., 2016). Even if it only implements mixtures of Gaussian distributions, it provides a fairly broad set of tools for summarizing and visualizing the results. The main function, `Mclust()` (note the capitalization), requires the raw data as the only mandatory argument. The number of clusters to consider in the analysis is chosen through the argument `G`, which is set by default to `1:9` (i.e., the function computes all the solutions with number of clusters from 1 to 9). The `Mclust()` function implements a list of 14 different Gaussian mixture models[53] that differ in

[53]The `mclust` package also includes features for performing classification through discriminant analysis based on Gaussian finite mixture modelling.

terms of their geometric characteristics. In the package these models are referred to with the acronyms shown in the first column of Table 2.2. The following code loads the package, fits the Gaussian mixtures to the simulated data we have been using so far and reports a summary of the results:

```
1  if (!require(mclust, quietly = TRUE)) install.packages("mclust")
2  library(mclust)
3  res_mclust <- Mclust(simdata_std, G = 1:8)
4  summary(res_mclust)
```

```
----------------------------------------------------
Gaussian finite mixture model fitted by EM algorithm
----------------------------------------------------

Mclust VVI (diagonal, varying volume and shape) model with 3
  components:

 log-likelihood  n df      BIC      ICL
      -45.12921 50 26 -191.971 -191.971

Clustering table:
 1  2  3
17 17 16
```

The summary shows that the best model (i.e., the model with the largest value of the BIC index[54]) is the VVI model with 3 components, which corresponds to the actual model we used to generate the data from.

The Mclust() function returns an object of class Mclust corresponding to a list containing the detailed results, for which the package includes plot and predict methods. The former produces different visualizations of the results, while the latter returns the class posterior probabilities based on which each observation is assigned to one of the groups. The visualizations available through the plot method are: (1) the plot of the BIC values for all the models as a function of the number of components, (2) the classification plot, that is the plot showing the cluster memberships for each observation, (3) the uncertainty plot, which shows the degree of uncertainty for the classification of each data point and (4) a plot of the estimated densities. The uncertainty plot for the previous fit is produced by the code below and it is shown in Figure 2.30.

```
1  plot(res_mclust, what = "uncertainty")
```

The plot shows that there is one datum, observation number 40, with an apparently uncertain classification. However, we must highlight that the actual uncertainty values are all very small with the largest one, associated with the observation above, being equal to 2.65×10^{-8}.

[54]Note that the BIC index in the mclust package is implemented with an opposite sign with respect to the definition we provide in equation (2.22).

TABLE 2.2: Characteristics of the models available for clustering in the `mclust` package.

Model	Distribution	Volume	Shape	Orientation
EII	Spherical	Equal	Equal	–
VII	Spherical	Equal	Variable	–
EEI	Diagonal	Equal	Equal	Coordinate axes
VEI	Diagonal	Equal	Variable	Coordinate axes
EVI	Diagonal	Variable	Equal	Coordinate axes
VVI	Diagonal	Variable	Variable	Coordinate axes
EEE	Ellipsoidal	Equal	Equal	Equal
EVE	Ellipsoidal	Variable	Equal	Equal
VEE	Ellipsoidal	Equal	Variable	Equal
VVE	Ellipsoidal	Variable	Variable	Equal
EEV	Ellipsoidal	Equal	Equal	Variable
VEV	Ellipsoidal	Equal	Variable	Variable
EVV	Ellipsoidal	Variable	Equal	Variable
VVV	Ellipsoidal	Variable	Variable	Variable

The `flexmix` package (Leisch, 2004) provides a general framework for finite mixture modelling, including mixtures with non-Gaussian components and mixtures of regression models (see Dolnicar et al., 2018, Chapter 7). Parameter estimation is performed through the EM algorithm (see Section 2.6 for a brief introduction). The main function in the package is `flexmix()`, which requires the following mandatory arguments:

- `formula`, the symbolic description of the model to fit
- `data`, the data frame containing the variables used in the model
- `k`, the number of clusters to use
- `model`, an object of `FLXM` providing the model family specification; this is the argument through which you can specify whether you want to use Gaussian or other types of components

As usual, other optional arguments are available to fine tune the performance of the EM algorithm. The `flexmix()` function returns an object of class `flexmix` which contains different components[55]. The following code fits a mixture of $K = 3$ multivariate Gaussian distributions to our simulated data and produces a plot showing the corresponding classification (see Figure 2.31):

[55]The `flexmix` package is implemented using the so called S4 system (Wickham, 2019). The components of S4 objects, called "slots", are extracted using the `@` operator instead of the `$` operator used for standard list objects. If you want to get the full list of slots available in a given S4 object, you can use the `slotNames()` function.

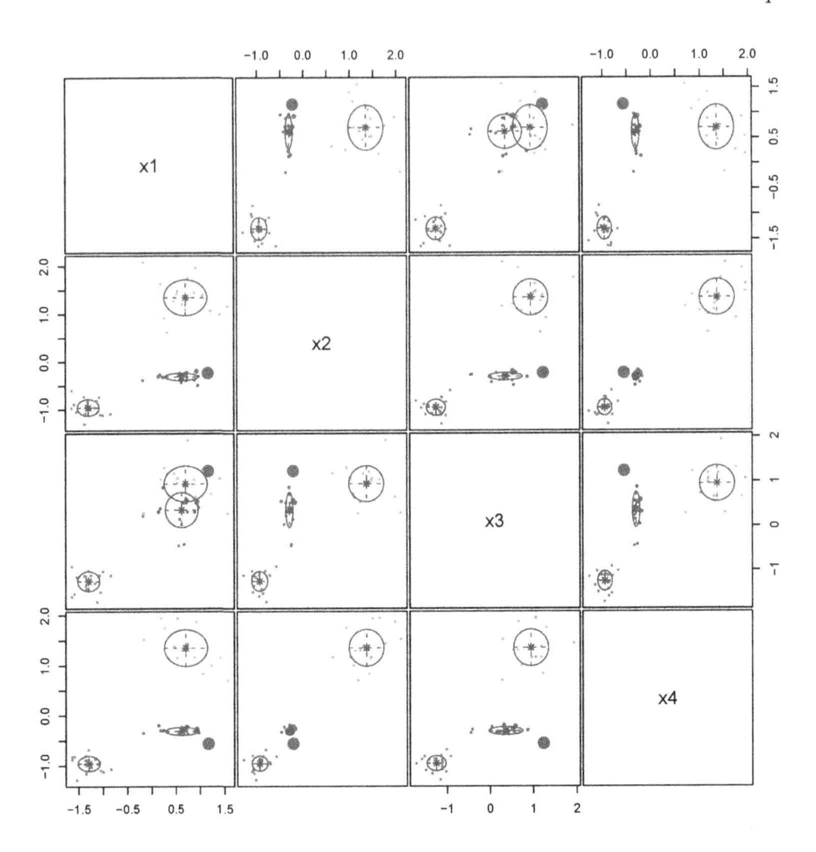

FIGURE 2.30: Uncertainty plot for the Gaussian mixture model fitted using the `Mclust()` function on simulated data.

```
if (!require(flexmix, quietly = TRUE)) install.packages("flexmix")
library(flexmix)

set.seed(301)    # for reproducibility
res_flexmix <- flexmix(formula = simdata_std ~ 1,
  data = data.frame(simdata_std), k = 3,
  model = FLXMCmvnorm())

par(mfrow = c(3, 2), mar = c(4, 4, 1, 1) + .1)
for (i in 1:(ncol(simdata_std) - 1)) {
  for (j in (i + 1):ncol(simdata_std)) {
    plotEll(res_flexmix, data = data.frame(simdata_std),
      which = c(i, j))
  }
}
```

As you can see, the specification of the model through the `formula` argument is not straightforward, because it has been developed to be as general as possible

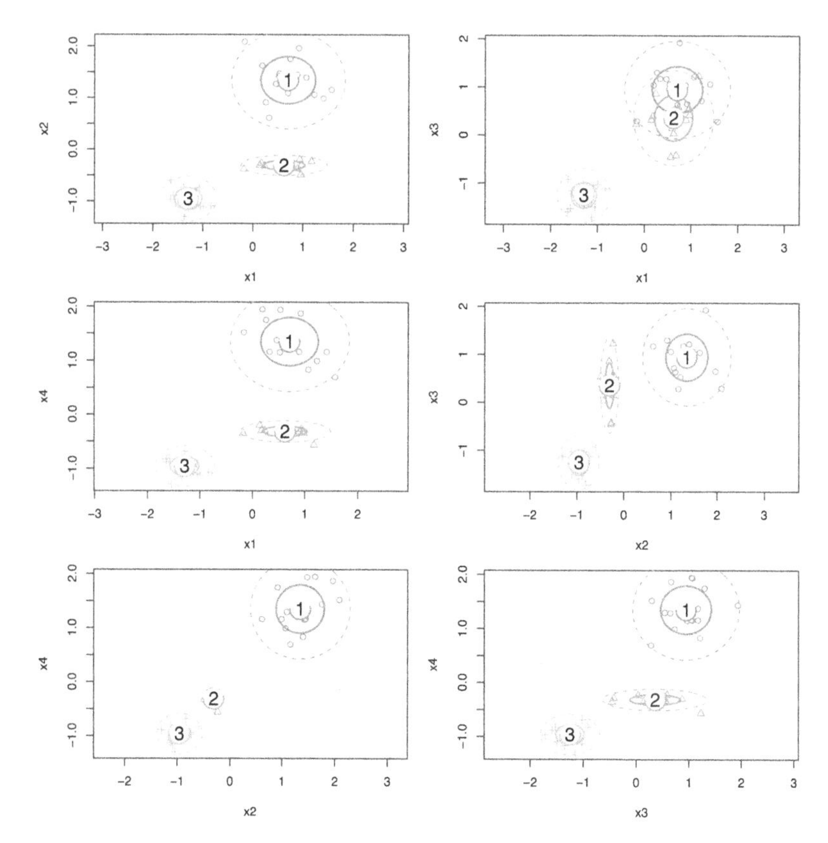

FIGURE 2.31: Classification produced by fitting a multivariate Gaussian mixture model using the `flexmix()` function on simulated data.

especially to fit mixtures of complex regression models. The `FLXMCmvnorm()` function represents the interface for fitting mixtures of multivariate Gaussian distributions, but others are available or can be provided directly by the user (see the documentation for more details).

The selection of the best model is performed using the standard information criteria AIC and BIC. To this end, the `flexmix` package also provides the `stepFlexmix()` function that allows to specify a range of values for the number of clusters to fit. The code below fits a mixture of multivariate Gaussian distributions to the same data using a number of components from 1 to 8, whose results are reported next:

```
1  set.seed(301)    # for reproducibility
2  res_flexmix_all <- stepFlexmix(formula = simdata_std ~ 1,
3    data = data.frame(simdata_std), k = 1:8,
```

```
4    model = FLXMCmvnorm())
5  print(res_flexmix_all)
```

```
Call:
stepFlexmix(formula = simdata_std ~ 1,
    data = data.frame(simdata_std),
    model = FLXMCmvnorm(), k = 1:8)

  iter converged k k0     logLik      AIC      BIC      ICL
1    2      TRUE 1  1 -281.78771 579.5754 594.8716 594.8716
2    9      TRUE 2  2 -139.73969 313.4794 345.9838 345.9838
3    8      TRUE 3  3  -45.31692 142.6338 192.3464 192.3464
4   30      TRUE 4  4  -36.49883 142.9977 209.9185 211.2668
5   28      TRUE 4  5  -36.49883 142.9977 209.9185 211.2668
6   14      TRUE 3  6  -45.31692 142.6338 192.3464 192.3464
7   33      TRUE 4  7  -37.12521 144.2504 211.1712 211.2719
8   40      TRUE 5  8  -29.82160 147.6432 231.7722 236.6819
```

We conclude with a brief mention to the `mixtools` package (Benaglia et al., 2009). Similar to `flexmix`, `mixtools` includes functions for fitting finite mixtures of Gaussian and non-Gaussian distributions as well as of more general regression models. In addition, `mixtools` also provides interfaces for fitting *non-parametric* mixtures, that is mixture models where the component distributions are left unspecified. We refer you to the package documentation for more details and examples.

Latent class analysis

As we already noted in Section 2.3.3, latent class analysis is a special type of finite mixtures where both the observed and latent (i.e., the cluster membership) variables are categorical. This implies that you can fit LCA models in `R` using the packages we discussed above, that is either `flexmix` or `mixtools` (`mclust` only allows for Gaussian components), specifying discrete distributions for the manifest variables. In addition to these general packages for FMMs, you may also use the `poLCA` package (Linzer and Lewis, 2011), which is entirely dedicated to LCA models. The `poLCA` package is fairly simple since it essentially consists of only one function, the `poLCA()` function, through which the model is specified, fitted and assessed by inspecting the results that are automatically printed.

Path analysis

Path models, known in econometrics as simultaneous equation models, can be estimated using a range of different methods depending on the structural and stochastic assumptions for the variables involved in the model (see for example Greene, 2018, Chapter 10). Some of these methods are implemented in the `systemfit` package (Henningsen and Hamann, 2007). The main function, `systemfit()`, requires the specification of the equations in the system as a list of `formula` objects. Optional arguments include the estimation method (`method`), the specification of the restrictions to impose on the model's coefficients (`restrict.matrix`) and a list of further control parameters (`control`). The `systemfit()` function returns an object of class `systemfit`, which is a list containing 10 different components with the detailed results of the analysis. The package includes different methods for summarizing and extracting portion of the results. The code snippet below installs the package and estimates the path model shown in Figure 2.27 using the data in the `ch2_hsb2.dta` file:

```
1  if (!require(foreign, quietly = TRUE)) install.packages("foreign")
2  library(foreign)
3  path_data <- ""                # write here the path for your data
4  hsb2 <- read.dta(file.path(path_data, "ch2_hsb2.dta"))

5  if (!require(systemfit, quietly = TRUE))
6    install.packages("systemfit")
7  library(systemfit)
8  eq_math <- math ~ read + write
9  eq_science <- science ~ math + read + write
10 eqs <- list(math = eq_math, science = eq_science)
11 res_systemfit <- systemfit(eqs, data = hsb2)
12 summary(res_systemfit)
```

```
systemfit results
method: OLS

         N  DF     SSR detRCov   OLS-R2 McElroy-R2
system 400 393 18220.4 2138.71 0.507202   0.507766

          N  DF     SSR     MSE    RMSE       R2   Adj R2
math    200 197 8465.52 42.9722 6.55532 0.515309 0.510388
science 200 196 9754.84 49.7696 7.05476 0.499944 0.492290

The covariance matrix of the residuals
               math     science
math    4.29722e+01 1.54256e-13
science 1.54256e-13 4.97696e+01

The correlations of the residuals
               math     science
math    1.00000e+00 3.33784e-15
science 3.33784e-15 1.00000e+00


OLS estimates for 'math' (equation 1)
Model Formula: math ~ read + write

              Estimate Std. Error t value    Pr(>|t|)
(Intercept) 12.8650676  2.8216198 4.55946 8.9956e-06 ***
read         0.4169486  0.0564838 7.38174 4.2886e-12 ***
write        0.3411219  0.0610982 5.58317 7.7567e-08 ***
---
Signif. codes:  0 '***' 0.001 '**' 0.01 '*' 0.05 '.' 0.1 ' ' 1

Residual standard error: 6.555315 on 197 degrees of freedom
```

```
Number of observations: 200 Degrees of Freedom: 197
SSR: 8465.515734 MSE: 42.972161 Root MSE: 6.555315
Multiple R-Squared: 0.515309 Adjusted R-Squared: 0.510388


OLS estimates for 'science' (equation 2)
Model Formula: science ~ math + read + write

             Estimate  Std. Error t value    Pr(>|t|)
(Intercept) 8.4073530  3.1927987 2.63322   0.0091325 **
math        0.3190094  0.0766753 4.16053 0.000047488 ***
read        0.3015317  0.0686815 4.39029 0.000018495 ***
write       0.2065257  0.0707644 2.91850   0.0039285 **
---
Signif. codes:  0 '***' 0.001 '**' 0.01 '*' 0.05 '.' 0.1 ' ' 1

Residual standard error: 7.054757 on 196 degrees of freedom
Number of observations: 200 Degrees of Freedom: 196
SSR: 9754.841936 MSE: 49.769602 Root MSE: 7.054757
Multiple R-Squared: 0.499944 Adjusted R-Squared: 0.49229
```

The coefficient estimates are the same we got with Stata (see Figure 2.28).

As we showed in Section 2.4 using Stata, path models may also be fitted using structural equation modelling software. One of the most popular choice in `R` is represented by the `sem` package (Fox et al., 2017), but other options are available (e.g., the `OpenMx` package; Boker et al., 2020). Here, we dedicate some space to another alternative, the `lavaan` package (Rosseel, 2012), which is increasingly gaining popularity among social scientists (for book-length presentations of `lavaan` see Beaujean, 2014; Finch and French, 2015; Gana and Broc, 2019). We choose to present `lavaan` because we will use its syntax also in other chapters of this book. The main function of the package is `lavaan()`, but other wrappers for more specific analyses (e.g., `cfa()` for confirmatory factor analysis or `sem()` for structural equation modelling) are also available. Model specification in `lavaan` must follow a set of strict rules which are summarized in Table 2.3 (you can find a more detailed presentation by executing the command `?model.syntax`).

The specification of the model is supplied to `lavaan()` through the `model` argument, which must be a text string enclosed within double (`"`) or single (`'`) quotes with each line corresponding to one equation in the path model. Optionally, the coefficients can be tagged with user-defined labels. As an example, the following code shows the specification of the path model described above for the HSB survey data where we name the coefficients as in Section 2.4:

```
1 hsb2_mod <- "
2   math ~ alpha_1*1 + gamma_11*read + gamma_12*write
3   science ~ alpha_2*1 + beta_21*math + gamma_21*read + gamma_22*write
4   math ~~ s2_math*math
5   science ~~ s2_science*science
6 "
```

The last two lines in the code above specify the variances of the endogenous

TABLE 2.3: Main syntax rules used by the `lavaan` package for the specification of a model.

Syntax	Description	Example
`~`	Regression onto	Regress `Y` onto `X`: `Y ~ X`
`~~`	Variance or covariance between two variables	Variance of `Y`: `Y ~~ Y`
		Covariance between `Y` and `X`: `Y ~~ X`
`~ 1`	Constant term/intercept to include in an equation	Regress `Y` onto `X` including a constant: `Y ~ 1 + X`
`*`	Labelling of a parameter	Label the coefficients in the regression of `Y` onto `X`: `Y ~ beta_0*1 + beta_1*X`
		Label the variance of the variable `Y`: `Y ~~ s2_Y*Y`
`:=`	Defining a non-model parameter	Define `beta_1_sq` to be the square of the `beta_1` coefficient of an already defined equation: `beta_1_sq := beta_1^2`

variables `math` and `science`. These are optional because the `lavaan()` function automatically creates an error term for each endogenous variable, but we added them because we wanted to provide labels for the two variances (i.e., `s2_math` and `s2_science`). Then, we can fit the model with the following command:

```
1 if (!require(lavaan, quietly = TRUE))
2   install.packages("lavaan")
3 library(lavaan)
4 res_lavann <- lavaan(model = hsb2_mod, data = hsb2)
```

The `lavaan()` function returns an object of class `lavaan` for which several methods are available, including a `summary` method, which accepts many optional arguments for printing additional results (for more information about the methods defined for `lavaan` objects, see the documentation with the command `?"lavaan-class"`):

```
1 summary(res_lavann, fit.measures = TRUE, ci = TRUE,
2   rsquare = TRUE, nd = 3L)
```

```
lavaan 0.6-5 ended normally after 22 iterations

  Estimator                                         ML
  Optimization method                           NLMINB
```

```
  Number of free parameters                          9

  Number of observations                           200

Model Test User Model:

  Test statistic                                 0.000
  Degrees of freedom                                 0

Model Test Baseline Model:

  Test statistic                               283.456
  Degrees of freedom                                 5
  P-value                                        0.000

User Model versus Baseline Model:

  Comparative Fit Index (CFI)                    1.000
  Tucker-Lewis Index (TLI)                       1.000

Loglikelihood and Information Criteria:

  Loglikelihood user model (H0)              -1330.839
  Loglikelihood unrestricted model (H1)      -1330.839

  Akaike (AIC)                                2679.679
  Bayesian (BIC)                              2709.364
  Sample-size adjusted Bayesian (BIC)         2680.851

Root Mean Square Error of Approximation:

  RMSEA                                          0.000
  90 Percent confidence interval - lower         0.000
  90 Percent confidence interval - upper         0.000
  P-value RMSEA <= 0.05                             NA

Standardized Root Mean Square Residual:

  SRMR                                           0.000

Parameter Estimates:

  Information                                 Expected
  Information saturated (h1) model          Structured
  Standard errors                             Standard

Regressions:
                   Estimate  Std.Err  z-value  P(>|z|) ci.lower ci.upper
  math ~
    read    (g_11)    0.417    0.056    7.438    0.000    0.307    0.527
    write   (g_12)    0.341    0.061    5.626    0.000    0.222    0.460
  science ~
    math    (b_21)    0.319    0.076    4.203    0.000    0.170    0.468
    read    (g_21)    0.302    0.068    4.435    0.000    0.168    0.435
    write   (g_22)    0.207    0.070    2.948    0.003    0.069    0.344

Intercepts:
                   Estimate  Std.Err  z-value  P(>|z|) ci.lower ci.upper
   .math    (al_1)   12.865    2.800    4.594    0.000    7.376   18.354
   .science (al_2)    8.407    3.161    2.660    0.008    2.212   14.602

Variances:
                   Estimate  Std.Err  z-value  P(>|z|) ci.lower ci.upper
   .math    (s2_m)   42.328    4.233   10.000    0.000   34.032   50.624
   .science (s2_s)   48.774    4.877   10.000    0.000   39.215   58.334
```

```
R-Square:
                   Estimate
    math              0.515
    science           0.500
```

It is possible to draw the path diagram using the `semPaths()` function in the `semPlot` package, but we do not provide here the details[56]. Rather, we focus now on the calculation of the indirect effects. As we already discussed in the chapter, an indirect effect represents the effect of an independent variable (endogenous or exogenous) on an endogenous variable that is mediated by at least another variable. Indirect effects are typically computed as products of direct effects. In our HSB data example, the `read` variable exerts an indirect effect on the `science` variable that is computed as `beta_21*gamma_11`, while the indirect effect of `write` on `science` is given by the product `beta_21*gamma_12` (there are no indirect effects of these two variables on `math`). To compute these effects with the `lavaan()` function, you need to explicitly define them within the model using the := operator. To this end, it is necessary that you already labelled the model's coefficients because you need to refer to them in the definition of the indirect effects. Similarly, you can also define the total effects as the sum of the direct and indirect effects. The code below modifies the previous model specification by adding the indirect and total effects of the `read` and `write` variables on `science`. The corresponding summary output reports the same results we showed above with the addition of the estimation of the new parameters:

```
1  hsb2_indirect <- "
2   math ~ alpha_1*1 + gamma_11*read + gamma_12*write
3   science ~ alpha_2*1 + beta_21*math + gamma_21*read + gamma_22*write
4   math ~~ s2_math*math
5   science ~~ s2_science*science
6   sc_read_ind := beta_21*gamma_11
7   sc_write_ind := beta_21*gamma_12
8   sc_read_tot := sc_read_ind + gamma_21
9   sc_write_tot := sc_write_ind + gamma_22
10 "
11 res_indirect <- lavaan(model = hsb2_indirect, data = hsb2)
12 summary(res_indirect, fit.measures = TRUE, ci = TRUE,
13   rsquare = TRUE, nd = 3L)
```

```
lavaan 0.6-5 ended normally after 22 iterations

# <Omitted output>

Defined Parameters:
                   Estimate  Std.Err  z-value  P(>|z|) ci.lower ci.upper
    sc_read_ind       0.133    0.036    3.659    0.000    0.062    0.204
    sc_write_ind      0.109    0.032    3.367    0.001    0.045    0.172
    sc_read_tot       0.435    0.063    6.922    0.000    0.312    0.558
    sc_write_tot      0.315    0.068    4.644    0.000    0.182    0.448
```

[56]Other alternatives to create path diagrams in `R` are represented by the `dagitty` package (`http://dagitty.net`) together with its `ggplot2` front-end `ggdag` (`https://ggdag.netlify.app`).

Appendix: Technical Details

More insights on the bootstrap

Suppose we are interested in estimating the unknown value of a parameter θ for a population with distribution F. To do that, we draw a random sample $\boldsymbol{x} = (x_1, x_2, \ldots, x_n)$ from F and compute the estimate $\widehat{\theta}$ (e.g., the sample mean). A critical issue in statistics is that of assessing the accuracy of $\widehat{\theta}$. This problem is typically solved by computing the standard error *SE* of the estimate. Ideally, the standard error is the standard deviation we would observe by drawing an infinite number of random samples of size n from the population and for each computing the estimate $\widehat{\theta}$. Clearly, this procedure is impossible to implement in practice, so we must rely on a different strategy. One possibility is to derive an analytical formula for the standard error. This is what is done in basic statistics courses for simple cases where the algebra is easy. For example, you know that the standard error of the sample mean is given by σ^2/n, where σ^2 is the variance of the population. Unfortunately, this approach is not available in most of the interesting cases because the algebra is too difficult. The bootstrap represents a general approach for calculating standard errors and confidence intervals. To perform the bootstrap we need to consider the original sample as if it were the population, and we then mimic the repeated sampling idea by drawing B different "bootstrap samples" from the observed sample. More specifically, the bth bootstrap sample $\boldsymbol{x}^{*b} = (x_1^{*b}, x_2^{*b}, \ldots, x_n^{*b})$ is formed by randomly drawing n observations with equal probability and with replacement from the original sample $\boldsymbol{x}$. Then, for each bootstrap sample we compute the corresponding *bootstrap replicate* $\widehat{\theta}^{*b}$, for $b = 1, 2, \ldots, B$. The bootstrap estimate of the standard error for $\widehat{\theta}$ is hence given by

$$SE_{\text{boot}} = \left[\frac{1}{B-1} \sum_{b=1}^{B} \left(\widehat{\theta}^{*b} - \overline{\theta}^{*} \right)^2 \right]^{1/2} \tag{2.5}$$

where $\overline{\theta}^{*} = 1/n \sum_{b=1}^{B} \widehat{\theta}^{*b}$ is the mean of the bootstrap replicates.

The collection of bootstrap replicates forms the bootstrap distribution of the statistic $\widehat{\theta}$. The bootstrap distribution is not centred around the true unknown parameter value θ like the sampling distribution of $\widehat{\theta}$ (assuming the statistics is unbiased), but it is approximately centred around the estimate $\widehat{\theta}$ itself. This is the reason why the bootstrap is not used to get better parameter estimates, rather it is useful to quantify the uncertainty of the same estimate. The *bias* of a statistic $\widehat{\theta}$ is defined as the difference between the average value of the statistic and the parameter value, that is

$$\text{Bias} = \text{E}[\widehat{\theta}] - \theta.$$

A statistic is *unbiased* if its bias is zero. Apart from simple cases, like the sample mean or the sample variance, the bias is generally unknown. However, thanks to the idea that the bootstrap "replaces" the population with the original sample, we can get a bootstrap estimate of the bias as

$$\text{Bias}_{\text{boot}} = \overline{\theta}^{*} - \widehat{\theta}, \tag{2.6}$$

the difference between the mean of the bootstrap distribution and the statistic computed for the original sample.

Together with standard errors and biases, the bootstrap distribution is also typically used to find more accurate confidence intervals than those based on the normal distribution assumption. There are different methods available, the most popular being:

- normal-based intervals,
- percentile intervals,
- bias-corrected percentile intervals.

Standard normal-based intervals are given by

$$\widehat{\theta} \pm z_{\frac{\alpha}{2}} \cdot SE,$$

and provide an approximate $100(1-\alpha)\%$ coverage. This formula depends on the asymptotic normality of the statistic sampling distribution and on a reliable estimate of its standard error. As we know, both these conditions are not often met in practice, so we can rely on bootstrap to refine this calculation.

A first refinement can be easily obtained by replacing the statistic standard error with its bootstrap estimate SE_{boot} as we defined it in (2.5), thus getting the first type of bootstrap confidence intervals listed above:

$$\widehat{\theta} \pm z_{\frac{\alpha}{2}} \cdot SE_{\text{boot}}. \tag{2.7}$$

A further refinement can be achieved by noting that, unless the sample size n is large enough, asymptotic normality of the sampling distribution is often not attained in practice. The percentile method allows to overcome this problem because it uses the bootstrap distribution and calculates the interval as

$$(\widehat{\theta}^{*}_{\frac{\alpha}{2}}, \widehat{\theta}^{*}_{1-\frac{\alpha}{2}}), \tag{2.8}$$

where the notation $\widehat{\theta}^{*}_{\ell}$ indicates the 100ℓth percentile of the bootstrap distribution.

Another refinement is provided by the so called *bias-corrected* (BC) percentile method, which also takes into account the fact that the bootstrap distribution is often not exactly centred around the statistic value $\widehat{\theta}$, that is the bootstrap estimate $\overline{\theta}^{*}$ is usually a biased estimator of $\overline{\theta}$ as measured by equation (2.6). The BC method provides an algorithm to correct for this bias. The algorithm is given below (for a justification see Section 11.3 in Efron and Hastie, 2016):

1. compute $p_0 = \#\left\{\overline{\theta}^{*b} \leq \overline{\theta}\right\} / B$, that is the proportion of bootstrap replicates lower than or equal to $\overline{\theta}$

2. find $z_0 = \Phi^{-1}(p_0)$, where Φ^{-1} denotes the inverse function of the standard normal cumulative distribution function

3. compute the quantities
$$p_{\alpha/2} = \Phi\left(2z_0 + z_{\alpha/2}\right)$$
and
$$p_{1-\alpha/2} = \Phi\left(2z_0 + z_{1-\alpha/2}\right).$$

Then, the BC percentile interval corresponds to the $100p_{\alpha/2}$th and $100p_{1-\alpha/2}$th percentiles of the bootstrap distribution. Note that if $p_0 = 0.5$ the bootstrap distribution is symmetric and $z_0 = 0$, so that the BC percentile interval reduces to percentile interval.

Finally, an even more accurate bootstrap interval is available, the so called bias-corrected and *accelerated* (BCa) percentile interval, whose definition is more involved, so we invite you to refer to the references for more details[57]. In Stata you can get it by adding the `bca` option to the `bootstrap` command.

The algebra of principal components analysis

Suppose that $\boldsymbol{X}$ is an $(n \times p)$ data matrix consisting of n observations over the p variables $X_1, X_2, \ldots, X_p$. We denote the generic value of the jth variable on the ith observation as x_{ij}, while $\boldsymbol{S}$ and $\boldsymbol{R}$ indicate the corresponding sample covariance and correlation matrices, respectively. We now illustrate the derivation of PCA from the covariance matrix, but a similar development applies to the correlation matrix.

The first principal component for the ith observation, z_{i1}, is defined as the linear combination
$$z_{i1} = a_{11}x_{i1} + a_{12}x_{i2} + \ldots a_{1p}x_{ip}$$
with the largest sample variance among the linear combinations for which the sum of squared coefficients $\sum_{j=1}^{p} a_{1j}^2$ is equal to one. The last restriction is needed because otherwise the variance of the first principal component can be increased without limit by increasing the values of the coefficients $(a_{11}, a_{12}, \ldots, a_{1p})$.

The second principal component for the ith observation is defined as the linear combination
$$z_{i2} = a_{21}x_{i1} + a_{22}x_{i2} + \ldots a_{2p}x_{ip}$$
with the largest variance subject to the conditions that the sum of squares of the coefficients $(a_{21}, a_{22}, \ldots, a_{2p})$, that is $\sum_{j=1}^{p} a_{2j}^2$, is equal to one and the inner product of the coefficients of the first and second components, that is $\sum_{j=1}^{p} a_{1j}a_{2j}$, is equal zero. The last requirement implies that the first two components must be uncorrelated.

Similarly, the generic kth principal component, for the ith observation is defined as the linear combination
$$z_{ik} = a_{k1}x_{i1} + a_{k2}x_{i2} + \ldots a_{kp}x_{ip}$$

[57] The BCa method requires the calculation of quantities that involve the jackknife approach. Hence, BCa is more computationally demanding.

with the largest variance subject to the same conditions as above, that is the sum of squared coefficients equal to one and no correlation with the previous components.

The coefficients that define the principal components are thus found by solving a series of constrained optimization problems (i.e., maximizing the variances subject to the conditions described above). It is possible to show that the solutions of these problems correspond to the eigenvectors of the sample covariance matrix $\boldsymbol{S}$. In particular, the kth set of coefficients $(a_{k1}, a_{k2}, \ldots, a_{kp})$ corresponds to the eigenvector of $\boldsymbol{S}$ associated with its kth largest eigenvalue λ_k. We remind that eigenvalues and eigenvectors define what is known in mathematics as the *spectral decomposition* of a square matrix defined as

$$\boldsymbol{S} = \boldsymbol{L}\boldsymbol{\Lambda}\boldsymbol{L}',$$

where $\boldsymbol{\Lambda}$ is a diagonal matrix containing the eigenvalues in decreasing order, while $\boldsymbol{L}$ is the matrix whose columns are the (normalized) eigenvectors[58].

The variance of the kth principal component is given by the kth largest eigenvalue λ_k. The total variance of the p principal components is equal to the sum of the variances of the original variables, that is

$$\sum_{k=1}^{p} \lambda_k = \text{tr}(\boldsymbol{S}),$$

where on the right hand side we used the trace operator defined as the sum of the elements on the main diagonal of $\boldsymbol{S}$, that is the variances. So, the proportion of the total variance explained by the kth component is given by

$$P_k = \frac{\lambda_k}{\text{tr}(\boldsymbol{S})}.$$

The covariance of the jth observed variable X_j with the kth principal component Z_h is given by

$$\text{Cov}(X_j, Z_k) = \lambda_k a_{kj}.$$

It then follows that the corresponding linear correlation index is

$$\begin{aligned} r_{X_j Z_k} &= \frac{\text{Cov}(X_j, Z_k)}{s_{X_j} \cdot s_{Z_k}} \\ &= \frac{\lambda_k a_{kj}}{s_{jj}^{1/2} \sqrt{\lambda_k}} \\ &= \frac{\sqrt{\lambda_k} a_{kj}}{s_{jj}^{1/2}}, \end{aligned}$$

where s_{jj} is the jth element on the main diagonal of $\boldsymbol{S}$. Clearly, if the original variables are standardized, that is we perform PCA on the correlation matrix, the previous

[58] A more general decomposition of a generic matrix (i.e., not necessarily squared) is the *singular value decomposition*, which is used by many software to perform PCA (see for example Rencher and Christensen, 2012, Chapter 2.)

expression reduces to

$$r_{X_j Z_k} = \sqrt{\lambda_k}\, a_{kj}. \tag{2.9}$$

Finally, the component score for the kth principal component of the ith observation is computed as

$$z_{ik} = a_{k1}(x_{i1} - \bar{x}_1) + a_{k2}(x_{i2} - \bar{x}_2) + \ldots a_{kp}(x_{ip} - \bar{x}_p),$$

where $\bar{x}_j$ denotes the sample mean of the jth observed variable.

Clustering stopping rules

Given an $(n \times p)$ data matrix $\boldsymbol{X}$, where n denotes the number of observations and p the number of continuous variables, we first define the *total* dispersion matrix as

$$\boldsymbol{T} = \sum_{g=1}^{K} \sum_{i=1}^{n_g} (\boldsymbol{x}_{gi} - \bar{\boldsymbol{x}})(\boldsymbol{x}_{gi} - \bar{\boldsymbol{x}})^{\top}, \tag{2.10}$$

where $\boldsymbol{x}_{gi}$ represents the p-dimensional vector of variable values for the ith observation in the gth cluster and $\bar{\boldsymbol{x}}$ corresponds to the p-dimensional vector containing the overall sample means for each variable[59]. The $\boldsymbol{T}$ matrix contains the overall sum of squares and cross-products for the p variables in $\boldsymbol{X}$. It is possible to show that the total dispersion matrix $\boldsymbol{T}$ can be decomposed into

$$\boldsymbol{T} = \boldsymbol{W} + \boldsymbol{B},$$

where W denotes the *within-cluster* dispersion matrix defined as

$$\boldsymbol{W} = \sum_{g=1}^{K} \sum_{i=1}^{n_g} (\boldsymbol{x}_{gi} - \bar{\boldsymbol{x}}_g)(\boldsymbol{x}_{gi} - \bar{\boldsymbol{x}}_g)^{\top}, \tag{2.11}$$

and B indicates the *between-cluster* dispersion matrix defined as

$$\boldsymbol{B} = \sum_{g=1}^{K} n_g (\bar{\boldsymbol{x}}_g - \bar{\boldsymbol{x}})(\bar{\boldsymbol{x}}_g - \bar{\boldsymbol{x}})^{\top}, \tag{2.12}$$

where $\bar{\boldsymbol{x}}_g$ represents the p-dimensional vector containing the group-specific sample means for the gth cluster.

For a fixed number of clusters g, the Caliński-Harabasz stopping rule is based on the following quantity, known as the **pseudo F index**,

$$CH_g = \frac{\mathrm{tr}(\boldsymbol{B})/(g-1)}{\mathrm{tr}(\boldsymbol{W})/(n-g)}. \tag{2.13}$$

Then, the optimal number of clusters is chosen as that with the maximum value of CH_g.

[59] Note that the quantity $\boldsymbol{T}$ defined here is a matrix, while the E quantity defined in Section 2.3.1.1 is a scalar. The two expressions provide the same result only for univariate data, that is when $p = 1$.

The Duda-Hart stopping rule provides a criterion for deciding whether to split a given cluster C_m into two subclusters C_{m_1} and C_{m_2} (so that $C_m = C_{m_1} \cup C_{m_2}$) by computing the index

$$DH_m = \frac{J_2(m)}{J_1(m)}, \tag{2.14}$$

where $J_2(m)$ is the sum of within-cluster sum of squared distances between the observations and the cluster centre when the data are partitioned into clusters C_{m_1} and C_{m_2}, while $J_1(m)$ is the within-cluster sum of squared distances when C_{m_1} and C_{m_2} are merged together. More technically,

$$J_1(m) = S_m$$

and

$$J_2(m) = S_{m_1} + S_{m_2},$$

where, for a generic cluster C_ℓ, S_ℓ is defined as

$$S_\ell = \sum_{i \in C_\ell} (\boldsymbol{x}_i - \bar{\boldsymbol{x}}_\ell)^\top (\boldsymbol{x}_i - \bar{\boldsymbol{x}}_\ell).$$

Large values of the Duda-Hart index (2.14) indicate that moving one step ahead in the dendrogram hierarchy (i.e., splitting one of the clusters in two subclusters) will produce a similar clustering solution, and so we should not go with it. A formal testing procedure for choosing the optimal number of clusters is provided by Gordon (1999), which consists in rejecting the null hypothesis that cluster C_m is homogeneous if

$$DH_m < DH_m^{\text{crit}} = 1 - \frac{2}{\pi p} - z_\alpha \sqrt{\frac{2(1 - 8/\pi^2 p)}{n_m p}}, \tag{2.15}$$

where z_α is the standard normal score for a significance level equal to α. Practically, we choose as optimal the smallest number of clusters such that (2.15) is not satisfied (i.e., the first one for which we do not reject the null hypothesis).

Duda et al. (2001) also propose a **pseudo t^2 index** defined as

$$t_m^2 = \frac{(1 - DH_m)}{DH_m} (n_{m_1} + n_{m_2} - 2), \tag{2.16}$$

also reported by Stata. In this case, small values of t_m^2 indicate that there is no need to proceed further in splitting the clusters. The corresponding testing procedure consists in rejecting the null hypothesis that cluster C_m is homogeneous if

$$t_m^2 > \left(\frac{1 - DH_m^{\text{crit}}}{DH_m^{\text{crit}}} \right) (n_{m_1} + n_{m_2} - 2). \tag{2.17}$$

So, according to this procedure, we choose as the optimal number of clusters the smallest value such that (2.17) is not satisfied. For a review of cluster validation measures we suggest to look at Chapter 17 of Zaki and Meira (2020).

Finite mixture models estimation and selection

Suppose our sample is made of n independent observations $\boldsymbol{x}_1,\ldots,\boldsymbol{x}_n$, where the generic observation $\boldsymbol{x}_i = (x_{i1},\ldots,x_{ip})$ includes the values of p different variables. We assume that the $\boldsymbol{x}_i$s follow a FMM whose density function[60] is defined as

$$f(\boldsymbol{x}_i) = \sum_{j=1}^{K} \pi_j f_j(\boldsymbol{x}_i;\boldsymbol{\theta}_j). \tag{2.18}$$

Then, the likelihood function for these data can be written as

$$L(\boldsymbol{\Theta},\boldsymbol{\pi}|\boldsymbol{x}_1,\ldots,\boldsymbol{x}_n) = \prod_{i=1}^{n}\sum_{j=1}^{K} \pi_j f_j(\boldsymbol{x}_i;\boldsymbol{\theta}_j), \tag{2.19}$$

where $\boldsymbol{\Theta} = (\boldsymbol{\theta}_1,\ldots,\boldsymbol{\theta}_K)$ and $\boldsymbol{\pi} = (\pi_1,\ldots,\pi_{K-1})$. The corresponding log-likelihood function is given by

$$\begin{aligned} \ell(\boldsymbol{\Theta},\boldsymbol{\pi}|\boldsymbol{x}_1,\ldots,\boldsymbol{x}_n) &= \log L(\boldsymbol{\Theta},\boldsymbol{\pi}|\boldsymbol{x}_1,\ldots,\boldsymbol{x}_n) \\ &= \sum_{i=1}^{n} \log\left(\sum_{j=1}^{K} \pi_j f_j(\boldsymbol{x}_i;\boldsymbol{\theta}_j)\right). \end{aligned} \tag{2.20}$$

Note that this is a rather complicated expression, since the logarithm is not acting directly on the mixture component densities.

The maximum likelihood estimates of the parameters $\boldsymbol{\Theta}$ and $\boldsymbol{\pi}$ are found by maximizing the log-likelihood function (2.20). This requires setting its derivatives to zero and solving the corresponding system of equations. Unfortunately, given the complexity of the problem, this procedure doesn't produce a closed-form solution and we must rely on numerical evaluations.

An elegant and powerful approach for finding the maximum likelihood estimates is the **expectation-maximization** (EM) algorithm. The EM algorithm is a two-stage iterative optimization technique that is guaranteed to improve the log-likelihood at each passage. Starting from some initial parameter values, at each iteration the algorithm involves two stages called *expectation step* (E-step) and *maximization step* (M-step). In the E-step, a lower bound of the log-likelihood function for the current parameter values is computed, which takes the same value as the log-likelihood itself. Moreover, the lower bound is defined to have the same slope as the log-likelihood. Then, the M-step entails finding new parameter values by maximizing the lower bound. This process is iterated until the maximum value of the log-likelihood is reached. Figure 2.32 provides a simplified graphical representation of what happens at each iteration of the EM algorithm.

The information criteria AIC and BIC are defined as

$$\text{AIC} = -2\ell(\widehat{\boldsymbol{\Theta}},\widehat{\boldsymbol{\pi}}|\boldsymbol{x}_1,\ldots,\boldsymbol{x}_n) + 2d \tag{2.21}$$

[60]For simplicity here we refer only to the case of continuous variables, but the same discussion also applies to the discrete case.

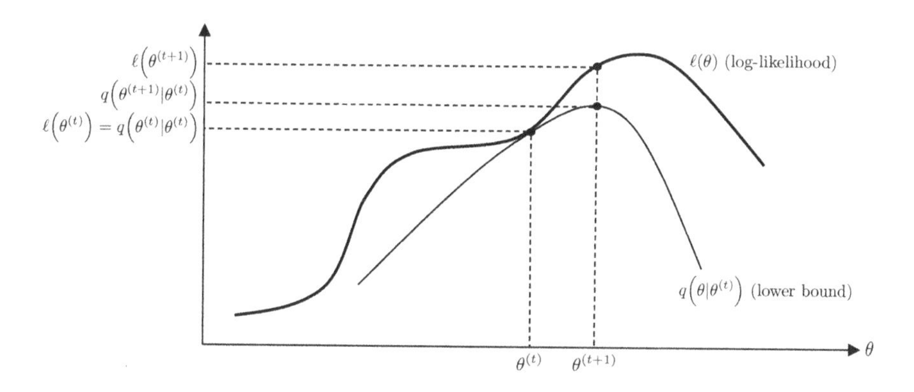

FIGURE 2.32: Graphical interpretation of a single iteration for the EM algorithm.

and

$$\text{BIC} = -2\ell(\widehat{\boldsymbol{\Theta}}, \widehat{\boldsymbol{\pi}}|\boldsymbol{x}_1, \ldots, \boldsymbol{x}_n) + d\log(n), \tag{2.22}$$

respectively. In the expressions above, the log-likelihood function is evaluated at the estimated parameter values, while d denotes the total number of parameters in the model. In both cases, the first term represents the lack of fit of the data, while the second one corresponds to a penalty for the complexity of the model. So, AIC and BIC only differ for the penalization of model's complexity. There is still no consensus in the literature about which one between AIC an BIC provides the best approach for model selection. Therefore, several authors suggest using multiple indicators as well as problem-specific considerations.

Path analysis using matrices

We provide here more technical details about PA. We indicate with p the number of endogenous variables and with q the number of exogenous variables in the model. The structural form of a general PA model is usually denoted as

$$\boldsymbol{y} = \boldsymbol{\alpha} + \boldsymbol{B}\boldsymbol{y} + \boldsymbol{\Gamma}\boldsymbol{x} + \boldsymbol{\zeta}, \tag{2.23}$$

where $\boldsymbol{y}$ is the $(p \times 1)$ vector of endogenous variables, $\boldsymbol{x}$ is the $(q \times 1)$ vector of exogenous variables, $\boldsymbol{\alpha}$ is the $(p \times 1)$ vector of structural intercepts, $\boldsymbol{B}$ is the $(p \times p)$ matrix that contains the coefficients relating the endogenous variables to each other, $\boldsymbol{\Gamma}$ is the $(p \times q)$ matrix of the coefficients relating the exogenous variables to the endogenous variables and $\boldsymbol{\zeta}$ is the $(p \times 1)$ vector of error terms. Moreover, we denote with $\boldsymbol{\Psi}$ the $(p \times p)$ covariance matrix of the error terms, while the corresponding means are assumed to be zero, and with $\boldsymbol{\Phi}$ the $(q \times q)$ covariance matrix of the exogenous variables. Therefore, the whole list of model's parameters is given by $\boldsymbol{\theta} = (\boldsymbol{B}, \boldsymbol{\Gamma}, \boldsymbol{\Psi}, \boldsymbol{\Phi})$, some of which will be fixed (most of the times set to zero), while the remaining ones are free and must be estimated.

The reduced form of the model is obtained by first rewriting (2.23) as

$$(\boldsymbol{I}-\boldsymbol{B})\boldsymbol{y}=\boldsymbol{\alpha}+\boldsymbol{\Gamma}\boldsymbol{x}+\boldsymbol{\zeta},$$

and then, after checking that the matrix $(\boldsymbol{I}-\boldsymbol{B})$ is non-singular, by rearranging to obtain

$$\begin{aligned}\boldsymbol{y} &= (\boldsymbol{I}-\boldsymbol{B})^{-1}\boldsymbol{\alpha}+(\boldsymbol{I}-\boldsymbol{B})^{-1}\boldsymbol{\Gamma}\boldsymbol{x}+(\boldsymbol{I}-\boldsymbol{B})^{-1}\boldsymbol{\zeta}\\ &= \boldsymbol{\Pi}_0+\boldsymbol{\Pi}_1\boldsymbol{x}+\boldsymbol{\zeta}^*, \end{aligned} \tag{2.24}$$

where $\boldsymbol{\Pi}_0=(\boldsymbol{I}-\boldsymbol{B})^{-1}\boldsymbol{\alpha}$ is the $(p\times 1)$ vector of reduced form intercepts, $\boldsymbol{\Pi}_1=(\boldsymbol{I}-\boldsymbol{B})^{-1}\boldsymbol{\Gamma}$ is the $(p\times q)$ matrix of reduced form slopes and $\boldsymbol{\zeta}^*=(\boldsymbol{I}-\boldsymbol{B})^{-1}\boldsymbol{\zeta}$ is the $(p\times 1)$ vector of reduced form error terms ($\boldsymbol{I}$ denotes the p-dimensional identity matrix).

Using the rules of matrix algebra, we can show that the model-implied covariance matrix for the endogenous and exogenous variables is given by

$$\begin{aligned}\boldsymbol{\Sigma}(\boldsymbol{\theta}) &= \begin{bmatrix}\boldsymbol{\Sigma}_{yy} & \boldsymbol{\Sigma}_{yx}\\ \boldsymbol{\Sigma}_{xy} & \boldsymbol{\Sigma}_{xx}\end{bmatrix}\\ &= \begin{bmatrix}\mathrm{E}(\boldsymbol{y}\boldsymbol{y}') & \mathrm{E}(\boldsymbol{y}\boldsymbol{x}')\\ \mathrm{E}(\boldsymbol{x}\boldsymbol{y}') & \mathrm{E}(\boldsymbol{x}\boldsymbol{x}')\end{bmatrix}\\ &= \begin{bmatrix}(\boldsymbol{I}-\boldsymbol{B})^{-1}(\boldsymbol{\Gamma}\boldsymbol{\Phi}\boldsymbol{\Gamma}'+\boldsymbol{\Psi})(\boldsymbol{I}-\boldsymbol{B})^{-1'} & (\boldsymbol{I}-\boldsymbol{B})^{-1}\boldsymbol{\Gamma}\boldsymbol{\Phi}\\ \boldsymbol{\Phi}\boldsymbol{\Gamma}'(\boldsymbol{I}-\boldsymbol{B})^{-1'} & \boldsymbol{\Phi}\end{bmatrix}. \end{aligned} \tag{2.25}$$

As we described in Section 2.4, to estimate the parameters we can use ML or WLS. The corresponding discrepancy functions are given by

$$F_{ML} = \log|\boldsymbol{\Sigma}(\boldsymbol{\theta})|+\mathrm{tr}\left[\boldsymbol{S}\boldsymbol{\Sigma}^{-1}(\boldsymbol{\theta})\right]-\log|\boldsymbol{S}|-(p+q) \tag{2.26}$$

$$F_{WLS} = [\boldsymbol{S}-\boldsymbol{\Sigma}(\boldsymbol{\theta})]'\boldsymbol{W}^{-1}[\boldsymbol{S}-\boldsymbol{\Sigma}(\boldsymbol{\theta})], \tag{2.27}$$

where $|\boldsymbol{A}|$ and $\mathrm{tr}(\boldsymbol{A})$ denote the determinant and the trace of a square matrix $\boldsymbol{A}$ respectively, while $\boldsymbol{W}^{-1}$ is a weight matrix for the differences $\boldsymbol{S}-\boldsymbol{\Sigma}(\boldsymbol{\theta})$. ULS corresponds to the case where $\boldsymbol{W}=\boldsymbol{I}$, the identity matrix, while the most common choice for GLS corresponds to the choice $\boldsymbol{W}=\boldsymbol{S}$, the sample covariance matrix of all variables.

According to the matrix representation above, one can show that the direct, indirect and total effects are given by the expressions reported in the following table (see Bollen, 1989):

Effect	Exogenous to Endogenous	Endogenous to Endogenous
Direct	$\boldsymbol{\Gamma}$	$\boldsymbol{B}$
Indirect	$(\boldsymbol{I}-\boldsymbol{B})^{-1}\boldsymbol{\Gamma}-\boldsymbol{\Gamma}$	$(\boldsymbol{I}-\boldsymbol{B})^{-1}-\boldsymbol{I}-\boldsymbol{B}$
Total	$(\boldsymbol{I}-\boldsymbol{B})^{-1}\boldsymbol{\Gamma}$	$(\boldsymbol{I}-\boldsymbol{B})^{-1}-\boldsymbol{I}$

3

PLS Structural Equation Modelling: Specification and Estimation

In this chapter we introduce the main topic of the book, the partial least squares structural equation modelling (PLS-SEM) methodology[1]. We anticipate that this chapter is longer than the other ones because our aim is to provide a comprehensive introduction to the subject. The formulas we show in the chapter are not strictly necessary to understand the overall logic, and so you can safely skip them at a first reading without compromising your understanding. For completeness, we place the more technical details in an appendix at the end of the chapter. What we instead suggest not to skip are the algorithmic details of the PLS-SEM approach, because they will allow you to better grasp the core ideas of the methodology.

3.1 Introduction

In Section 2.4 we reviewed path analysis, which represents an approach for estimating the relationships between a set of *observed* variables described by a system of equations. One of the underlying assumptions of path analysis is that we are able to accurately measure the quantities we need to study. However, often this is not easy in practice because, due to their complex nature, the quantities involved in the study are not directly observable. A classic example from marketing is brand image, that is the collective consumer perception of what a company's products or services represent. Brand image is a "concept" that can't be measured directly, but instead it must be assessed indirectly using instruments like customer surveys or social media listening tools. An equivalent way to refer to the same issue is that in path analysis, as well as in linear regression, principal components and the other more traditional statistical analyses, it is assumed that the study quantities are measured *without error*. Measurement error is usually defined as the difference between a measured quantity and its true value. Measurement error can originate from many different sources, the most common being poor wording in survey items, misunderstanding of scales

[1] An alternative terminology for PLS-SEM is partial least squares path modelling (PLS-PM). Both the terms are used in the literature and it is mainly a matter of personal preferences which one to opt for. In this book we adopt the PLS-SEM parlance because we think it better signals the fact that this methodology falls within the realm of SEM.

and misinterpretation of the questions as they were conceived by the researcher. The consequences of measurement error are potentially biased and inaccurate coefficient estimates. As a first example, let's consider the simple situation of two observed variables x and y both of which are measured with error, that is

$$x = \lambda_1 \xi + \delta \tag{3.1}$$

and

$$y = \lambda_2 \eta + \varepsilon. \tag{3.2}$$

In the equations above, ξ and η represent the unobserved true scores of which x and y are only imprecise measurements, while δ and ε are the corresponding measurement errors. For simplicity, we assume that λ_1 and λ_2 are both equal to one[2]. We are interested in estimating the structural relationship between ξ and η described by the following equation

$$\eta = \gamma \xi + \zeta. \tag{3.3}$$

Suppose that to estimate γ (the slope coefficient between η and ξ) we fit the model

$$y = \gamma^* x + \zeta^*, \tag{3.4}$$

thus disregarding the fact that x and y are both measured with error. Under suitable conditions on the different error terms, it can be shown (see Bollen, 1989, Chapter 5) that

$$\gamma^* = \frac{\mathrm{Cov}(x,y)}{\mathrm{Var}(x)} = \gamma \left[\frac{\mathrm{Var}(\xi)}{\mathrm{Var}(x)} \right], \tag{3.5}$$

where the quantity within brackets, called the *reliability coefficient* of x, represents the proportion of total variability of x that is accounted for by ξ; therefore it is a number in between zero and one. The consequence of the last equation is that, in presence of measurement error in the predictor variable x, the slope coefficient γ^* underestimates the actual association between the unobserved true scores[3]. The only case where this effect doesn't occur is when x is measured without error (i.e., when $x = \xi$, or equivalently when $\delta = 0$), so that the reliability of x is equal to one (i.e., x is perfectly reliable as a measurement of ξ). If an estimate of the reliability coefficient of x is available, we may correct (or "disattenuate") the estimated value of γ^* by reversing equation (3.5). Otherwise, one should at least assess the sensitivity of the estimate to different hypothetical values of the x reliability. In presence of several covariates, the effects of measurement error are less clear cut, but they still apply. The same issue produces an even more critical consequence, that is the bias does not vanish if one increases the sample size, a situation that in statistics is known as *inconsistency*.

[2]When λ_1 and λ_2 take values different from one, the calculations are more involved, but the final conclusion still holds. Note also that we are assuming mean-centred variables since we didn't include intercepts in the equations.

[3]Another terminology that is often used in this context is that the slope coefficient γ is *attenuated*. Also notice that for the conclusion to hold, it must be that x is measured with error, but the response does not necessarily be. Therefore, if y only is measured with error, this doesn't imply any bias in the assessment of the association between y and x, but only an increased uncertainty in the estimate variability.

The main methodological innovation introduced by the structural equation modelling (SEM) framework is allowing for the possibility that the observed variables are measured with error. Within this approach, in fact, it is assumed that the observed variables are imprecise measurements of the concepts we are actually interested in studying. Observed variables are usually referred to as *manifest variables* or *indicators*, while the unobserved ones are called *latent variables*, *constructs* or sometimes simply *proxies*. So, every SEM model is composed of two parts, one that regards the relationships between the manifest and latent variables, which is called the **measurement** or **outer model**, and the other that represents the relationships between the latent variables, the so called **structural** or **inner model**.

As we introduced in the first chapter, there are fundamentally two different approaches for SEM, the more traditional **covariance-based SEM** (CB-SEM) introduced by Jöreskog (1969), and the **variance-based** or **partial least squares SEM** (PLS-SEM) approach proposed by Wold (1975)[4]. Both these methodologies share a common terminology and the general aim, that is assessing the association between (unobserved) quantities described by a set of simultaneous structural equations while accounting for the fact that these quantities are measured by imprecise instruments. The main difference between the two approaches consists in the estimation process. CB-SEM is typically based on maximum likelihood. In CB-SEM the idea is to look for parameter values that provide (model-implied) covariances between the manifest variables which reproduce as closely as possible those actually observed. PLS-SEM, instead, estimates the parameters by maximizing the explained variance of the exogenous latent variables. Moreover, in CB-SEM, multi-item variables are incorporated into the model using the factor analytic technique[5]. In PLS-SEM instead multi-item variables are used to directly generate weighted composites[6].

As we already discussed in Chapter 1, even if the two approaches appeared approximately at the same time in the literature, CB-SEM has seen a more rapid and wide diffusion compared to PLS-SEM. One of the motivations for the increased popularity of CB-SEM can be certainly attributed to the early availability of software packages, in particular the highly popular LISREL package[7] (Jöreskog et al., 2016). Other software for CB-SEM analysis that contributed to the spread of the method are AMOS[8], EQS[9] and Mplus[10]. More recently, Stata also introduced its own CB-SEM toolset with the `sem` and `gsem` commands. Similarly, SAS provides the comprehensive CALIS procedure. On the side of open-source implementations, the only platform that currently includes functions for CB-SEM is `R` with the comprehensive

[4]There are also other approaches to SEM that have been developed so far in the literature, such as *generalized structured component analysis* (Hwang and Takane, 2014), but currently they are less popular so we do not describe them here.

[5]A SEM model including only the measurement model part is usually referred to as confirmatory factor analysis.

[6]As a consequence of this difference, it is also said that CB-SEM is a factor-based method, while PLS-SEM is composite-based.

[7]`https://ssicentral.com/index.php/products/lisrel/`.

[8]`https://www.ibm.com/products/structural-equation-modeling-sem`.

[9]`http://www.mvsoft.com/eqs60.htm`.

[10]`https://www.statmodel.com`.

lavaan, OpenMx and sem packages. For what regards PLS-SEM, the first available software, LVPLS, has been introduced in 1984 by Lohmöller, more than 10 years later than the first release of LISREL. Then, other tools like PLS-Graph and PLS-GUI appeared, but they are no longer available or maintained. Currently, the most popular commercial software packages for PLS-SEM are ADANCO[11], SmartPLS[12], WarpPLS[13] and XLSTAT-PLSPM[14]. For what regards open-source implementations, the following R packages are available: cSEM (Rademaker and Schuberth, 2020), matrixpls (Rönkkö, 2020), plspm (Sanchez, 2013) and semPLS (Monecke and Leisch, 2012). Recently, we developed our own open-source implementation for Stata, the plssem package (Venturini and Mehmetoglu, 2019). As we will describe in the following, plssem is able to perform all the analyses illustrated in this book.

In the next sections we present the basics of PLS-SEM, and in particular we provide the details on how to specify and estimate a PLS-SEM model. Before proceeding, let us spend some words about notation: in this book we adopt the standard PLS-SEM notation (e.g., Esposito Vinzi et al., 2010). This notation is similar but not perfectly overlapped with that used in LISREL. In particular, we do not explicitly distinguish between exogenous and endogenous variables. We are aware that this may be a little uncomfortable for those who are accustomed to LISREL, but we think that it allows moving back and forth from the literature more easily. Nonetheless, we may occasionally modify the notation later in the book whenever this simplifies our discussion.

3.2 Model Specification

Figure 3.1 shows the path diagram for a hypothetical model that involves nine manifest variables $x_{11}, x_{21}, \ldots, x_{33}$ and three constructs ξ_1, ξ_2 and ξ_3. The diagram shows that the constructs form blocks with the manifest variables they are connected to, which are represented as dashed boxes in the picture. In particular, ξ_1 is connected to the three indicators x_{11}, x_{21} and x_{31}, ξ_2 to x_{12}, x_{22} and x_{32}, and ξ_3 forms a block with the x_{13}, x_{23} and x_{33} variables. However, we can see a difference between ξ_2 and the other two constructs, because the arrows in the ξ_2 block point towards the latent variable instead of outwards, while for ξ_1 and ξ_3 we have the opposite situation.

As for classical CB-SEM, PLS-SEM models too are typically specified starting from a theoretical model that is either already established in the scientific literature or in the common practice of a specific field, or it corresponds to a new set of hypotheses regarding the relationships between the quantities involved in the phenomenon under

[11]https://www.composite-modeling.com.
[12]https://www.smartpls.com.
[13]http://warppls.com.
[14]https://www.xlstat.com/en/solutions/features/pls-path-modelling.

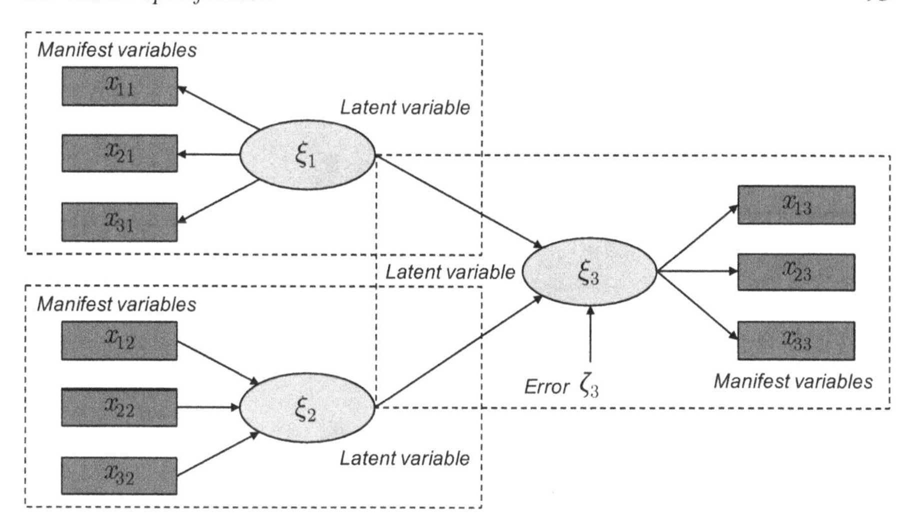

FIGURE 3.1: Path diagram for an hypothetical model with nine manifest variables, x_{11} to x_{33}, and three constructs, ξ_1, ξ_2 and ξ_3.

investigation. Indeed, we remark that a PLS-SEM model is only the analytic step that comes after the theoretical model has been carefully devised.

3.2.1 Outer (measurement) model

The outer model is the part of a PLS-SEM model establishing the relationships between each construct and the corresponding manifest variables. Figure 3.2 shows the outer model for the hypothetical example we introduced above, represented in the diagram by the set of connections in the shaded boxes.

The first step in defining the outer model is the conceptualization of the construct variables. As already stated, these are unobserved quantities which refer to concepts representing intangible but real phenomena (e.g., intelligence, happiness, perceptions, behaviours). Without delving into the details of measurement theory and scale development[15], here it suffices to say that the design of valid and reliable measures is a time-consuming but critical premise for any SEM analysis. In practice, researchers typically rely on scales published and already validated by others[16], or developed directly on their own. Since PLS-SEM is based on the estimation of variances, it works best with continuous (i.e., metric) data.

There are two different approaches to specify blocks in the measurement model[17], the **reflective** and the **formative** models. In a reflective block, indicators

[15]For a book-length technical treatment of these topics you can see Bandalos (2018) and the classic Nunnally and Bernstein (1994).

[16]An example for marketing studies is Bearden et al. (2011).

[17]In the literature you also find a third type of measurement model, the so called *multiple indica-*

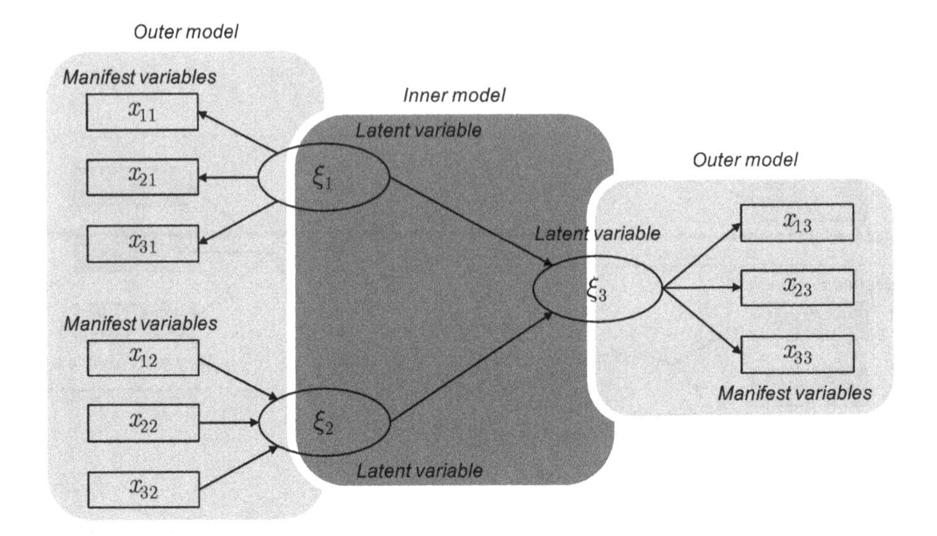

FIGURE 3.2: Path diagram for an hypothetical model with nine manifest variables, x_{11} to x_{33}, and three constructs, ξ_1, ξ_2 and ξ_3. The inner and outer models are highlighted with dark and light-shaded boxes respectively.

related to a given construct are assumed to measure a common underlying concept. More technically, in a reflective model the observed variability of a block of indicators is assumed to be fully explained by means of a single unobserved measure (i.e., the construct) and the indicator-specific error terms. From the definition, it follows that the indicators in a given reflective block are expected to be strongly linearly correlated with each other. For this reason, reflective measures should be checked for homogeneity and unidimensionality (more on these assessments will be provided in Chapter 4). In path diagrams reflective blocks are represented with arrows originating from the construct and pointing towards the corresponding manifest variables.

In the **formative** model, instead, each manifest variable represents a different dimension of the underlying concept. Therefore, unlike the reflective model, the formative model assumes neither homogeneity nor unidimensionality of the block. For this reason, the formative approach does not impose any restriction on the covariances between the indicators of the same construct. In other words, the formative model doesn't assume the existence of a common factor explaining the association between indicators in the block[18]. In path diagrams formative blocks are represented with arrows going from the indicators towards the construct. The technical appendix

tors and multiple indicator causes (*MIMIC*) model, which corresponds to a mixture of the reflective and formative approaches within the same block. Since MIMIC is not frequently used in practice, we don't provide further details here (see for example Tenenhaus et al., 2005).

[18]In reality, the correlation between the indicators for a formative construct may become a problem. As we will see later, this issue is related to multicollinearity (see Section A.2.6).

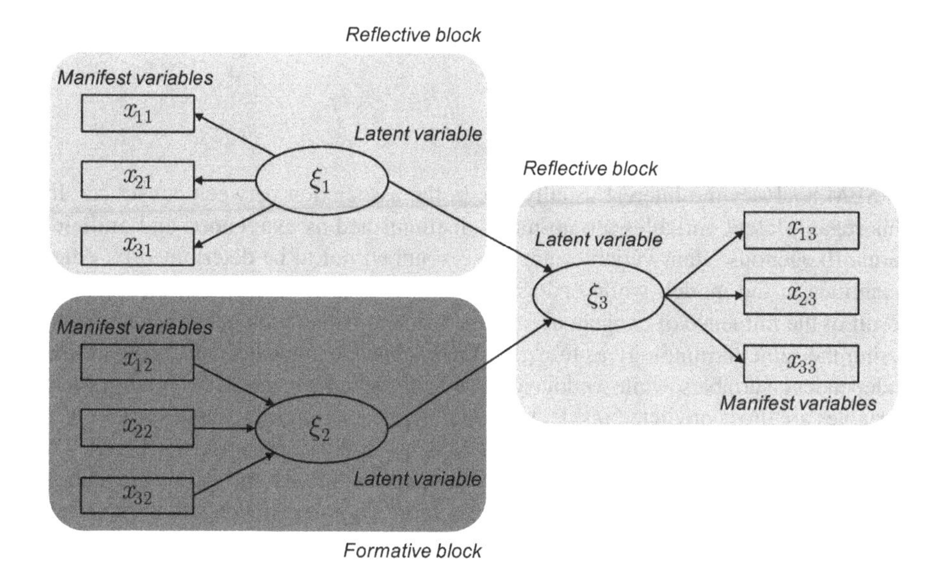

FIGURE 3.3: Path diagram for an hypothetical model with nine manifest variables, x_{11} to x_{33}, and three constructs, ξ_1, ξ_2 and ξ_3. The reflective and formative approaches are highlighted with light and dark-shaded boxes respectively.

at the end of the chapter provides a more formal definition of the different measurement frameworks. Figure 3.3 shows the same hypothetical model introduced above but highlighting that the constructs ξ_1 and ξ_3 are measured using a reflective approach, while the ξ_2 block involves a formative measurement model.

The choice of the model for measuring a given latent variable, either reflectively or formatively, is a modelling decision and there is no clear answer as to when to use one or the other. In general, we can say that formative measures should be used when the indicators precede in time the construct itself. Moreover, in the formative case, the indicators should provide the full list of motivations for the construct to vary. In some cases, it is possible that a single construct can be conceptualized both as reflective and formative. In these contexts it is the research objective that should guide you in the selection of the appropriate representation. Hair et al. (2017) provide some guidelines for choosing the measurement model for each construct (see in particular Exhibit 2.9 on page 52).

As it was implicit in our discussion, every latent variable must have indicators connected to it. A situation that occurs quite frequently is when researchers measure a construct using a single indicator (so called **single-item measures**). For this kind of construct, it does not matter the distinction between reflective and formative, since there is a substantial overlap between the "cause" and the "effect" in this case. Single-item measures are appealing because they require less efforts in the design and less money in the data collection. However, unless it is strictly necessary or requested by the theoretical model we want to estimate, we discourage measuring constructs in this way.

3.2.2 Inner (structural) model

In SEM analyses the interest usually lies in the association between constructs. In this regard, latent variables are normally distinguished as **exogenous** and **endogenous**. Exogenous latent variables are those assumed not to be determined by other quantities in the model, while endogenous variables are those that instead are the result of the influence of exogenous, and maybe some other endogenous, constructs. Using the same terminology as in regression analysis, exogenous constructs are the independent variables while endogenous are the dependent ones. So, endogenous variables are those predicted inside the model, while exogenous variables are predictors of some endogenous constructs, but they are not predicted themselves. Figure 3.4 shows through a different shading that in our simple model ξ_1 and ξ_2 are exogenous while ξ_3 is the only endogenous variable. In more complex models, it is very often the case that a latent variable is exogenous in a given relationship, but it becomes endogenous in another one. An example is provided in Figure 3.5, where we modified the previous diagram by adding the path from ξ_1 to ξ_2. Note that now ξ_2 is endogenous in the ξ_2 versus ξ_1 relationship, while it becomes exogenous when we focus on the relationship between ξ_3 and ξ_2. In this last example, we are assuming that ξ_2 is acting as a mediator of the association between ξ_1 and ξ_3. Briefly speaking, this means that ξ_1 has a direct effect on ξ_3, represented by the arrow connecting these two circles in the diagram. However, we also see that there is another path with which ξ_1 influences ξ_3, the path that goes through ξ_2. This further "compound" intervention corresponds to the indirect effect of ξ_1 on ξ_3, that is the effect that is mediated by the ξ_2 construct. So, the total effect of ξ_1 on ξ_3 is given by sum of the direct and indirect effects. Mediating effects can be very complicated and they can span a sequence of more than one direct path between two latent variables. Chapter 5 is dedicated to discuss the nuts and bolts of mediation analysis. Chapter 6 will instead introduce moderation, which corresponds to the situation where a third variable can directly intervene on the relationship between exogenous and endogenous variables modifying the strength or even the direction of the corresponding association.

The specification of the structural relationships is dictated by the theoretical model we intend to investigate. As we already mentioned, a study typically involves the formulation of a new theory for a given phenomenon or the modification of an existing one. The theory establishes a set of implications, that technically correspond to associations between the concepts involved in the study. In practice, most published studies include a collection of *research questions* regarding the relationships between the variables, which represent the source of the structural part specification. Since a model is nothing else than a simplification of a complex real phenomenon, to be useful it should be "parsimonious", in the sense that it shouldn't be too much detailed. In other terms, the model should include the least number of connections that allows answering the research questions. The overparameterization of a model, that is the inclusion of too many connections, implies a more complicated interpre-

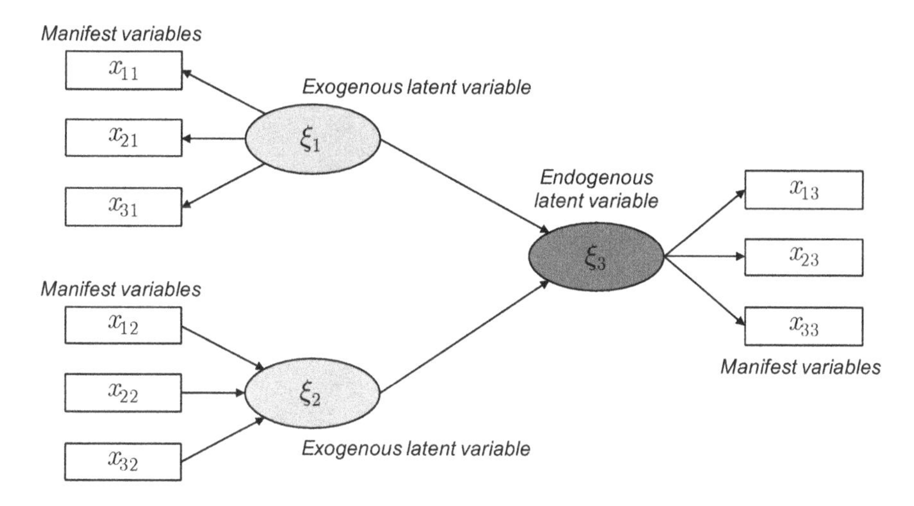

FIGURE 3.4: Path diagram for an hypothetical model with nine manifest variables, x_{11} to x_{33}, and three constructs, ξ_1, ξ_2 and ξ_3. Exogenous variables ξ_1 and ξ_2 are highlighted with light-shaded boxes, while the only endogenous variable, ξ_3, is shown using a darker shading.

tation of the findings, and it also requires the collection of a larger sample to reliably estimate the parameters in the model.

Even if the inner model is theoretically distinct from the outer model, they do not represent two independent parts of the analysis. This means that the practical relevance and the implications one can get from the structural model would be useless without a valid and reliable measurement model. In Chapter 4 we will provide a systematic strategy for assessing the goodness of both the measurement and structural parts of a PLS-SEM model.

3.2.3 Application: Tourists satisfaction

In this section we introduce an example taken from the first author's research agenda. We will refer to this application throughout the book. In particular we start here considering a simple model, whose aim is basically to serve as a practical implementation of the concepts we are introducing. The data for this application are contained in the `ch3_MotivesActivity.dta` file that is available on the GitHub repository of the book. This dataset is the result of a survey on a sample of 1000 domestic and international tourists visiting a destination in Norway in the summer of 2010. The respondents were asked to indicate:

- how important each of the following reasons/motives was for choosing to travel to the current destination using an ordinal scale from 1 (not important) to 5 (very important),
  - to gain new energy (`spm1_6`)

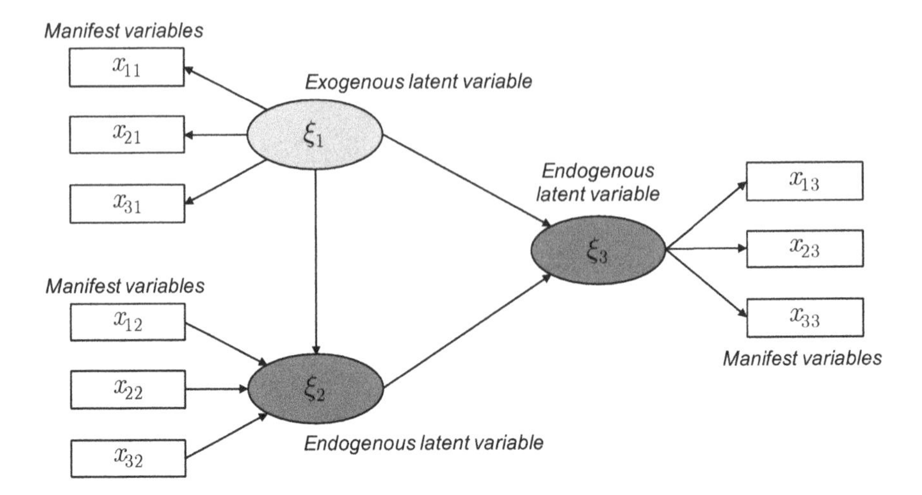

FIGURE 3.5: Path diagram for an hypothetical model with nine manifest variables, x_{11} to x_{33}, and three constructs, ξ_1, ξ_2 and ξ_3. ξ_1, the only exogenous variable here, is highlighted with a light-shaded box, while the endogenous ones, ξ_2 and ξ_3, are shown using a darker shading. In this model, ξ_2 is both exogenous and endogenous.

  - to get away from everyday life (`spm1_7`)
  - to avoid boredom (`spm1_8`)
  - to experience excitement (`spm1_9`)

- whether they had participated in the following activities during their stay in the current destination using a nominal scale, 0 (no) or 1 (yes),
  - to visit amusement/theme/family parks (`spm3_2`)
  - to visit the town (`spm3_6`)
  - to be in and see nature (`spm3_8`)
  - to catch fish (`spm3_12`)
- the respondents were also asked to indicate to what extent they agreed with the following statements regarding the current destination using an ordinal scale from 1 (totally disagree) to 5 (totally agree),
  - "overall I am satisfied with my holiday here" (`spm15_7`)
  - "my holiday here exceeded my expectations" (`spm15_8`)
  - "I am going to recommend my friends to spend a holiday here" (`spm15_3`).

The dataset contains some missing data that we ignore for the moment, that is we only use observations for which values on all variables are available. We will discuss different approaches to deal with missing values in Section 3.6.

```
    variable |         N      mean       p50        sd       p25       p75
-------------+------------------------------------------------------------
      energy |       983  3.343845         3  1.263534         3         4
     getaway |       997  3.985958         4  1.112093         3         5
     boredom |       993  2.897281         3  1.418314         2         4
    exciting |       991  3.123108         3  1.250851         2         4
   entertain |       991  .3733602         0  .4839408         0         1
   visittown |       991  .6377397         1  .4808962         0         1
      nature |       991  .5590313         1  .4967538         0         1
     fishing |       991  .1150353         0  .3192257         0         0
   recommend |       934  3.891863         4  .9668005         3         5
      satisf |       928   4.15625         4  .7843835         4         5
     expecta |       919  3.314472         3   .981372         3         4
--------------------------------------------------------------------------
```

FIGURE 3.6: Tourists satisfaction application. Summary statistics for the manifest variables in the model shown in Figure 3.7.

With the following code, we first load the data and rename the columns to get more informative outputs. Then, to start getting familiar with the data we compute some summary statistics, that are reported in Figure 3.6:

```
use ch3_MotivesActivity, clear

rename spm1_6 energy
rename spm1_7 getaway
rename spm1_8 boredom
rename spm1_9 exciting
rename spm3_2 entertain
rename spm3_6 visittown
rename spm3_8 nature
rename spm3_12 fishing
rename spm15_3 recommend
rename spm15_7 satisf
rename spm15_8 expecta

tabstat energy getaway boredom exciting entertain visittown ///
        nature fishing recommend satisf expecta, ///
        statistics(n mean median sd p25 p75) columns(statistics)
```

The first example we present is shown in Figure 3.7. According to the diagram, the structural model involves three constructs, namely SATISFACTION, MOTIVES and ACTIVITY. These represent respectively the overall satisfaction, the extent of motivation and experience intensity of the visit. SATISFACTION is assumed to be endogenous with its values that are predicted by the exogenous constructs MOTIVES and ACTIVITY. For what regards the measurement model, MOTIVES is modelled as reflective with indicators `energy`, `getaway`, `boredom` and `exciting`. For ACTIVITY it is assumed instead a formative model with indicators `entertain`, `visittown`, `nature` and `fishing`. Finally, SATISFACTION is reflective with indicators `recommend`, `satisf` and `expecta`. A couple of remarks are needed here. We note that in this example ACTIVITY is correctly modelled as formative

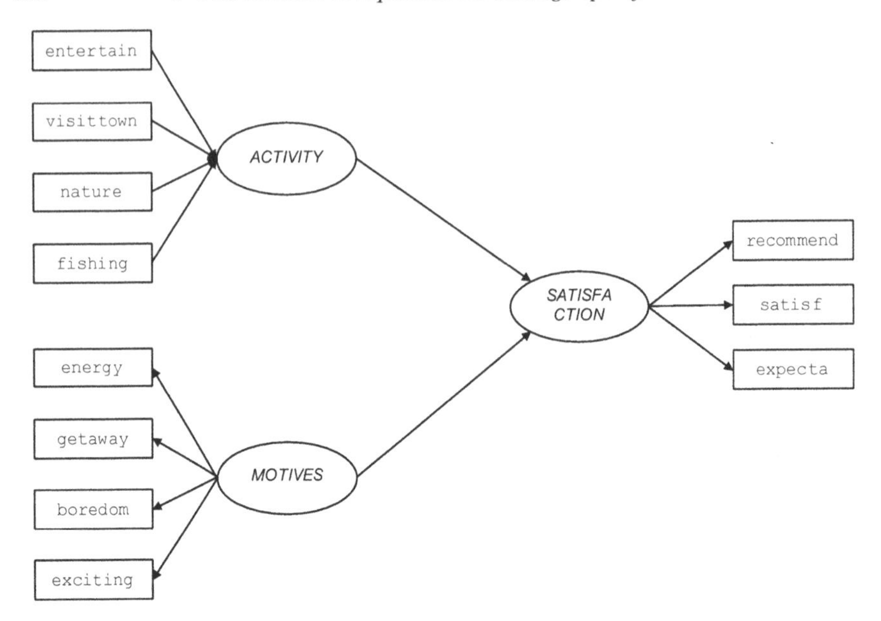

FIGURE 3.7: Tourists satisfaction application. The path diagram shows the assumed measurement and structural models for the first model we discuss in this chapter. SATISFACTION is the only endogenous latent variable, while ACTIVITY and MOTIVES are the two exogenous constructs that are assumed to predict the unobserved tourists satisfaction. MOTIVES is a reflective construct measured by the four indicators `energy` (originally called `spm1_6`), `getaway` (`spm1_7`), `boredom` (`spm1_8`) and `exciting` (`spm1_9`), while ACTIVITY is modelled as a formative measure using the observed variables `entertain` (`spm3_2`), `visittown` (`spm3_6`), `nature` (`spm3_8`) and `fishing` (`spm3_12`). SATISFACTION forms a reflective block with the indicators `recommend` (`spm15_3`), `satisf` (`spm15_7`) and `expecta` (`spm15_8`).

both because the indicators precede in time the construct itself and because they represent the possible explanations for the values taken by the construct itself. Modelling ACTIVITY as reflective here would therefore be mistaken. On the other side, it is appropriate to measure MOTIVES and SATISFACTION as reflective substantially for the opposite reasons: these are unobserved cognitive constructs that are assumed to originate the observed values for the indicators they are linked to. As such, the indicators of MOTIVES and SATISFACTION are expected to be strongly correlated among themselves.

The full description of the model is provided by the following equations[19]:

$$\text{SATISFACTION}_i = \beta_{03} + \beta_{13}\,\text{MOTIVES}_i + \beta_{23}\,\text{ACTIVITY}_i + \zeta_i \qquad (3.6)$$

[19]The notation we use in this example appears to be overly complicated, but it follows the conventions described in the technical appendix at the end of this chapter.

for the structural part, and

$$\text{ACTIVITY}_i = \pi_{11}\,\texttt{entertain}_i + \pi_{21}\,\texttt{visittown}_i + \pi_{31}\,\texttt{nature}_i + \pi_{41}\,\texttt{fishing}_i + \delta_i \quad (3.7)$$

$$\texttt{energy}_i = \lambda_{120} + \lambda_{12}\,\text{MOTIVES}_i + \varepsilon_{i12} \quad (3.8)$$
$$\texttt{getaway}_i = \lambda_{220} + \lambda_{22}\,\text{MOTIVES}_i + \varepsilon_{i22} \quad (3.9)$$
$$\texttt{boredom}_i = \lambda_{320} + \lambda_{32}\,\text{MOTIVES}_i + \varepsilon_{i32} \quad (3.10)$$
$$\texttt{exciting}_i = \lambda_{420} + \lambda_{42}\,\text{MOTIVES}_i + \varepsilon_{i42} \quad (3.11)$$

$$\texttt{recommend}_i = \lambda_{130} + \lambda_{13}\,\text{SATISFACTION}_i + \varepsilon_{i13} \quad (3.12)$$
$$\texttt{satisf}_i = \lambda_{230} + \lambda_{23}\,\text{SATISFACTION}_i + \varepsilon_{i23} \quad (3.13)$$
$$\texttt{expecta}_i = \lambda_{330} + \lambda_{33}\,\text{SATISFACTION}_i + \varepsilon_{i33} \quad (3.14)$$

for the measurement part.

3.3 Model Estimation

In this section we provide a detailed presentation of the PLS-SEM estimation process using the tourist satisfaction data. The tools for assessing the goodness of the measurement and structural parts will be the subject of Chapter 4 instead. The quantities that need to be estimated in a PLS-SEM model are:

1. the outer model parameters,
2. the inner model parameters,
3. the latent variable scores.

Before illustrating the steps for estimation, we introduce additional notation that will be useful later. The measurement model can be described by an adjacency matrix $\boldsymbol{M}$ whose entries m_{pq} take value 1 if indicator x_p belongs to the block that defines the qth latent variable $\boldsymbol{\xi}_q$, and 0 otherwise, with $p = 1, \ldots, P$ and $q = 1, \ldots, Q$. The adjacency matrix of the measurement model for the example in Figure 3.7 is provided in Table 3.1. Note that the matrix $\boldsymbol{M}$ does not convey any information about whether a construct is measured in a reflective or formative way.

Similarly, the structural model can be summarized by an adjacency matrix $\boldsymbol{S}$ whose entries s_{kq} take value 1 if the latent variable $\boldsymbol{\xi}_k$ is a predecessor of the latent variable $\boldsymbol{\xi}_q$ in the model, and 0 otherwise, with $k, q = 1, \ldots, Q$. The adjacency matrix of the structural model for the example in Figure 3.7 is reported in Table 3.2. Note that matrix $\boldsymbol{S}$ allows to recover the information about whether a latent variable

TABLE 3.1: Measurement model adjacency matrix $\boldsymbol{M}$ for the example shown in Figure 3.7. The generic element m_{pq} of the matrix is set to 1 if indicator $\boldsymbol{x}_p$ belongs to the block that defines latent variable $\boldsymbol{\xi}_q$, and 0 otherwise.

	ACTIVITY	MOTIVES	SATISFACTION
`entertain`	1	0	0
`visittown`	1	0	0
`nature`	1	0	0
`fishing`	1	0	0
`energy`	0	1	0
`getaway`	0	1	0
`boredom`	0	1	0
`exciting`	0	1	0
`recommend`	0	0	1
`satisf`	0	0	1
`expecta`	0	0	1

is exogenous or endogenous. More specifically, if the column corresponding to the latent variable $\boldsymbol{\xi}_q$ contains only zeros, that indicates that $\boldsymbol{\xi}_q$ is exogenous. In other words, contrary to the matrix $\boldsymbol{M}$ for the measurement model, $\boldsymbol{S}$ accounts for the directionality of the relationships among the latent variables.

TABLE 3.2: Structural model adjacency matrix $\boldsymbol{S}$ for the example shown in Figure 3.7. The generic element s_{kq} of the matrix is set to 1 if the latent variable $\boldsymbol{\xi}_k$ is a predecessor of the latent variable $\boldsymbol{\xi}_q$ in the model, and 0 otherwise.

	ACTIVITY	MOTIVES	SATISFACTION
ACTIVITY	0	0	1
MOTIVES	0	0	1
SATISFACTION	0	0	0

3.3.1 The PLS-SEM algorithm

The basic PLS-SEM algorithm (Lohmöller, 1989) involves the estimation of the so called **outer weights** by means of an iterative procedure in which the latent variable scores are obtained by alternating the outer and inner estimation steps. More specifically, the estimation algorithm consists of three sequential stages. In the first stage, latent variable scores are iteratively estimated for each case in the sample. Using these scores, in the second stage measurement model parameters (so called **outer coefficients** and **outer loadings**) are computed. In the same manner, in the third stage structural parameters (also called **path coefficients**) are finally estimated. The first stage is what makes PLS-SEM a novel method in that the second and third stages are

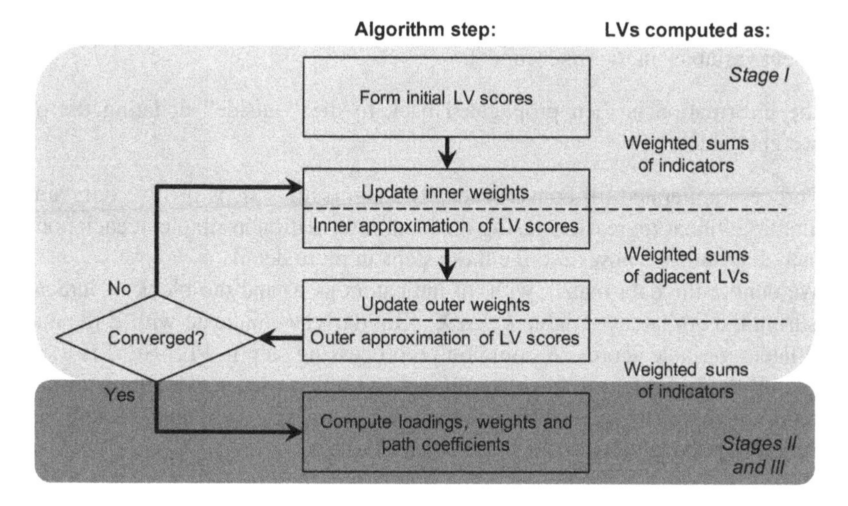

FIGURE 3.8: Diagrammatic representation of the basic PLS-SEM estimation algorithm stages (adapted from Rigdon (2013); LV means latent variables).

about conducting a series of standard OLS regressions[20]. To help grasping the whole process, we summarize it in Figure 3.8. We now provide more details on each stage.

3.3.2 Stage I: Iterative estimation of latent variable scores

The first stage of the algorithm involves an iterative procedure consisting of the following steps, whose aim is the estimation of the latent variable scores:

Step 0: Initialization of the latent variable scores

Step 1: Estimation of the inner weights

Step 2: Inner approximation of the latent variable scores

Step 3: Estimation of the outer weights

Step 4: Outer approximation of the latent variable scores

Step 5: Convergence checking

The basic idea behind these steps is as follows:

- after the latent variable scores have been initialized, the information flows from the "outside" to the "inside" by using the outer model information,

[20]The algorithm that we describe in this section allows to fit only recursive models. To estimate non-recursive models, you should use other estimation methods such as two-stage least squares (see Dijkstra and Henseler, 2015a,b).

- then, the latent scores are revised taking into account the association between the latent variables in the inner model,
- the information is then propagated back to the "outside" updating the outer weight values.

This process is iterated till convergence. The algorithm's steps involve only simple and multiple linear regressions using the local information available at each node of the path diagram. We now describe these steps in more details.

We denote the data matrix with all indicators as $\boldsymbol{X}$ and the block of indicators measuring the qth latent variable $\boldsymbol{\xi}_q$ as $\boldsymbol{X}_q$. Similarly, we indicate with Ξ the matrix of all latent variable scores. A common preprocessing step in PLS-SEM is to standardize all the indicators to have zero mean and unit variance. This transformation permits to disregard the scale differences among the manifest variables. Additionally, after each step the latent variables are scaled likewise.

Step 0: Initialization of the latent variable scores. In general we estimate the latent variable scores as a weighted sum of the indicators in the corresponding block. In the very first step, each latent variable is initialized setting all weights equal to one. In other terms, initially we compute the scores as

$$\widehat{\boldsymbol{\xi}}_q = \sum_{p=1}^{P_q} m_{pq}\boldsymbol{x}_{pq}. \tag{3.15}$$

This equation implies that the scores of each latent variable are initially set equal to the algebraic sum of the indicators in the corresponding block, irrespective of whether the construct is measured reflectively or formatively. This is the most popular initialization method, but it is not the only one available. Indeed, in some software it is possible to choose a different method that consists in assigning a weight equal to 1 to all indicators in a block except the last one, which instead receives a weight of −1. This approach was frequently used in the past because it permits a faster convergence. However, it also has the drawback of producing unexpected sign flipping of the parameter estimates.

Step 1: Estimation of the inner weights. Inner weights are calculated for each latent variable to reflect how strongly the other latent variables are connected to it. The most common schemes for computing the inner weights are the *centroid scheme*, originally proposed by Wold (1982), the *factorial scheme* and the *path scheme*, both introduced by Lohmöller (1989). We provide below a brief description of each one assuming that the inner weights are collected in a matrix denoted as $\boldsymbol{E}$ with generic element $e_k q$.

Centroid scheme: this scheme produces weights e_{kq} based on the sign of $r_{kq} = \text{Cor}(\boldsymbol{\xi}_k, \boldsymbol{\xi}_q)$, the linear correlation coefficient between the latent variables $\boldsymbol{\xi}_k$

and $\boldsymbol{\xi}_q$ resulting from the outer approximation (see Step 4 below[21]), assuming they are neighbours. In particular, if $\boldsymbol{\xi}_k$ and $\boldsymbol{\xi}_q$ are adjacent, the weight e_{kq} is set to $+1$ if the correlation is positive and to -1 if the correlation is negative. If $\boldsymbol{\xi}_k$ and $\boldsymbol{\xi}_q$ are not adjacent, that is there is no relationship between them, e_{kq} is set to 0. More formally, for $k,q = 1,\ldots,Q$,

$$e_{kq} = \begin{cases} \text{sign}(r_{kq}) & \text{if } c_{kq} = 1 \\ 0 & \text{otherwise} \end{cases}, \qquad (3.16)$$

where c_{kq} denotes the (k,q)th element of the matrix $\boldsymbol{C} = \boldsymbol{S} + \boldsymbol{S}^{\top}$ and $\boldsymbol{S}$ is the adjacency matrix of the structural model introduced in Section 3.3. Thus, $\boldsymbol{C}$ is a symmetric matrix whose element c_{kq} takes value 1 if the latent variables $\boldsymbol{\xi}_k$ and $\boldsymbol{\xi}_q$ are neighbours in the structural model, and 0 otherwise.

Note that, as implied by Equation (3.16), correlations that are very close to zero may cause the weights to take a non-zero value, which may lead to instability. Moreover, the centroid scheme should not be used when the model contains higher order constructs (see Section 3.10).

Factorial scheme: in this scheme the correlation value between each pair of latent variables is directly used as the weight, that is

$$e_{kq} = \begin{cases} r_{kq} & \text{if } c_{kq} = 1 \\ 0 & \text{otherwise} \end{cases}, \qquad (3.17)$$

with the same interpretation of the notation as above. Note that the aim of this scheme is to take into consideration the strength of the association between the two latent variables $\boldsymbol{\xi}_k$ and $\boldsymbol{\xi}_q$, and not simply its sign as in the centroid scheme, but irrespective of the *direction* of the relationship.

Path scheme: in this scheme two types of weights are produced depending on the relationship between the latent variables. When a latent variable, say $\boldsymbol{\xi}_k$, is "causing" another latent variable $\boldsymbol{\xi}_q$ (so called *successor*), the weight value corresponds to the linear correlation coefficient $r_{kq} = \text{Cor}(\boldsymbol{\xi}_k, \boldsymbol{\xi}_q)$. If instead the latent variable $\boldsymbol{\xi}_k$ is "caused" by another latent variable $\boldsymbol{\xi}_q$ (so called *predecessor*), the weight is determined using a multiple regression model. In particular, the estimated linear regression coefficient on the predecessor will be used as the weight. More formally, according to the path scheme the weights are computed as follows

$$e_{kq} = \begin{cases} \hat{\gamma}_q & \text{for } q \in \boldsymbol{\xi}_k^{\text{pred}} \\ r_{kq} & \text{for } q \in \boldsymbol{\xi}_k^{\text{succ}} \\ 0 & \text{otherwise} \end{cases}, \qquad (3.18)$$

where $\boldsymbol{\xi}_k^{\text{pred}}$ indicates the set of predecessors of $\boldsymbol{\xi}_k$ and $\boldsymbol{\xi}_k^{\text{succ}}$ represents the corresponding set of successors. The coefficient $\hat{\gamma}_q$ provides the estimate of the $\boldsymbol{\xi}_q$

[21] At the first iteration of the algorithm the outer proxies of the latent variable scores correspond to the initial values computed in Step 0.

coefficient in the linear regression model

$$\boldsymbol{\xi}_k = \boldsymbol{\xi}_k^{\text{pred}} \boldsymbol{\gamma} + \varepsilon_k,$$

assuming $\boldsymbol{\xi}_q$ belongs to the predecessor set of $\boldsymbol{\xi}_k$. The path scheme is usually the default weighting scheme used in most PLS-SEM software because it allows to account for the largest amount of variance for the endogenous latent variables[22].

Step 2: Inner approximation of the latent variable scores. Here, the latent variable scores $\widehat{\boldsymbol{\xi}}_1, \ldots, \widehat{\boldsymbol{\xi}}_Q$ obtained in the previous iteration (usually referred to as the *outer approximation* of the latent variable scores) are updated getting new scores, $\widetilde{\boldsymbol{\xi}}_1, \ldots, \widetilde{\boldsymbol{\xi}}_Q$ (called the *inner approximation* of the latent variable scores), which are computed as a weighted sum of their respective adjacent latent variables. More specifically, the inner approximation of the latent variable scores is computed as

$$\widetilde{\boldsymbol{\xi}}_q = \sum_{k=1}^{Q} e_{kq} \widehat{\boldsymbol{\xi}}_k, \tag{3.19}$$

where the e_{kq}s are the inner weights as obtained from Step 1.

Step 3: Estimation of the outer weights. So far we did not make any distinction between reflective and formative measures. Now, we need to take this difference into account to properly estimate the parameters of the measurement model. Therefore, we recalculate the latent variable scores obtained from Step 2 using yet another weighting update. In the classical algorithm, there are two possible choices for updating the outer weights, usually referred to as *mode A* and *mode B*, which typically refer to reflective and formative models respectively.

With mode A, for each indicator in a given block, we fit a simple linear regression of the indicator versus the corresponding latent variable. Since both the indicators and the latent variables obtained from Step 2 are standardized[23], the regression coefficients computed here correspond to linear correlation coefficients, that is

$$\begin{aligned} \widehat{\boldsymbol{w}}_q^\top &= \left(\widetilde{\boldsymbol{\xi}}_q^\top \widetilde{\boldsymbol{\xi}}_q\right)^{-1} \widetilde{\boldsymbol{\xi}}_q^\top \boldsymbol{X}_q \\ &= \operatorname{Cor}(\widetilde{\boldsymbol{\xi}}_q, \boldsymbol{X}_q). \end{aligned} \tag{3.20}$$

In mode B we regress each latent variable against the indicators in its block. The weights will then correspond to the partial coefficients, that is[24]

$$\begin{aligned} \widehat{\boldsymbol{w}}_q &= \left(\boldsymbol{X}_q^\top \boldsymbol{X}_q\right)^{-1} \boldsymbol{X}_q^\top \widetilde{\boldsymbol{\xi}}_q \\ &= \operatorname{Var}(\boldsymbol{X}_q)^{-1} \operatorname{Cor}(\boldsymbol{X}_q, \widetilde{\boldsymbol{\xi}}_q). \end{aligned} \tag{3.21}$$

[22] Intuitively, this is due to the fact that, differently from the factorial scheme where the correlation between any two latent variables is used irrespective of the direction of the relationship, the path scheme differentiates between the possible directions.

[23] If the indicators are not standardized a priori, then we get covariances instead of correlations.

[24] The second equality below still refers to the situation where the indicators have been standardized.

Step 4: Outer approximation of the latent variable scores. In this step, we estimate the latent variable scores using the outer weights $\widehat{\boldsymbol{w}}_q$ obtained from Step 3 above by computing

$$\widehat{\boldsymbol{\xi}}_q = \sum_{p=1}^{P_q} \widehat{w}_{pq} \boldsymbol{x}_{pq}, \tag{3.22}$$

which is usually referred to as the *weight relation.*

Step 5: Convergence checking. The process from Step 1 through Step 4 is then repeated until a convergence criterion is met. The most common criterion used is the maximum relative difference between the outer weights from iteration to iteration. When the criterion falls below the chosen tolerance value (e.g., 10^{-5}), the algorithm stops. More formally, the procedure ends when

$$\max_{\substack{p=1,\ldots,P \\ q=1,\ldots,Q}} \left| \frac{\widehat{w}_{pq}^{\text{old}} - \widehat{w}_{pq}^{\text{new}}}{\widehat{w}_{pq}^{\text{new}}} \right| < \text{tolerance}. \tag{3.23}$$

However, other convergence criteria may be used, such as the maximum of the squared differences between the outer weights from two consecutive iterations

$$\max_{\substack{p=1,\ldots,P \\ q=1,\ldots,Q}} \left(\widehat{w}_{pq}^{\text{old}} - \widehat{w}_{pq}^{\text{new}} \right)^2 < \text{tolerance}, \tag{3.24}$$

or the sum of the absolute weight differences

$$\sum_{\substack{p=1,\ldots,P \\ q=1,\ldots,Q}} \left| \widehat{w}_{pq}^{\text{old}} - \widehat{w}_{pq}^{\text{new}} \right| < \text{tolerance}. \tag{3.25}$$

3.3.3 Stage II: Estimation of measurement model parameters

Having estimated the latent variable scores, in the second stage of the PLS-SEM algorithm the loadings for reflective constructs and coefficients for formative constructs are computed. To do that, we use the final latent variables scores ($\widehat{\boldsymbol{\Xi}}$) to compute the outer loadings (and cross-loadings) as the linear correlation between $\boldsymbol{X}$ and $\widehat{\boldsymbol{\Xi}}$, and the outer coefficients by regressing $\widehat{\boldsymbol{\Xi}}$ on $\boldsymbol{X}$.

3.3.4 Stage III: Estimation of structural model parameters

In this stage, using the final latent variable scores, we estimate the path coefficients (i.e., the structural model parameters) for each endogenous latent variable using OLS according to the specified PLS-SEM model. In particular, for each latent variable ($\widehat{\boldsymbol{\xi}}_q$) in the model, the path coefficients are computed as the regression coefficients of its predecessors ($\widehat{\boldsymbol{\xi}}_q^{\text{pred}}$), that is

$$\begin{aligned}\widehat{\boldsymbol{\beta}}_q &= \left(\widehat{\boldsymbol{\xi}}_q^{\text{pred}\top}\widehat{\boldsymbol{\xi}}_q^{\text{pred}}\right)^{-1}\widehat{\boldsymbol{\xi}}_q^{\text{pred}\top}\widehat{\boldsymbol{\xi}}_q \\ &= \text{Cor}\left(\widehat{\boldsymbol{\xi}}_q^{\text{pred}},\widehat{\boldsymbol{\xi}}_q^{\text{pred}}\right)^{-1}\text{Cor}\left(\widehat{\boldsymbol{\xi}}_q^{\text{pred}},\widehat{\boldsymbol{\xi}}_q\right). \end{aligned} \tag{3.26}$$

For convenience, we summarize the whole procedure in Algorithm 3.2. Before moving to the practical illustration of the estimation algorithm, we want to make a final remark about the convergence properties of the algorithm. In practice, the algorithm usually converges in few iterations. However, it has been shown that in some situations the algorithm may not converge (Hanafi, 2007; Henseler, 2010). This statement should not be taken as a motivation for not using PLS-SEM, since as Henseler (2010, page 118) puts it:

> "It should be noted that the non-convergence of PLS path modelling does not mean that researchers in behavioural, social, and business science should not use PLS any more. Many of the most important psychometric methods, as for instance common factor analysis or covariance-based structural equation modelling, face the issue of non-convergence. It is just that users must learn how to deal with it."

3.4 Bootstrap-based Inference

As we have seen in the previous section, the estimation procedure in PLS-SEM involves a sequence of stages and iterative steps. As a consequence, the path coefficient estimates we get at the end of the procedure cannot be expressed as an explicit function of the indicators data. For this reason, it is not possible to derive the (exact) sampling distributions of the corresponding estimators. It follows that the only viable way to perform inference (i.e., compute p-values and confidence intervals) for a PLS-SEM model is through (non-parametric) bootstrap (Davison and Hinkley, 1997). As we discussed in Section 2.1, the bootstrap involves randomly drawing with replacement a large number of subsamples from the observed data. So, if we denote with $\boldsymbol{X}$ the original data matrix containing the observed values for the full set of manifest variables, for each subsample b, with $b = 1, \ldots, B$, the bootstrap procedure generates a new matrix $\boldsymbol{X}^{(b)}$ whose rows are drawn randomly (with repetitions allowed) from $\boldsymbol{X}$. Then, for each subsample $\boldsymbol{X}^{(b)}$ a complete PLS-SEM analysis is performed and the corresponding parameter estimates are stored. After completion of the B bootstrap iterations, we end up with a list of B values for each parameter in the model that represents its bootstrap distribution. The bootstrap distributions can then be used to perform the required inferential analyses.

Algorithm 3.2 The PLS-SEM estimation algorithm.

1: Given data $\boldsymbol{X}$ on indicators, measurement and structural model adjacency matrices $\boldsymbol{M}$ and $\boldsymbol{S}$. Choose the latent variables measured in reflective (mode A) and formative (mode B) way. Set the outer weights initial values $\widehat{\boldsymbol{W}}^{\text{old}}$ to the zero matrix. Fix the tolerance tol and the maximum number of iterations $t_{\max}$.
2: Scale the indicators to have zero mean and unit variance.
3: Set the scores initial value to

$$\widehat{\boldsymbol{\Xi}} = \boldsymbol{XM}.$$

4: Scale the latent variables scores to have zero mean and unit variance.
5: Set the iteration counter to zero ($t \leftarrow 0$) and the maximum relative difference of the outer weights δ to 1 ($\delta \leftarrow 1$).
6: **while** $\delta \geq tol$ and $t < t_{\max}$ **do**
7: Estimate the inner weights using either (3.16), (3.17) or (3.18) and form matrix $\boldsymbol{E}$.
8: Compute the inner approximation of the latent variable scores as

$$\widetilde{\boldsymbol{\Xi}} = \widehat{\boldsymbol{\Xi}}\boldsymbol{E}.$$

9: Scale the latent variables scores to have zero mean and unit variance.
10: **for** $q \leftarrow 1, Q$ **do**
11: **if** $\boldsymbol{\xi}_q$ is in the set of mode A latent variables **then**
12: Compute the outer weights as

$$\widehat{\boldsymbol{w}}_q^\top = \left(\widetilde{\boldsymbol{\xi}}_q^\top \widetilde{\boldsymbol{\xi}}_q\right)^{-1} \widetilde{\boldsymbol{\xi}}_q^\top \boldsymbol{X}_q.$$

13: **else if** $\boldsymbol{\xi}_q$ is in the set of mode B latent variables **then**
14: Compute the outer weights as

$$\widehat{\boldsymbol{w}}_q = \left(\boldsymbol{X}_q^\top \boldsymbol{X}_q\right)^{-1} \boldsymbol{X}_q^\top \widetilde{\boldsymbol{\xi}}_q.$$

15: **end if**
16: **end for**
17: Compute the outer approximation of the latent variable scores as

$$\widehat{\boldsymbol{\Xi}} = \boldsymbol{X}\widehat{\boldsymbol{W}},$$

where $\widehat{\boldsymbol{W}}$ is a diagonal matrix collecting the estimated weights $\widehat{\boldsymbol{w}}_q$.
18: Scale the latent variables scores to have zero mean and unit variance.
19: Compute

$$\delta = \max_{\substack{p=1,\ldots,P \\ q=1,\ldots,Q}} \left| \frac{\widehat{w}_{pq}^{\text{old}} - \widehat{w}_{pq}^{\text{new}}}{\widehat{w}_{pq}^{\text{new}}} \right|.$$

20: Increase the iteration counter ($t \leftarrow t+1$).
21: **end while**

22: **for** $q \leftarrow 1, Q$ **do**
23: **if** $\boldsymbol{\xi}_q$ is in the set of mode A latent variables **then**
24: Compute the cross loadings as

$$\widehat{\boldsymbol{\lambda}}_q^{\text{cross}} = \text{Cor}(\boldsymbol{X}, \widehat{\boldsymbol{\xi}}_q).$$

25: Compute the outer loadings as

$$\widehat{\lambda}_{pq}^{\text{outer}} = \begin{cases} \widehat{\lambda}_{pq}^{\text{cross}} & \text{if } m_{pq} = 1 \\ 0 & \text{otherwise} \end{cases}.$$

26: **else if** $\boldsymbol{\xi}_q$ is in the set of mode B latent variables **then**
27: Compute the outer coefficients as

$$\widehat{\boldsymbol{w}}_q = \left(\boldsymbol{X}_q^\top \boldsymbol{X}_q\right)^{-1} \boldsymbol{X}_q^\top \widehat{\boldsymbol{\xi}}_q.$$

28: **end if**
29: Compute the path coefficients (i.e., the structural model parameters) as

$$\widehat{\boldsymbol{\beta}}_q = \left(\widehat{\boldsymbol{\xi}}_q^{\text{pred}\top} \widehat{\boldsymbol{\xi}}_q^{\text{pred}}\right)^{-1} \widehat{\boldsymbol{\xi}}_q^{\text{pred}\top} \widehat{\boldsymbol{\xi}}_q.$$

30: **end for**

A last issue we must mention regarding bootstrapping PLS-SEM models is related to the sign indeterminacy of the latent variable scores, an unpleasant feature which is shared with other factorial techniques (see for example Brown, 2015). From our point of view, score indeterminacy implies the possibility that during resampling any of the parameter sign does change unexpectedly. The main consequence of this situation is that the bootstrap distribution of the corresponding parameter becomes more dispersed thus influencing the estimate of the bootstrap standard error, which in turn will produce higher p-values and wider confidence intervals. Even if different strategies to deal with this situation have been proposed (see Hair et al., 2017, pages 153–154), the general suggestion is to take no action for it and accept the potential negative effects of the latent scores sign changes on the results.

3.5 The `plssem` Stata Package

As we described in the introductory section, we developed an open-source package for Stata called `plssem`, which implements all the analyses we present in this book. The package is freely available as a GitHub repository at `https://github.`

com/sergioventurini/plssem, where you also find the instructions for installing it in your computer[25]. Among the features included in plssem there are:

- Model specification using an equation-like style.
- Standard and bootstrap based estimation of PLS-SEM models.
- Mediation analysis through estimation and inference (including bootstrap) for up to five indirect effects (see Chapter 5).
- Moderation analysis through the inclusion of interactions among latent variables in the structural model (see Chapter 6).
- Multi-group analysis of outer loadings/coefficients and path coefficients for dealing with observed heterogeneity; in particular, it allows the comparison of an arbitrary number of groups using either normal-based, bootstrap or permutation tests.
- Potential to estimate higher order construct models (see Section 3.10).
- Postestimation commands to deal with unobserved heterogeneity (see Chapter 7).
- A range of graphical and postestimation commands for representing and inspecting the results of a fitted PLS-SEM model.

We provide now a description of the basic plssem characteristics, while more advanced features will be presented throughout the rest of the book.

3.5.1 Syntax

The syntax of plssem reflects the measurement and structural part of a PLS-SEM model, and accordingly requires the user to specify both of these parts simultaneously. Since a full PLS-SEM model typically includes a structural part, we need to define at least two latent variables in the measurement model. For example, assuming the model involves only two latent variables LV1 and LV2, the plssem syntax requires to specify the measurement part as

```
plssem (LV1 > varlist1) (LV2 > varlist2).
```

To specify reflective measures we need to use the greater-than sign (>) between a latent variable and its associated indicators (e.g., LV1 > *varlist1*), while the less-than sign (<) is required to include latent variables measured in a formative way (e.g., LV1 < *varlist1*).

The specification of the structural part[26] requires that the user must provide each endogenous/dependent latent variable (say, LV2) followed by the exogenous ones it depends upon (say, LV1) as shown in the following example

[25] The package works on Stata 15 or later.

[26] While the measurement part is mandatory, the plssem package allows to fit PLS-SEM models that do not include the structural part.

```
plssem (LV1 > varlist1) (LV2 > varlist2), ///
  structural(LV2 LV1).
```

One may specify more than one structural relationship by separating them using commas. For example, suppose we have two additional latent variables in the model, `LV3` and `LV4`, both measured reflectively, with `LV4` endogenous and `LV3` exogenous. Then, the syntax for the structural part would be

```
plssem (LV1 > varlist1) (LV2 > varlist2) ///
       (LV3 > varlist3) (LV4 > varlist4), ///
       structural(LV2 LV1, LV4 LV3 LV1).
```

In addition, in line with most Stata commands, we can fit a full PLS-SEM model by sub-setting the data directly in the syntax using the `if` and `in` qualifiers.

The `plssem` command can also be used with the `by` prefix, which causes Stata to repeat the analysis on subsets of the sample data corresponding to the values of the variable specified with `by`.

3.5.2 Options

The `plssem` command allows setting many options, in particular:

`wscheme(weighting_scheme)` provides the choice for the weighting scheme. The default is `path` for the path scheme as given in equation (3.18). Alternative choices are `factorial` or `centroid`.

`boot(#)` sets the number of bootstrap replications.

`seed(#)` sets the seed number for the bootstrap calculations. This option allows to make the results reproducible.

`tol(#)` sets the tolerance value used for checking convergence attainment (see Step 5 in Stage I described in Section 3.3). The default tolerance value is 1e-7.

`maxiter(#)` indicates the maximum number iterations the algorithm runs. The default is 100 iterations. Note that usually the algorithm requires a very limited number of iterations to reach convergence, typically less than 10.

`init(init_method)` lets the user choose between two options for initialization. These are `indsum`, the default, which sets the initial values of the weights to 1s for all indicators, and `eigen`, which instead initializes the latent scores using the first eigenvector of the factor analysis for the corresponding block.

`loadpval` shows the table of loadings' p-values.

`correlate(mv lv cross, cutoff(#))` lets the user ask for correlations among the indicators or manifest variables (`mv`), latent variables (`lv`) as well as cross-loadings (`cross`) between the indicators and latent variables[27]. When

[27]These correlations are computed using the original indicators and the estimated latent variable scores.

doing so, the user can also set a certain cut-off value for the correlations to be displayed by using the suboption `cutoff(#)`. For instance, `cutoff(0.3)` will display correlations that are larger than 0.3 in absolute terms.

`noscale` if chosen, the manifest variables are not standardized before running the algorithm.

`convcrit(convergence_criterion)` the convergence criterion to use. Alternative choices are `relative` (default) or `square`. The former corresponds to (3.23) while the latter to (3.24).

Other options will be presented in the rest of the book.

3.5.3 Stored results

Most of the estimation results and intermediate calculations performed by the `plssem` command are stored in memory and can be listed using the `ereturn list` command. Some of the stored results are:

`e(iterations)` a scalar that provides the number of iterations performed to reach convergence,

`e(mvs)` a macro containing the list of manifest variables used in the analysis,

`e(lvs)` a macro containing the list of latent variables used in the analysis,

`e(loadings)` the matrix with the estimated outer loadings and coefficients,

`e(pathcoef)` the matrix containing the estimated path coefficients,

`e(adj_meas)` the measurement model adjacency matrix $\boldsymbol{M}$,

`e(adj_struct)` the structural model adjacency matrix $\boldsymbol{S}$.

Most importantly, the command always saves in the active dataset the final estimates of the construct scores as new columns which are labelled as in the command call.

As usual in Stata, the stored results can be accessed using the `matrix list` (or `matlist`) command for matrix results and `display` for scalars and macros. The advantage of having a rich set of stored results is that one could use them for further analyses or for building custom graphical representations.

3.5.4 Application: Tourists satisfaction (cont.)

We now use the `plssem` command to estimate the model we introduced in Section 3.2.3, whose path diagram is reported in Figure 3.7. To fit the model we run the following code:

```
1 plssem (ACTIVITY < entertain visittown nature fishing) ///
2        (MOTIVES > energy getaway boredom exciting) ///
3        (SATISFACTION > recommend satisf expecta), ///
4        structural(SATISFACTION ACTIVITY MOTIVES)
```

As you can notice, we conventionally write a latent variable using uppercase letters, while for indicators we use lowercase (or capitalized) words, but this is not strictly required. The first part of the command, up to the comma, specifies the measurement model given by equations (3.7)–(3.14). Since we want to model ACTIVITY as a formative construct, the `plssem` syntax requires to specify it using the lower-than sign, that is

```
(ACTIVITY < entertain visittown nature fishing)
```

which is a way to mimic the fact that in this case the indicators are predictive of the latent variable values. The other two constructs, MOTIVES and SATISFACTION, are reflective, which we specify using the greater-than sign instead:

```
(MOTIVES > energy getaway boredom exciting)
(SATISFACTION > recommend satisf expecta)
```

The single option we included in the previous code, that is `structural()`, specifies the inner model. In `plssem` the inner model is provided using the standard approach in Stata for specifying regression models, that is as a list of column names with the dependent variable followed by its predictors list. In our first example the inner model is made of a single relationship given by equation (3.6). Assuming we have many equations, we should separate them using a comma. The output of the analysis is reported in Figure 3.9.

The first part of the output provides the iterations performed with the corresponding attained value of the convergence criterion. In this example, the algorithm took 13 iterations to converge. The default value for the tolerance in `plssem` is 10^{-7}, but it can be overridden with the option `tol()`. For example, if you try rerunning the same code with the addition of the `tol(1e-9)` option, it means you want to run the algorithm till the convergence criterion has attained a value smaller than 10^{-9}. In this case the algorithm will need 16 iterations to converge and stop.

The second part of the output summarizes the algorithm settings and reports some goodness-of-fit indices. In particular, we see that we used the path scheme for estimating the outer parameters, which represents the default in `plssem`. If you want to use another scheme, you need to add either the `wscheme(centroid)` or the `wscheme(factorial)` option. Table 3.3 reports the parameter estimates using the three weighting schemes. As we see, they are very similar. The default method for initializing the latent variable scores is the sum of the indicators in the block (referred to as `Initialization: indsum` in the output). Finally, we see that only 882 observations have been used even if 1000 were available in the sample because of missing data. The goodness-of-fit measures will be presented in the next chapter.

Next, the output in Figure 3.9 reports the table of standardized loadings. These are the parameters related to the outer model. The columns of the table are labelled

```
Iteration 1: outer weights rel. diff. = 6.11e+01
Iteration 2: outer weights rel. diff. = 5.23e-01
Iteration 3: outer weights rel. diff. = 1.78e-01
Iteration 4: outer weights rel. diff. = 2.04e-02
Iteration 5: outer weights rel. diff. = 1.21e-02
Iteration 6: outer weights rel. diff. = 9.11e-04
Iteration 7: outer weights rel. diff. = 6.60e-04
Iteration 8: outer weights rel. diff. = 4.43e-05
Iteration 9: outer weights rel. diff. = 3.44e-05
Iteration 10: outer weights rel. diff. = 2.22e-06
Iteration 11: outer weights rel. diff. = 1.77e-06
Iteration 12: outer weights rel. diff. = 1.12e-07
Iteration 13: outer weights rel. diff. = 9.14e-08

Partial least squares SEM                       Number of obs        =      882
                                                Average R-squared    =      0.09930
                                                Average communality  =      0.57438
Weighting scheme: path                          Absolute GoF         =      0.23882
Tolerance: 1.00e-07                             Relative GoF         =      0.93839
Initialization: indsum                          Average redundancy   =      0.06820

Measurement model - Standardized loadings
-----------------------------------------------------------
             |  Formative:    Reflective:    Reflective:
             |    ACTIVITY        MOTIVES   SATISFACTION
-------------+---------------------------------------------
   entertain |       0.248
   visittown |       0.145
      nature |       0.719
     fishing |       0.680
      energy |                      0.734
     getaway |                      0.707
     boredom |                      0.680
    exciting |                      0.678
   recommend |                                     0.850
      satisf |                                     0.838
     expecta |                                     0.797
-------------+---------------------------------------------
    Cronbach |                      0.666          0.772
          DG |                      0.793          0.868
       rho_A |       1.000          0.674          0.772
-----------------------------------------------------------

Discriminant validity - Squared interfactor correlation vs. Average variance
extracted (AVE)
-----------------------------------------------------------
             |    ACTIVITY        MOTIVES   SATISFACTION
-------------+---------------------------------------------
    ACTIVITY |       1.000          0.025          0.024
     MOTIVES |       0.025          1.000          0.087
SATISFACTION |       0.024          0.087          1.000
-------------+---------------------------------------------
         AVE |                      0.490          0.687
-----------------------------------------------------------

Structural model - Standardized path coefficients
-----------------------------
    Variable | SATISFACTION
-------------+---------------
    ACTIVITY |       0.111
             |     (0.001)
     MOTIVES |       0.278
             |     (0.000)
-------------+---------------
        r2_a |       0.097
-----------------------------
 p-values in parentheses
```

FIGURE 3.9: Tourists satisfaction application. Output of the `plssem` command for the model reported in Figure 3.7.

TABLE 3.3: Comparison of parameter estimates using different weighting schemes for the model in Figure 3.7.

	Centroid scheme	Factorial scheme	Path scheme
Measurement model			
`entertain`	0.24614	0.24748	0.24792
`visittown`	0.15004	0.14608	0.14477
`nature`	0.71576	0.71802	0.71876
`fishing`	0.68459	0.68110	0.67994
`energy`	0.73367	0.73380	0.73385
`getaway`	0.70590	0.70677	0.70706
`boredom`	0.68051	0.68005	0.67990
`exciting`	0.67885	0.67816	0.67794
`recommend`	0.85264	0.85102	0.85046
`satisf`	0.82978	0.83592	0.83792
`expecta`	0.80195	0.79810	0.79682
Structural model			
ACTIVITY	0.11180	0.11114	0.11093
MOTIVES	0.27787	0.27794	0.27796

with the construct names and they also indicate whether the construct is reflective or formative. For example, the fitted measurement model for ACTIVITY is

$$\widehat{\text{ACTIVITY}}_i = 0.248\,\texttt{entertain}_i + 0.145\,\texttt{visittown}_i + 0.719\,\texttt{nature}_i + 0.680\,\texttt{fishing}_i. \tag{3.27}$$

These quantities are "standardized" because they are computed using the standardized indicators. In the last three rows of the outer parameters table we find some indices that are useful for assessing the goodness of this part of the model. Similarly, the next table, titled `Discriminant validity`, contains further tools for the same aim. As we already said, we will dedicate Chapter 4 to the interpretation of these measures.

Finally, the last table provides the path coefficient estimates and the corresponding p-values. We can therefore conclude that both ACTIVITY and MOTIVES seem to positively and significantly affect the tourists satisfaction for the visit. The p-values reported here are those based on the standard normal-theory (i.e., no bootstrap is used). To make inferences using the bootstrap, we need to add the `boot()` option as shown in the code below:

```
1  plssem (ACTIVITY < entertain visittown nature fishing) ///
2         (MOTIVES > energy getaway boredom exciting) ///
3         (SATISFACTION > recommend satisf expecta), ///
```

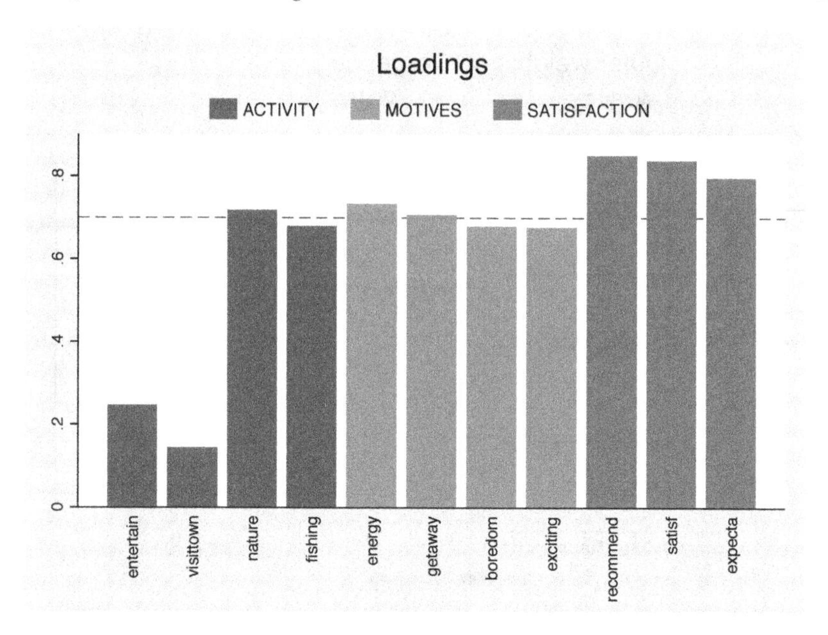

FIGURE 3.10: Tourists satisfaction application. Bar chart of the estimated outer parameters for the model reported in Figure 3.7.

```
4        structural(SATISFACTION ACTIVITY MOTIVES) ///
5        boot(1000) seed(1406)
```

In this example we used 1000 bootstrap replications and the output, not reported here, appears to be identical to the results of the standard `plssem` output (Figure 3.9). However, if we compare the standard errors of the outer and inner model parameters, which are available in the stored matrices `e(loadings_se)` and `e(struct_se)`, we conclude that there are some differences. In particular, the bootstrap standard errors of the ACTIVITY outer coefficients are much higher.

The `plssem` package also includes commands for visualizing the results. This is possible with the `plssemplot` postestimation command. For example, the following code produces: 1) a bar chart of the estimated outer coefficients (Figure 3.10), 2) a graph of the outer weights convergence paths (Figure 3.11) and 3) a scatterplot matrix with the estimated latent scores (Figure 3.12):

```
1  quietly plssem (ACTIVITY < entertain visittown nature fishing) ///
2                 (MOTIVES > energy getaway boredom exciting) ///
3                 (SATISFACTION > recommend satisf expecta), ///
4                 structural(SATISFACTION ACTIVITY MOTIVES)
5  plssemplot, loadings
6  plssemplot, outerweights
7  plssemplot, scores
```

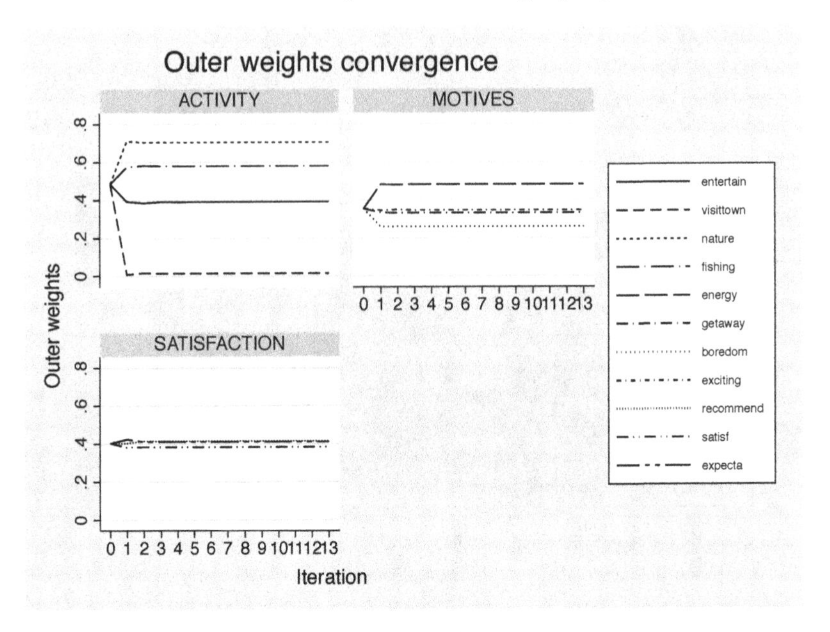

FIGURE 3.11: Tourists satisfaction application. Evolution upon convergence of the outer parameters for the model reported in Figure 3.7.

The graph of the outer weights evolution shows that the algorithm reached convergence and it did so very quickly since the estimates stabilized practically after the first two iterations. In addition, the scatter plot matrix of the estimated construct scores does not provide evidence of outliers or other unusual observations.

3.6 Missing Data

Working with real data it is very common that some of the values in the data table are *missing*, that is they are not available. Data may miss for many reasons such as subjects dropping out from a study before the end, non-response in sample surveys, refusal to answer particular questions in a questionnaire or simply an inadvertent loss of information. There is a vast literature on how to deal with missing values in different context and it is not our intention to provide a thorough presentation here[28]. It suffices to say that in some circumstances missing data can be quite dangerous in the sense that ignoring them may produce biased and inaccurate results.

[28]Good references are van Buuren (2018) or Molenberghs et al. (2015). You may also find an effective and brief review in Hair et al. (2018a, Chapter 2).

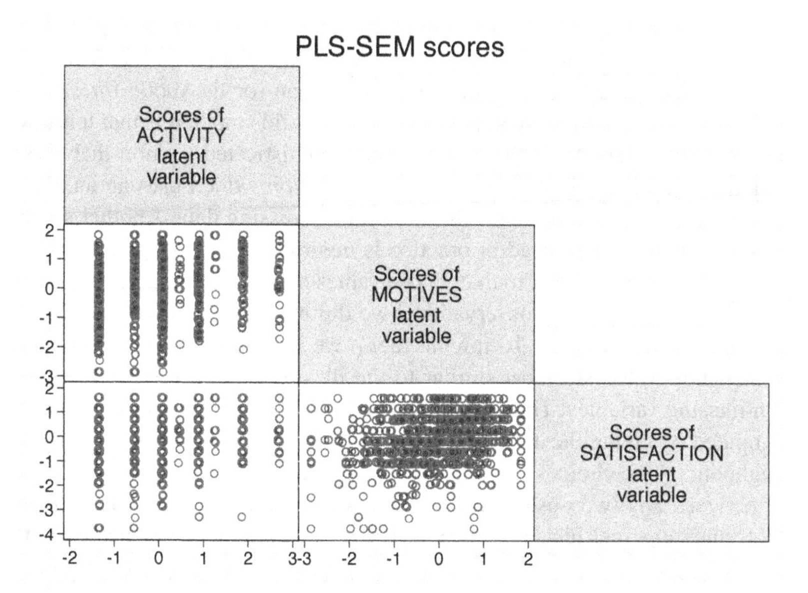

FIGURE 3.12: Tourists satisfaction application. Scatterplot matrix of the estimated latent scores for the model reported in Figure 3.7.

The most common approaches to deal with missing data in practice are:

- *Complete-case analysis* (also called *casewise* or *listwise deletion*), which consists in omitting any case with a missing value on any of the variables. This is the default strategy adopted in most statistical software packages (Stata included). Unless one can prove that the missing data are a true random subset of the original full sample[29], this option may dramatically reduce the effective number of observations used in the analysis providing less accurate results. Additionally, using this approach it is possible that entire groups of observations may be discarded thus leading to severe biases in the estimates.

- *Available-case analysis* (sometimes called *pairwise deletion*) uses all available data to estimate parameters of the model. For example, when interest focuses on bivariate associations for a set of variables assessed using covariances or correlation coefficients, then available-case analysis consists of a complete-case approach but performed separately for each pair of variables[30]. Even if this method seems to make a better use of the data, the set of observations used is different for each analysis and the resulting covariance or correlation matrices may even be numerically inadequate[31].

[29]This situation is referred to in the literature as data *missing completely at random* (MCAR).

[30]For example, this is the approach implemented in the `pwcorr` command in Stata.

[31]More technically, the covariance or correlation matrices may not be positive definite, which is a basic requirement for most multivariate analyses.

- *Single imputation*, which consists in "filling in" missing data with plausible values computed according to some rule. A popular choice is *mean imputation* which replaces the missing values with the mean (or the mode for categorical variables) of the available data. One drawback of this method is that it artificially reduces the variability in the data. A more sophisticated version that allows to partially reduce this issue is *regression imputation*, which uses an auxiliary regression model to predict the values to replace missing data. Another single imputation method often used in practice is *nearest neighbours*, which uses the k observations most similar to that whose values need to be imputed. For example, suppose that for a given observation i we did not observe all the variables, but 2 out of 10 are missing. To impute them, we first look for the k observations in the sample that are most similar to the ith observation for what regards the non-missing variables. Then, the values to impute for the 2 missing variables are computed by taking the averages of the corresponding columns for the identified neighbours. Two choices must be taken in this case: how to measure the similarity between any two observations and how many neighbours to use. Regarding the former, any metric can be used but the most common choice is the Euclidean distance (also called L_2 norm), which has shown to provide good performance results in many situations (Tutz and Ramzan, 2015). As for the number of neighbours, k, this quantity behaves as a smoothing parameter and there is no general rule for choosing it: smaller values (1 or 2) typically produce unstable results, while larger values tend to approach the mean imputation case. A reasonable choice for k is usually in between 3 and 7, with 5 often used as the default. The advantages of this approach are that it deals with both discrete (i.e., categorical) and continuous data and it is non-parametric in the sense that it does not require the specification of a predictive model. The main drawback is that it is computationally more demanding than the other methods, an issue that may become critical for large databases.
- *Multiple imputation* consists in generating multiple copies (e.g., 10) of the dataset where the missing data are simulated according to an external imputation model. Each of the simulated complete dataset is then analysed using the method of interest. Finally, the results are combined to produce estimates that also take into account the uncertainty that is due to the presence of missing data. Thanks to its flexibility, multiple imputation is currently considered one of the best options for dealing with missing data. We refer to the literature for more details.

In general, complete-case analysis may be considered acceptable only when missing data represent at most 5% of the entire sample. Otherwise, one of the strategies discussed above should be adopted. Nevertheless, as a general advice we suggest to perform a thorough exploration of the data to understand whether those missing are systematically different from those available. Additionally, it is often useful to try different strategies to assess the robustness of the findings. The most popular options to deal with missing data in PLS-SEM are mean imputation and nearest neighbours, which are also implemented in the plssem command. We now show an example using the tourists satisfaction application.

3.6.1 Application: Tourists satisfaction (cont.)

As we noted in the previous analyses, the tourists satisfaction data contain a non-negligible number of missing data, as the following code proves (output not reported):

```
1 misstable summarize energy-expecta
2 misstable pattern energy-expecta
```

This first analysis shows that around 12% of the observations contain at least one missing value. Thus, the results we reported in our previous examples, which were using the complete-case approach, may be biased. To address this issue, we now repeat the same analysis using different imputation methods. In particular, we use mean imputation and nearest neighbours with $k = 1,5$ and 10. The plssem command allows to impute missing values through the missing() option[32]. More specifically:

- to perform mean imputation you need to specify missing(mean),
- to use nearest neighbours imputation you must provide missing(knn) and the additional k() option for setting the number of neighbouring observations to use (if not specified, k() is automatically fixed at 5).

The following code fits the model under these different scenarios using the path weighting scheme[33]:

```
1  set seed 1404

2  plssem (ACTIVITY < entertain visittown nature fishing) ///
3         (MOTIVES > energy getaway boredom exciting) ///
4         (SATISFACTION > recommend satisf expecta), ///
5         structural(SATISFACTION ACTIVITY MOTIVES)
6  estimates store complete_case
7  label variable ACTIVITY "Scores of ACTIVITY (complete)"
8  label variable MOTIVES  "Scores of MOTIVES (complete)"
9  label variable SATISFACTION "Scores of SATISFACTION (complete)"
10 rename ACTIVITY ACTIVITY_cc
11 rename MOTIVES MOTIVES_cc
12 rename SATISFACTION SATISFACTION_cc

13 plssem (ACTIVITY < entertain visittown nature fishing) ///
```

[32] Imputation occurs before standardizing the manifest variables.

[33] Note that we set the random seed because, in case of ties (i.e., when there are more than one observation at a given distance from that whose values are to be imputed), the nearest neighbours method randomly selects those to use for computing the average. Therefore, setting the seed allows to reproduce the same numbers we report here.

```
14          (MOTIVES > energy getaway boredom exciting) ///
15          (SATISFACTION > recommend satisf expecta), ///
16          structural(SATISFACTION ACTIVITY MOTIVES) ///
17          missing(mean)
18    estimates store mean_imputation
19    label variable ACTIVITY "Scores of ACTIVITY (mean)"
20    label variable MOTIVES  "Scores of MOTIVES (mean)"
21    label variable SATISFACTION "Scores of SATISFACTION (mean)"
22    rename ACTIVITY ACTIVITY_mean
23    rename MOTIVES MOTIVES_mean
24    rename SATISFACTION SATISFACTION_mean

25    plssem (ACTIVITY < entertain visittown nature fishing) ///
26          (MOTIVES > energy getaway boredom exciting) ///
27          (SATISFACTION > recommend satisf expecta), ///
28          structural(SATISFACTION ACTIVITY MOTIVES) ///
29          missing(knn) k(1)
30    estimates store knn_imputation_1
31    label variable ACTIVITY "Scores of ACTIVITY (k-NN - k=1)"
32    label variable MOTIVES  "Scores of MOTIVES (k-NN - k=1)"
33    label variable SATISFACTION "Scores of SATISFACTION (k-NN - k=1)"
34    rename ACTIVITY ACTIVITY_knn1
35    rename MOTIVES MOTIVES_knn1
36    rename SATISFACTION SATISFACTION_knn1

37    plssem (ACTIVITY < entertain visittown nature fishing) ///
38          (MOTIVES > energy getaway boredom exciting) ///
39          (SATISFACTION > recommend satisf expecta), ///
40          structural(SATISFACTION ACTIVITY MOTIVES) ///
41          missing(knn) k(5)
42    estimates store knn_imputation_5
43    label variable ACTIVITY "Scores of ACTIVITY (k-NN - k=5)"
44    label variable MOTIVES  "Scores of MOTIVES (k-NN - k=5)"
45    label variable SATISFACTION "Scores of SATISFACTION (k-NN - k=5)"
46    rename ACTIVITY ACTIVITY_knn5
47    rename MOTIVES MOTIVES_knn5
48    rename SATISFACTION SATISFACTION_knn5

49    plssem (ACTIVITY < entertain visittown nature fishing) ///
50          (MOTIVES > energy getaway boredom exciting) ///
51          (SATISFACTION > recommend satisf expecta), ///
52          structural(SATISFACTION ACTIVITY MOTIVES) ///
53          missing(knn) k(10)
54    estimates store knn_imputation_10
55    label variable ACTIVITY "Scores of ACTIVITY (k-NN - k=10)"
56    label variable MOTIVES  "Scores of MOTIVES (k-NN - k=10)"
57    label variable SATISFACTION "Scores of SATISFACTION (k-NN - k=10)"
58    rename ACTIVITY ACTIVITY_knn10
59    rename MOTIVES MOTIVES_knn10
60    rename SATISFACTION SATISFACTION_knn10
```

The parameter estimates are reported in Table 3.4, from which we notice some differences for both the outer and inner model parameters. In particular, the extent of the differences with respect to the complete-case method provides some evidence of the bias produced from disregarding missing values.

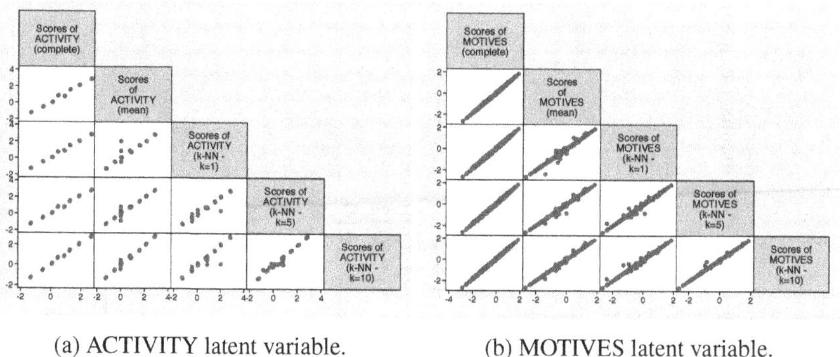

(a) ACTIVITY latent variable. (b) MOTIVES latent variable.

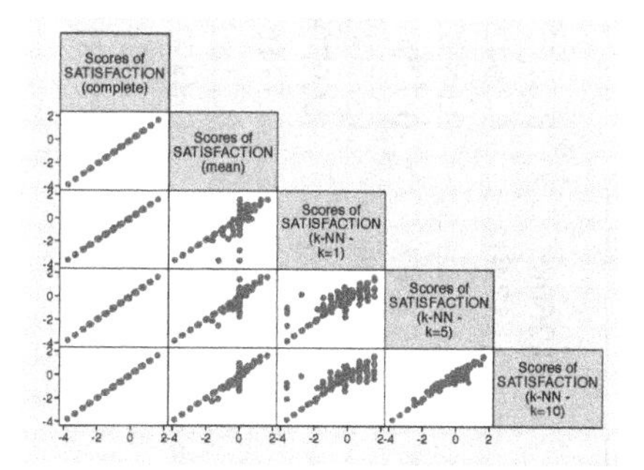

(c) SATISFACTION latent variable.

FIGURE 3.13: Tourists satisfaction application. Scatterplot matrix of the estimated latent scores for the model reported in Figure 3.7 under different missing data imputation schemes.

A further comparison is provided by the graphs in Figure 3.13, which report the estimated latent scores obtained after applying the different methods. The construct that seems to be most affected by the choice of the missing data approach is SATISFACTION, because its indicators (i.e., `recommend`, `satisf` and `expecta`) are those containing the highest number of missing values.

3.7 Effect Decomposition

As we have seen in Section 2.4 for path analysis, a useful way to summarize and interpret the results of a PLS-SEM analysis is through the computation of the direct, indirect and total effects. We recall that direct effects correspond to the impacts of one latent variable on another without the intervention of other variables along the causal path that goes from the former to the latter. On the contrary, indirect effects are

TABLE 3.4: Comparison of parameter estimates using different imputation methods for the model in Figure 3.7 (k-NN means k nearest neighbours imputation).

	Complete-case	Mean imputation	k-NN ($k = 1$)	k-NN ($k = 5$)	k-NN ($k = 10$)
Measurement model					
`entertain`	0.24792	0.22429	0.26089	0.21738	0.22911
`visittown`	0.14477	0.16486	0.13740	0.19424	0.19456
`nature`	0.71876	0.66646	0.67466	0.68552	0.70413
`fishing`	0.67994	0.74623	0.71565	0.72942	0.70350
`energy`	0.73385	0.72921	0.72118	0.73060	0.72759
`getaway`	0.70706	0.69853	0.69228	0.69396	0.69950
`boredom`	0.67990	0.70895	0.71705	0.71161	0.71042
`exciting`	0.67794	0.71546	0.72613	0.71893	0.71910
`recommend`	0.85046	0.84407	0.84535	0.84725	0.84708
`satisf`	0.83792	0.82223	0.83383	0.82245	0.82435
`expecta`	0.79682	0.80984	0.81908	0.81818	0.81644
Structural model					
ACTIVITY	0.11093	0.09982	0.10472	0.10159	0.10371
MOTIVES	0.27796	0.24208	0.25291	0.25924	0.26368

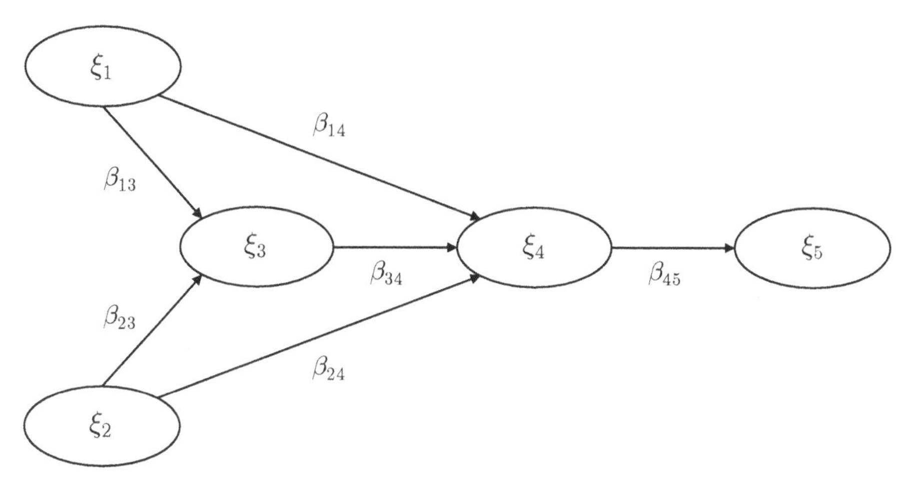

FIGURE 3.14: Hypothetical path diagram for a model showing the assumed structural relationships between five latent variables $\xi_1,\ldots,\xi_5$ together with the corresponding path coefficients.

the influences that are mediated by at least one other variable[34]. Total effects are the summation of direct and indirect effects. In practice, direct effects correspond to the path coefficients. Indirect effects for a variable ξ_A on another variable ξ_B are computed by considering all paths going from ξ_A to ξ_B, multiplying the corresponding coefficients and finally summing up the results.

As an example, consider the hypothetical structural model shown in Figure 3.14 (we skip the measurement model because it is not involved in the definition of the direct and indirect effects), which involves five latent variables $\xi_1,\ldots,\xi_5$. The corresponding path coefficients β_{mj} are also reported in the diagram. For illustration purposes, we focus on two of the relationships, namely the effect of ξ_1 on ξ_4 and that of ξ_2 on ξ_5. Regarding the former, we have:

- the direct effect of ξ_1 on ξ_4 is given by the corresponding path coefficient β_{14},
- the indirect effect of ξ_1 on ξ_4 is instead represented by the influence that ξ_1 has on ξ_4 that is mediated by ξ_3, the only variable that lies along the indirect path going from ξ_1 to ξ_4. This indirect effect is computed as the product of the corresponding path coefficients, that is $(\beta_{13} \times \beta_{34})$. Since there are no other indirect paths linking ξ_1 to ξ_4, this value represents the final measure of the indirect effect,
- the total effect is then given by $\beta_{14} + (\beta_{13} \times \beta_{34})$.

For what regards the relationship between ξ_2 and ξ_5:

- there is no direct effect of ξ_2 on ξ_5 because there is no path linking directly ξ_2 to ξ_5,

[34]For the moment we rely on the common intuition you already have about the notion of mediation, but in Chapter 5 we provide a more detailed discussion of this topic.

- the indirect effect of ξ_2 on ξ_5 is more articulated because there are now two paths going from ξ_2 to ξ_5, that is the $\xi_2 \rightarrow \xi_4 \rightarrow \xi_5$ path, which is the effect mediated by the single ξ_4 variable, and the $\xi_2 \rightarrow \xi_3 \rightarrow \xi_4 \rightarrow \xi_5$ path, that is instead the indirect effect mediated by two variables. The first indirect path produces an effect equal to $(\beta_{24} \times \beta_{45})$, while the second one is given by $(\beta_{23} \times \beta_{34} \times \beta_{45})$ so that the overall indirect effect of ξ_2 on ξ_5 is equal to their summation,
- since there is no direct effect, the total effect is simply equal to the overall indirect effect, that is $(\beta_{24} \times \beta_{45}) + (\beta_{23} \times \beta_{34} \times \beta_{45})$.

Indirect and total effects are useful for providing a more detailed view of the relationships among the latent variables in a structural model. In some cases, in fact, a direct effect can be misleading because its value can be offset by that of the corresponding indirect effect producing thus a total effect that has the opposite sign. In these situations, the total effects give a complete view of the overall net effect of one variable on another. In other cases, one may be interested not only in the net effect of a given change in an independent variable on an endogenous variable but also in the process that leads to the final change. In these situations, the indirect effects would provide the full answer.

The `plssem` Stata package includes a series of postestimation commands that produce additional output after a PLS-SEM has been estimated. As it is common in Stata, these further analyses are available via the `estat` command. More specifically, to get the decomposition of the total effects into direct and indirect effects for a fitted PLS-SEM model you can use the `estat total` command. This command returns a table containing the estimates for all the effects. Moreover, adding the `plot` option also produces a graphical representation of the same values, which is particularly useful for complex models that involve many latent variables and a complicated structural model. As an illustrative example, we consider again the model shown in Figure 3.7, but in which we also add a path going from ACTIVITY to MOTIVES so that on top of the direct effects now we also have an indirect effect of ACTIVITY on SATISFACTION. The code below runs the analysis (output not reported) and computes the effects (see Figure 3.15):

```
1  plssem (ACTIVITY < entertain visittown nature fishing) ///
2         (MOTIVES > energy getaway boredom exciting) ///
3         (SATISFACTION > recommend satisf expecta), ///
4         structural(SATISFACTION ACTIVITY MOTIVES, ///
5                    MOTIVES ACTIVITY)
6
7  estat total
```

The results show that in this hypothetical model ACTIVITY has both a direct and indirect effect on SATISFACTION, which is mediated by MOTIVES. In particular, the magnitude of the indirect effect is larger than the direct effect producing thus a total effect which is more than twice that provided by the direct effect alone. This allows to fully assess the impact of the ACTIVITY construct on SATISFACTION.

```
Direct, Indirect (overall) and Total Effects
---------------------------------------------------------------------
                   Effect |     Direct      Indirect          Total
-------------------------+-------------------------------------------
       ACTIVITY -> MOTIVES |      0.236                          0.236
  ACTIVITY -> SATISFACTION |      0.056         0.064            0.120
   MOTIVES -> SATISFACTION |      0.273                          0.273
---------------------------------------------------------------------
```

FIGURE 3.15: Illustrative example of computation of the direct, indirect and total effects for the model reported in Figure 3.7 with the addition of a path going from ACTIVITY to MOTIVES.

3.8 Sample Size Requirements

One of the features of PLS-SEM that is often advanced as a justification for its use is a higher efficiency compared to CB-SEM methods in presence of small samples (Reinartz et al., 2009). This means that, everything else equal, PLS-SEM requires a smaller sample to achieve a given accuracy in the results. Even if papers have been published showing that this is not true (see for example Rönkkö and Evermann, 2013 and Goodhue et al., 2012), we agree with the perspective reported by Hair et al. (2018b) according to which some authors have abused in the past of the advantageous sample size requirements of PLS-SEM by fitting complex models using samples that were too small. This situation has clearly contributed to discredit the reputation of PLS-SEM, which, as any other statistical technique, is not a magic wand. Therefore, without adding more material to this debate, in the following we briefly describe the approaches developed so far for determining the minimum sample size in a PLS-SEM study.

We first remind that PLS-SEM is composed of separate OLS regression models which use the local information available for each construct. This implies that, at least as a first step, we may use the sample size requirements usually adopted in multiple linear regression (Cohen, 1992; see also Exhibit 1.7 in Hair et al., 2017). Moreover, since the OLS regressions involved in PLS-SEM are distinct, the overall complexity of the structural model is not an element that affects sample size requirements. More generally, when deciding the minimum sample size for whatever analysis, the following elements should be taken into consideration:

- The *statistical power* we want to achieve; the power of a test in statistics corresponds to the probability of rejecting the null hypothesis of no effect when in the population there is a non-null effect. Conventionally, an analysis is deemed as reasonable if its power is at least 80%.

- The *effect size* to detect, that is the extent of the effect of a variable on another one. Examples of effect sizes are the mean difference between two groups, the correlation index between two variables and a coefficient in linear regression.

Within the context of PLS-SEM, one of the most common measure of effect size often reported is the Cohen's f^2 (Cohen, 1988), which is defined as

$$f^2 = \frac{R^2_{A,B} - R^2_A}{1 - R^2_{A,B}}, \tag{3.28}$$

where R^2_A denotes the R-squared index for a linear regression involving a set of predictors A, while $R^2_{A,B}$ refers to the same quantity for a linear regression involving both predictors in sets A and B. In other terms, f^2 is a relative measure of the contribution provided by the predictors in B to the explanation of the response variability in addition to that already provided by the predictors in A. Conventionally, f^2 values of 0.02, 0.15 and 0.35 are referred to as small, moderate and large respectively.

- The *significance* of the test, usually denoted as α, which represents the risk of wrongly rejecting a true null hypothesis (so called type I error). Typically, 5% is taken as the default reference value for α, while 1% and 10% are used to describe a more or less conservative situation compared to the standard respectively.

Thus, for a given value of the power, effect size and significance, one can determine the corresponding minimum sample size.

In PLS-SEM, sample size requirements are typically provided with reference to the so called **10-times rule**, according to which the minimum sample size should be larger than 10 times the maximum number of paths pointing to any latent variable in both the inner and outer models (Hair et al., 2017, p. 24)[35]. So, for example, regarding the model represented in Figure 3.7, the minimum sample size is equal to $10 \times \max\{4, 2\} = 40$, where 4 refers to the number of indicators pointing to ACTIVITY and 2 to the number of predictors of SATISFACTION. The clear advantage of this method is its simplicity. However, it does not require any of the elements we described above so that in our example 40 should be the recommendation to provide irrespective to any considerations about power, effect size and significance. Moreover, a number of studies (Kock and Hadaya, 2018; Goodhue et al., 2012) have shown that the 10-times rule leads to a substantial underestimation of the minimum sample size.

Other methods that solve some of the limitations of the 10-times rule have been proposed in the literature, such as the *inverse square root* and the *gamma-exponential* methods (Kock and Hadaya, 2018), which are both based on approximating the standard error of a regression coefficient. However, the gold standard approach for determining the minimum sample size in PLS-SEM is based on **Monte Carlo simulation**. The Monte Carlo simulation is a general method for assessing the finite sample properties of statistical procedures and it is typically used to discover features that are too difficult to derive analytically. The first step in a Monte Carlo study is the design of the generating model, that is the model that is used to generate the observations to analyse afterwards. Then, a large number of samples (e.g., 1000) of fixed size (e.g.,

[35]This definition implies that to apply the 10-times rule to the outer model, we must focus only on the formatively measured constructs.

$n = 150$) are repeatedly drawn from the model thus mimicking the sampling process that characterizes all real-life statistical analysis. Finally, for each generated sample, the model parameters are estimated using PLS-SEM (or whatever other method of interest) and some of the corresponding characteristics are studied. The whole process is repeated for different values of the sample size (e.g., $30, 50, 100, 200$ and 500). For determining the minimum sample size, after fixing the power target value, the researcher must compute the percentage of significant effects obtained during the simulation process. Monte Carlo studies are time-consuming both in terms of design and implementation, but they provide a precise way to determine the minimum sample size in PLS-SEM. For a complete review of Monte Carlo simulation in the SEM context we suggest to see Paxton et al. (2001) and Muthén and Muthén (2002).

3.9 Consistent PLS-SEM

As any other statistical technique, PLS-SEM is not a perfect tool. One of its most critical limitations is that in some cases it yields *inconsistent* estimates. Consistency is a desirable statistical property of an estimator according to which a parameter estimate should approach its true unknown value more closely as the sample size increases indefinitely[36]. For PLS-SEM this means that when the data are assumed to originate from a common factor model, the method tends to overestimate the loadings of reflectively measured constructs and to underestimate (i.e., attenuate) the path coefficients in the structural model (Gefen et al., 2011; Dijkstra, 1983). The justification for the inconsistency comes from the fact that PLS-SEM is composite-based, that is the constructs are defined as linear combinations of the manifest variables, which are typically measured with error. Therefore, the construct scores themselves will also be affected by the same measurement error. As we discussed in Section 3.1, when you run a linear regression between variables that are measured with error, the coefficient estimates are attenuated and the bias does not go away if we increase the sample size.

Even if the original aim of PLS-SEM was not parameter estimation, inconsistent estimates can considerably deceive the interpretation of the results of a PLS-SEM analysis thus leading to wrong conclusions about the significance of a given effect. For this reasons, the inconsistency issue has always attracted a lot of interest in the literature and it has culminated in the development of a new approach called **consistent PLS-SEM** (usually referred to as PLSc-SEM) by Dijkstra and Henseler (2015a,b). The aim of the PLSc-SEM algorithm is to correct the estimates of the standard PLS-SEM approach to produce consistent and asymptotically normal estimates of loadings for reflective constructs and of correlations among latent variables. We briefly describe it in the rest of this section.

The main idea in PLSc-SEM is to adjust the traditional PLS-SEM estimates using

[36] A further property, that is instead held by PLS-SEM, is *consistency at large*, which states that the estimates converge to the correct values as long as both the sample size *and* the number of indicators in each block increase indefinitely (Schneeweiss, 1993; Hui and Wold, 1982).

a "correction for attenuation" using a newly developed reliability coefficient, ρ_A, that provides a consistent estimate of construct reliability[37]. More specifically, the steps involved in PLSc-SEM are as follows (for more details see the technical appendix at the end of the chapter):

Step 1. The traditional PLS-SEM algorithm is used to estimate all the unknown quantities described so far in this chapter, including latent variable scores, correlations and weights. The choice of the weighting scheme in the PLS-SEM's inner approximation step (see Section 3.3.2) does not affect the consistency of the PLSc-SEM procedure.

Step 2. For each reflective construct, compute the new reliability coefficient ρ_A mentioned above. For formative constructs, ρ_A is conventionally set to 1.

Step 3. For every pair of latent variable scores obtained from the traditional PLS-SEM algorithm, the corresponding correlation is corrected for attenuation. The correction is also needed if one of the two latent variables is formative, but not if both are formative.

Step 4. Compute the consistent path coefficients using the latent variable correlations from the previous step as described in equation (3.26).

Step 5. Get the consistent loading estimates using the values of the weights and reliability coefficients ρ_A computed in the previous steps.

The suggested approach to perform inference in PLSc-SEM is through bootstrap as we discussed in Section 3.4 (Aguirre-Urreta and Rönkkö, 2018).

3.9.1 The `plssemc` command

The `plssem` Stata package includes another command called `plssemc` for estimating the model's parameters using the PLSc-SEM approach. The syntax of `plssemc` is the same as that of `plssem` described in Section 3.5.

As an illustration of PLSc-SEM, we now present an example taken from Sanchez (2013, Section 5.1.2). The data come from a survey among students of an American college and the aim of the study was assessing the satisfaction for the education they received. The data are available in the `education.dta` file[38]:

```
1 use education, clear
```

The latent variables in the model and the corresponding indicators used to measure them are reported in Table 3.5, while the assumed structural relationships are shown in Figure 3.16. We now fit the model using both PLS-SEM and PLSc-SEM and we compare the results. To avoid having indicators that point in different direction in each block, we apply the trick suggested in Sanchez (2013, Section 5.4), that

---

[37]We will present the reliability coefficients commonly used in PLS-SEM in Chapter 4.

[38]The original dataset can be retrieved at `https://www.gastonsanchez.com`.

TABLE 3.5: Content of the latent and manifest variables for the satisfaction in education example described in Section 3.9.1.

Latent variable	**Manifest variable**	**Content**
Support	suphelp	I feel comfortable asking for help from the program's staff
	supunder	I feel underappreciated in the program
	supsafe	I can find a place where I feel safe in the program
	supconc	I go to the program when I have concerns about school
Advising	advcomp	Competence of advisors
	advacces	Access to advisors
	advcomm	Communication skills of advisors
	advqual	Overall quality of advising
Tutoring	tutprof	Proficiency of tutors
	tutsched	Tutoring schedules
	tutstud	Variety of study groups
	tutqual	Overall quality of tutoring
Value	valdevel	Helpfulness in my personal development
	valdeci	Helpfulness in personal decision-making
	valmeet	Facilitating meeting people and contacts
	valinfo	Accessibility to support and information
Satisfaction	satglad	I'm glad to be a member of the program
	satexpe	The program meets my expectations
	satover	Overall, I'm very satisfied with the program
Loyalty	loyproud	I'm proud to tell others I'm part of the program
	loyrecom	I would recommend the program to my colleagues
	loyasha	I often feel ashamed of being a member of the program
	loyback	I'm interested in giving something back to the program

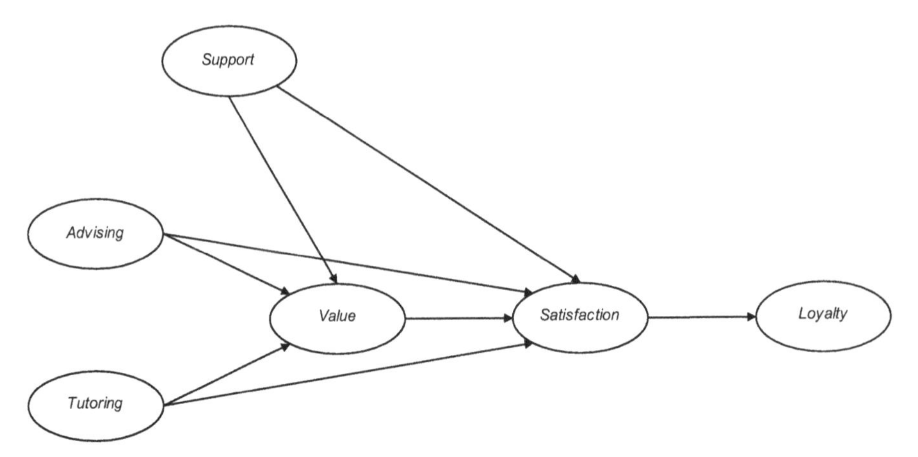

FIGURE 3.16: Satisfaction in education application (Sanchez, 2013, Section 5.1.2). The diagram shows the assumed structural relationships between the latent variables in the model.

is we invert the scale of the `supunder` and `loyasha` manifest variables renaming them as `supappre` and `loypleas` respectively:

```
generate supappre = 8 - supunder
generate loypleas = 8 - loyasha
drop supunder loyasha
```

The estimates using the two approaches are obtained by running the following code, whose results are reported in Table 3.6:

```
set seed 101

plssem (Support > sup*) (Advising > adv*) ///
       (Tutoring > tut*) (Value > val*) ///
       (Satisfaction > sat*) (Loyalty > loy*), ///
       structural(Value Support Advising Tutoring, ///
       Satisfaction Support Advising Tutoring Value, ///
       Loyalty Satisfaction) ///
       wscheme("centroid") tol(1e-6) boot(1000)

plssemc (Support > sup*) (Advising > adv*) ///
        (Tutoring > tut*) (Value > val*) ///
        (Satisfaction > sat*) (Loyalty > loy*), ///
        structural(Value Support Advising Tutoring, ///
        Satisfaction Support Advising Tutoring Value, ///
        Loyalty Satisfaction) ///
        wscheme("centroid") tol(1e-6) boot(1000)
```

TABLE 3.6: Comparison of parameter estimates for the satisfaction in education example using the PLS-SEM and PLSc-SEM algorithms.

	PLS-SEM	PLSc-SEM
Measurement model		
`Support` → `suphelp`	0.84742	0.76432
`Support` → `supsafe`	0.75250	0.62668
`Support` → `supconc`	0.74660	0.67225
`Support` → `supappre`	0.65386	0.54136
`Advising` → `advcomp`	0.90610	0.84302
`Advising` → `advacces`	0.85651	0.81764
`Advising` → `advcomm`	0.91652	0.91523
`Advising` → `advqual`	0.94908	0.92278
`Tutoring` → `tutprof`	0.87721	0.77808
`Tutoring` → `tutsched`	0.84663	0.81170
`Tutoring` → `tutstud`	0.77459	0.76556
`Tutoring` → `tutqual`	0.83875	0.72833
`Value` → `valdevel`	0.89512	0.78534
`Value` → `valdeci`	0.88892	0.82450
`Value` → `valmeet`	0.90446	0.81850
`Value` → `valinfo`	0.86108	0.94306
`Satisfaction` → `satglad`	0.90436	0.84530
`Satisfaction` → `satexpe`	0.91276	0.84706
`Satisfaction` → `satover`	0.92704	0.91492
`Loyalty` → `loyproud`	0.87260	0.80308
`Loyalty` → `loyrecom`	0.90350	0.84617
`Loyalty` → `loyback`	0.78421	0.71491
`Loyalty` → `loypleas`	0.65612	0.58359
Structural model		
`Support` → `Value`	0.67111	0.96335
`Advising` → `Value`	0.13045	-0.01623
`Tutoring` → `Value`	0.10386	-0.01189
`Support` → `Satisfaction`	0.14222	0.24685
`Advising` → `Satisfaction`	0.36425	0.37261
`Tutoring` → `Satisfaction`	0.12952	0.11927
`Value` → `Satisfaction`	0.32428	0.24853
`Satisfaction` → `Loyalty`	0.7765	0.88882

To conclude this section, a natural question may arise: since the PLS-SEM estimates are inconsistent, why should one not use PLSc-SEM every time? As you can expect, the answer is not clear-cut (see Chapter 1 for details). First, we need to recall that PLSc-SEM provides consistent parameter estimates only when the data have been generated by an underlying common factor model, as in standard CB-SEM. Clearly, this is an assumption that is not guaranteed to hold all the times. On the other hand, PLSc-SEM should be used in cases where the model is under-identified or when convergence problems arise with CB-SEM (Hair et al., 2019). In general, it must be said that PLSc-SEM has provided a valid solution to one of the most disputed limitations of traditional PLS-SEM. In the rest of this book we will continue to present only the "standard" PLS-SEM results, but you may replicate the examples by replacing `plssem` with the `plssemc` command[39].

3.10 Higher Order Constructs

So far we considered PLS-SEM models containing only **first-order latent variables** (FOLVs), that is latent variables that are directly related to the corresponding indicators. However, sometimes researchers are interested in applying further levels of abstraction assuming that the FOLVs are in turn related to other latent variables, which are thus called **higher order latent variables** (HOLVs). This structure can be motivated in particular when the FOLVs are strongly correlated so that their association can be explained by higher order constructs. The application of HOLVs is mostly limited to second-order latent variables (SOLVs), but some complex models may justify also the use of third or even higher order constructs. These models are also referred to in the literature as *hierarchical models* (Lohmöller, 1989). A typical example of HOLVs from psychology is represented by models for cognitive ability, which state the existence of a general cognitive ability construct that in turn determines more specific abilities like verbal, numerical or spatial ability (see Figure 3.17a, where we use the convention to represent HOLVs using grey-shaded ovals). Another example from the marketing literature is about customer satisfaction, which can be assumed to be determined by different dimensions like satisfaction for the quality, satisfaction for the price and satisfaction for the service provided by the personnel (see Figure 3.17b).

The main advantage from using HOLVS is that they allow to reduce the model's complexity making it more parsimonious by reducing the number of structural relationships. Moreover, when the FOLVs are highly correlated, using them in the structural equations may produce collinearity issues (see Section A.2.6), a problem that can be weakened by introducing a higher order construct.

We can distinguish two types of HOLVs according to the direction of the relationships between the latent variables at the different levels, which are called the **molecular** and the **molar** model respectively[40]. The molecular model states that the HOLVs are the "cause" of the lower order constructs, such as for the example in

[39]The GitHub repository of the book includes the code for all the examples using both `plssem` and `plssemc`.

[40]This jazzy terminology has been introduced by Chin and Gopal (1995), but other terms are also used in the literature (see for example Wetzels et al., 2009 or Hair et al., 2018c, Chapter 2).

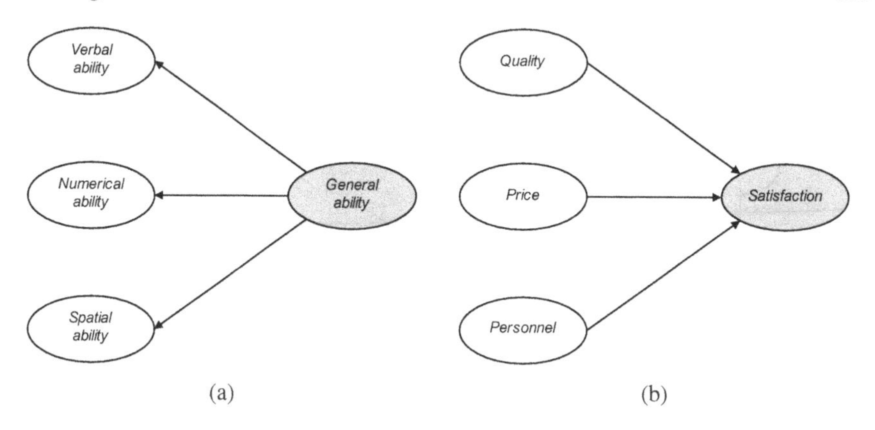

FIGURE 3.17: Examples of higher order constructs. Higher order constructs are shown as grey-shaded ovals.

Figure 3.17a. The molar model instead assumes that the higher order constructs are instead the "effects" of the latent variables at the lower levels, an example being the model reported in Figure 3.17b.

As for the measurement part, two different approaches have been suggested:

- the **repeated indicators approach**, the most popular in practice, which consists in using all the indicators of the lower order constructs as the manifest variables for the corresponding HOLVs. For example, suppose that in your model you have four FOLVs, each with four indicators, and the FOLVs are determined by a second-order variable. Then, the second-order construct must use sixteen manifest variables corresponding to all the indicators of the FOLVs (see Figure 3.18).
- the **two-step approach**, which first requires getting the FOLVs scores through an auxiliary technique such as principal component extracting only one component for each block. Then, the second step consists of a PLS-SEM analysis with the FOLVs scores obtained at the previous step playing the role of the manifest variables for the HOLVs.

As you may have noticed, in the example reported in Figure 3.18 we implicitly assumed that both the FOLVs $\xi_1, \ldots, \xi_4$ and the second-order construct ξ_5 have been modelled reflectively. However, other choices are possible (see for example Wetzels et al., 2009 and Sarstedt et al., 2019). The repeated indicators approach has pros and cons, but overall it is suggested in the literature as the method of reference when using higher order constructs within PLS-SEM.

We now show with a simple example how to fit models that include HOLVs using Stata. In particular, the `plssem` command allows to reuse the same indicators multiple times in different parts of the measurement model, so that you can choose the same manifest variables that measure the FOLVs also for defining the measurement model of the higher order constructs. For illustration, we use the data available in

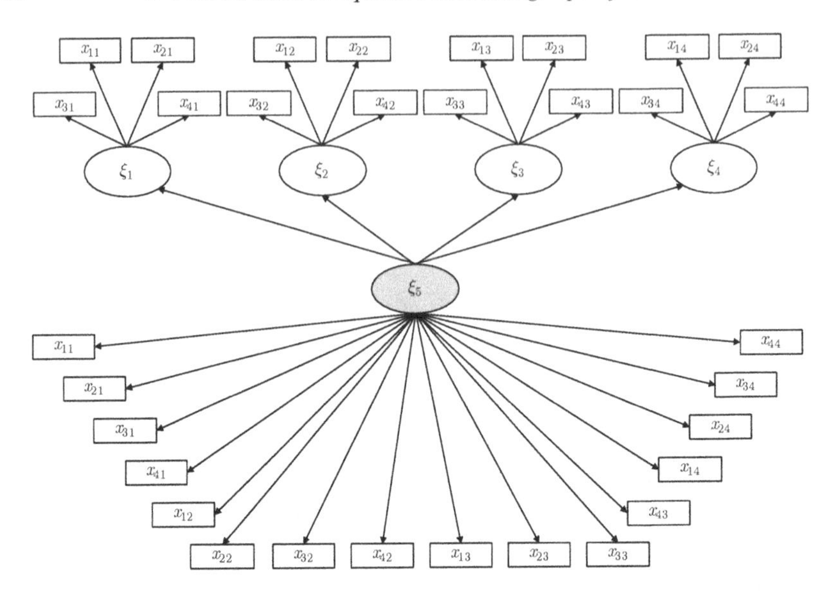

FIGURE 3.18: Hypothetical example to show the application of the repeated indicators approach to deal with higher order constructs.

the file `ch3_Curiosity.dta`. These data are about studying the determinants of the propensity to travel abroad for holiday. More specifically, we assume that there are two types of curiosity, perceptual and epistemic. The former represents curiosity which leads to increased perception of stimuli evoked by visual, auditory, or tactile stimulation, and the latter represents a drive to know that is aroused by conceptual puzzles and gaps in knowledge (Litman and Spielberger, 2003). The relevant literature also established that curiosity has a positive impact on the people's propensity to travel abroad for holiday. The measurement of perceptual curiosity employed an ordinal scale (from 1 = totally disagree to 5 = totally agree), and the respondents in the study were asked to reveal to what extent they agreed with the following statements:

1. "I like to discover new places to go to", and

2. "I like to travel to places I have never been to before".

Epistemic curiosity was measured in a similar way by requesting respondents to respond to the following statements:

1. "I like to learn about subjects that are unfamiliar to me", and

2. "I become fascinated when learning new information".

Finally, the interest for holiday was measured by the following items:

1. "I like holidaying", and

2. "I like holidaying abroad".

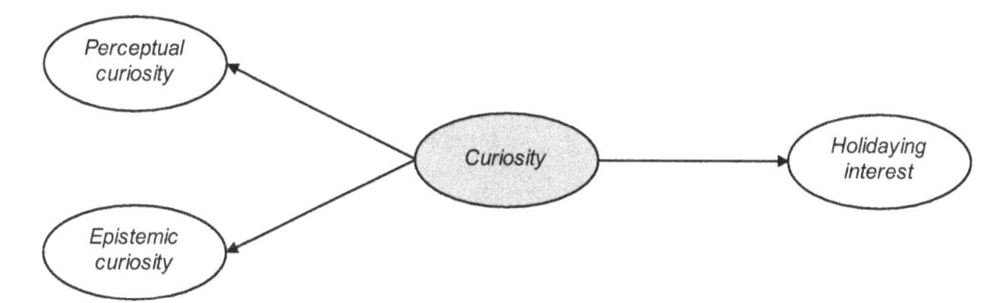

FIGURE 3.19: Curiosity example. The diagram shows the assumed structural relationships between the latent variables in the model. `Curiosity` is a second-order latent variable.

The items considered here and the corresponding FOLVs are reported in Table 3.7. The structural model considered in this example is fairly simple and it is reported

TABLE 3.7: Content of the latent and manifest variables for the curiosity example in Section 3.10.

Latent variable	**Manifest variable**	**Content**
`PerceptCur`	`v2a`	I like to discover new places to go to
	`v2b`	I like to travel to places I have never been to before
`EpistemCur`	`v2e`	I like to learn about subjects that are unfamiliar to me
	`v2f`	I become fascinated when learning new information
`HolidayInt`	`v3a`	I like holidaying
	`v3b`	I like holidaying abroad

in Figure 3.19, where `Curiosity` acts as a second-order construct that causes `PerceptCur` and `EpistemCur`.

We first apply the repeated indicators approach, which uses all the indicators of both `PerceptCur` and `EpistemCur` as manifest variables for the second-order construct `Curiosity`. The code to fit the model with `plssem` is reported below and the corresponding results are shown in Figure 3.20:

```
Partial least squares SEM                  Number of obs          =       1000
                                           Average R-squared      =    0.51586
                                           Average communality    =    0.72095
Weighting scheme: path                     Absolute GoF           =    0.60984
Tolerance: 1.00e-07                        Relative GoF           =    0.83107
Initialization: indsum                     Average redundancy     =    0.42674

Measurement model - Standardized loadings
-------------------------------------------------------------------------
             |  Reflective:    Reflective:    Reflective:    Reflective:
             |   PerceptCur     EpistemCur      Curiosity     HolidayInt
-------------+-----------------------------------------------------------
         v2a |        0.927
         v2b |        0.919
         v2e |                       0.911
         v2f |                       0.873
         v2a |                                      0.775
         v2b |                                      0.738
         v2e |                                      0.788
         v2f |                                      0.666
         v3a |                                                     0.917
         v3b |                                                     0.929
-------------+-----------------------------------------------------------
    Cronbach |        0.825          0.745          0.728          0.827
          DG |        0.920          0.886          0.831          0.920
       rho_A |        0.827          0.760          0.733          0.830
-------------------------------------------------------------------------

Discriminant validity - Squared interfactor correlation vs. Average variance extracted (AVE)
-------------------------------------------------------------------------
             |   PerceptCur     EpistemCur      Curiosity     HolidayInt
-------------+-----------------------------------------------------------
  PerceptCur |        1.000          0.119          0.673          0.077
  EpistemCur |        0.119          1.000          0.672          0.208
   Curiosity |        0.673          0.672          1.000          0.202
  HolidayInt |        0.077          0.208          0.202          1.000
-------------+-----------------------------------------------------------
         AVE |        0.851          0.796          0.553          0.852
-------------------------------------------------------------------------

Structural model - Standardized path coefficients
---------------------------------------------------------
    Variable |   PerceptCur     EpistemCur     HolidayInt
-------------+-------------------------------------------
   Curiosity |        0.821          0.820          0.449
             |      (0.000)        (0.000)        (0.000)
-------------+-------------------------------------------
        r2_a |        0.673          0.672          0.201
---------------------------------------------------------
 p-values in parentheses
```

FIGURE 3.20: Curiosity example. Output of the `plssem` command for the model reported in Figure 3.19 using the repeated indicators approach.

```
1 use ch3_Curiosity.dta, clear

2 /* Repeated indicators approach */
3 plssem (PerceptCur > v2a v2b) ///
4        (EpistemCur > v2e v2f) ///
5        (Curiosity > v2a v2b v2e v2f) ///
6        (HolidayInt > v3a v3b), ///
7        structural(PerceptCur Curiosity, ///
8                   EpistemCur Curiosity, ///
9                   HolidayInt Curiosity)
```

In the next chapter you will learn how to assess the goodness of the results from HOLVs models, but for the moment we can say that Curiosity is a significant driver of the propensity to travel abroad for holiday.

We now apply the two-step approach to the same data. In the first step we get temporary scores for the FOLVs, that is for PerceptCur and EpistemCur. We can do that using principal component analysis (PCA). We define as PerceptCur_s the score of the first principal component on the PerceptCur's items. Similarly, EpistemCur_s provides the same quantity for the EpistemCur's items (the results of the two PCAs are not reported):

```
1  use ch3_Curiosity.dta, clear

2  /* Two-step approach */
3  /* -- Step 1 -- */
4  quietly {
5    pca v2a v2b, components(1)
6    predict PerceptCur_s, score

7    pca v2e v2f, components(1)
8    predict EpistemCur_s, score
9  }
```

Then, the second step involves a PLS-SEM analysis where we treat the scores obtained in the previous step as indicators of the second-order construct Curiosity (see Figure 3.21 for the corresponding results):

```
1  /* -- Step 2 -- */
2  plssem (Curiosity > PerceptCur_s EpistemCur_s)///
3         (HolidayInt > v3a v3b), ///
4         structural(HolidayInt Curiosity)
```

As we can see by comparing the results of the two approaches, they produce similar path coefficient estimates, that is 0.449 using repeated indicators and 0.471 using the two-step approach.

3.11 Summary

In this chapter we introduced the basics of the PLS-SEM approach. After illustrating the fundamental ideas of PLS-SEM, in parallel we presented the plssem Stata package, which will be used in the rest of the book. Furthermore, we presented some strategies to use in case some of the data are missing. Then, we discussed sample size requirements for PLS-SEM and introduced the innovative PLSc-SEM algorithm that allows to get consistent parameter estimates. Finally, we presented the options

```
Partial least squares SEM                            Number of obs          =      1000
                                                     Average R-squared      =   0.22146
                                                     Average communality    =   0.75525
Weighting scheme: path                               Absolute GoF           =   0.40897
Tolerance: 1.00e-07                                  Relative GoF           =   0.97864
Initialization: indsum                               Average redundancy     =   0.18868

Measurement model - Standardized loadings
------------------------------------------
              |  Reflective:   Reflective:
              |   Curiosity    HolidayInt
--------------+---------------------------
PerceptCur_s  |       0.706
EpistemCur_s  |       0.905
          v3a |                     0.917
          v3b |                     0.929
--------------+---------------------------
     Cronbach |       0.504         0.827
           DG |       0.792         0.920
        rho_A |       0.589         0.830
------------------------------------------

Discriminant validity - Squared interfactor correlation vs. Average variance extracted (AVE)
------------------------------------------
              |   Curiosity    HolidayInt
--------------+---------------------------
    Curiosity |       1.000         0.221
   HolidayInt |       0.221         1.000
--------------+---------------------------
          AVE |       0.659         0.852
------------------------------------------

Structural model - Standardized path coefficients
---------------------------
     Variable |  HolidayInt
--------------+------------
    Curiosity |       0.471
              |     (0.000)
--------------+------------
         r2_a |       0.221
---------------------------
 p-values in parentheses
```

FIGURE 3.21: Curiosity example. Output of the `plssem` command for the model reported in Figure 3.19 using the two-step approach.

available to deal with higher order constructs. In the next chapter we will provide more details on how to assess the goodness of a PLS-SEM analysis for both the measurement and the structural parts.

Appendix: R Commands

Different packages are available to perform PLS-SEM in `R`. Currently, these are `matrixpls`, `semPLS`, `plspm` and `cSEM`. They mainly differ for the interface to use for specifying the model. In particular, some of them require to specify the model through suitably structured matrices, while others allow for more user-friendly interfaces. Here, we briefly present the two packages that will be used in the rest of the book, that is `plspm` and `cSEM`.

The plspm package

The plspm package (Sanchez et al., 2017) is the first R package that has been developed for fitting PLS-SEM models and it currently provides many functionalities including those for dealing with observed and unobserved heterogeneity (see Chapters 6 and 7).

The specification of the model in plspm is not easy since it requires creating a set of separate objects representing the different components of the model. The structural model must be specified through a lower triangular Boolean matrix, that is a square matrix whose elements below the main diagonal can only be either zero or one. The ones in the matrix correspond to connections in the path diagram. The next code chunk shows how to set up this matrix for the tourists satisfaction example we presented in Section 3.2.3:

```
1 ACTIVITY <- c(0, 0, 0)
2 MOTIVES <- c(0, 0, 0)
3 SATISFACTION <- c(1, 1, 0)
4 tour_path <- rbind(ACTIVITY, MOTIVES, SATISFACTION)
5 colnames(tour_path) <- rownames(tour_path)
```

The tour_path matrix corresponds to the (transposed) structural model's adjacency matrix. The structural model can also be plotted using the innerplot() function.

Next, we must specify the measurement model by creating a list object with one numerical vector for each block, where each vector indicates the column indices in the dataset of the manifest variables forming the block. For the tourists satisfaction example this corresponds to:

```
1 tour_blocks <- list(5:8, 1:4, 9:11)
```

This means, for example, that the ACTIVITY latent variable, the first one appearing in the model, is measured through the variables in the fifth, sixth, seventh and eighth columns of the dataset.

Finally, we need to specify whether each construct is reflective or formative. We can do that by creating a character vector with elements either "A" or "B", where the former indicates a reflective construct and the latter a formative one:

```
1 tour_modes <- c("B", "A", "A")
```

Then, we can estimate the model using the plspm() function, which includes the following arguments:

- Data, a matrix or data frame containing the observed values of the manifest variables,
- path_matrix, a square lower triangular Boolean matrix representing the inner model,

- `blocks`, a list of vectors with column indices or column names from the `Data` indicating the sets of manifest variables forming each block,
- `modes`, character vector indicating the type of measurement for each block,
- `scheme`, length-one character vector indicating the type of inner weighting scheme; possible values are `"centroid"`, `"factorial"` or `"path"`,
- `scaled`, a logical argument indicating whether the manifest variables should be standardized,
- `tol`, the tolerance value,
- `maxiter`, an integer indicating the maximum number of iterations,
- `boot.val`, whether bootstrap validation should be performed (default to FALSE),
- `br`, the number of bootstrap replications (default to 100).

The `plspm()` function returns an object of class `plspm` containing all the results. We can get a full summary by applying the corresponding `summary` method. The next code estimates and summarizes the tourists satisfaction model:

```
1 tour_res <- plspm(Data = tour_data_nomiss,
2   path_matrix = tour_path,
3   blocks = tour_blocks,
4   modes = tour_modes,
5   scheme = "path", tol = 1e-7)
6 summary(tour_res)
```

```
PARTIAL LEAST SQUARES PATH MODELING (PLS-PM)

----------------------------------------------------------
MODEL SPECIFICATION
1   Number of Cases       882
2   Latent Variables      3
3   Manifest Variables    11
4   Scale of Data         Standardized Data
5   Non-Metric PLS        FALSE
6   Weighting Scheme      path
7   Tolerance Crit        0.0000001
8   Max Num Iters         100
9   Convergence Iters     4
10  Bootstrapping         FALSE
11  Bootstrap samples     NULL

----------------------------------------------------------
BLOCKS DEFINITION
          Block         Type   Size   Mode
1      ACTIVITY    Exogenous      4      B
2       MOTIVES    Exogenous      4      A
3  SATISFACTION   Endogenous      3      A

----------------------------------------------------------
BLOCKS UNIDIMENSIONALITY
```

```
               Mode  MVs  C.alpha  DG.rho  eig.1st  eig.2nd
ACTIVITY          B    4    0.000   0.000     1.32    0.986
MOTIVES           A    4    0.666   0.800     2.01    0.920
SATISFACTION      A    3    0.772   0.868     2.06    0.557

-----------------------------------------------------------
OUTER MODEL
               weight  loading  communality  redundancy
ACTIVITY
  1 entertain  0.3913    0.248       0.0613      0.0000
  1 visittown  0.0142    0.145       0.0210      0.0000
  1 nature     0.7046    0.719       0.5167      0.0000
  1 fishing    0.5802    0.680       0.4625      0.0000
MOTIVES
  2 energy     0.4820    0.734       0.5384      0.0000
  2 getaway    0.3323    0.707       0.4999      0.0000
  2 boredom    0.2594    0.680       0.4624      0.0000
  2 exciting   0.3465    0.678       0.4598      0.0000
SATISFACTION
  3 satisf     0.3831    0.838       0.7021      0.0697
  3 expecta    0.4104    0.797       0.6349      0.0630
  3 recommend  0.4138    0.850       0.7233      0.0718

-----------------------------------------------------------
CROSSLOADINGS
               ACTIVITY  MOTIVES  SATISFACTION
ACTIVITY
  1 entertain    0.2476   0.1801        0.0384
  1 visittown    0.1449   0.0707        0.0224
  1 nature       0.7188   0.0216        0.1113
  1 fishing      0.6801   0.1226        0.1053
MOTIVES
  2 energy       0.1183   0.7338        0.2687
  2 getaway      0.1715   0.7070        0.1853
  2 boredom      0.0431   0.6800        0.1445
  2 exciting     0.0942   0.6781        0.1931
SATISFACTION
  3 satisf       0.0940   0.2438        0.8379
  3 expecta      0.1439   0.2440        0.7968
  3 recommend    0.1444   0.2462        0.8505

-----------------------------------------------------------

INNER MODEL
$SATISFACTION
             Estimate   Std. Error     t value   Pr(>|t|)
Intercept   -2.11e-16       0.0320   -6.60e-15   1.00e+00
ACTIVITY     1.11e-01       0.0324    3.42e+00   6.50e-04
MOTIVES      2.78e-01       0.0324    8.57e+00   4.45e-17

-----------------------------------------------------------
CORRELATIONS BETWEEN LVs
              ACTIVITY  MOTIVES  SATISFACTION
ACTIVITY         1.000    0.158         0.155
MOTIVES          0.158    1.000         0.295
SATISFACTION     0.155    0.295         1.000

-----------------------------------------------------------
SUMMARY INNER MODEL
                    Type      R2  Block_Communality  Mean_Redundancy    AVE
ACTIVITY       Exogenous  0.0000              0.265           0.0000  0.000
MOTIVES        Exogenous  0.0000              0.490           0.0000  0.490
SATISFACTION  Endogenous  0.0993              0.687           0.0682  0.687
```

```
----------------------------------------------------------
GOODNESS-OF-FIT
[1]  0.2142

----------------------------------------------------------
TOTAL EFFECTS
              relationships  direct  indirect  total
1       ACTIVITY -> MOTIVES   0.000         0  0.000
2  ACTIVITY -> SATISFACTION   0.111         0  0.111
3   MOTIVES -> SATISFACTION   0.278         0  0.278
```

To enable bootstrap inference, we need to set the `boot.val` argument to `TRUE`:

```
1 set.seed(1406)
2 tour_boot <- plspm(Data = tour_data_nomiss,
3   path_matrix = tour_path,
4   blocks = tour_blocks,
5   modes = tour_modes, scheme = "path",
6   tol = 1e-7, boot.val = TRUE, br = 500)
7 # summary(tour_boot)
8 tour_boot$boot
```

```
$weights
                            Original Mean.Boot  Std.Error    perc.025  perc.975
ACTIVITY-entertain        0.39125832 0.3730243 0.19496051 -0.07993971 0.7338443
ACTIVITY-visittown        0.01419865 0.0107126 0.20875906 -0.41388384 0.4027639
ACTIVITY-nature           0.70462254 0.6449775 0.16584501  0.28185340 0.9130454
ACTIVITY-fishing          0.58018360 0.5522470 0.16810493  0.19649775 0.8527681
MOTIVES-energy            0.48198101 0.4841079 0.05276579  0.37554123 0.5872616
MOTIVES-getaway           0.33234335 0.3313903 0.04516998  0.24096101 0.4154432
MOTIVES-boredom           0.25942601 0.2560793 0.04328139  0.16844640 0.3406964
MOTIVES-exciting          0.34653196 0.3447588 0.04604970  0.25985363 0.4431503
SATISFACTION-satisf       0.38313747 0.3838447 0.02984427  0.32351970 0.4405398
SATISFACTION-expecta      0.41041384 0.4094214 0.03558346  0.34086305 0.4825443
SATISFACTION-recommend 0.41381980 0.4139794 0.02900239  0.35697009 0.4727902

$loadings
                          Original Mean.Boot  Std.Error   perc.025  perc.975
ACTIVITY-entertain       0.2476225 0.2447812 0.19703834 -0.1928380 0.6075208
ACTIVITY-visittown       0.1449344 0.1342296 0.20631034 -0.2925520 0.5164586
ACTIVITY-nature          0.7187989 0.6437762 0.20672191  0.2585761 0.8998321
ACTIVITY-fishing         0.6800879 0.6297343 0.19950004  0.2630712 0.9021197
MOTIVES-energy           0.7337576 0.7339382 0.04145715  0.6450601 0.8101335
MOTIVES-getaway          0.7070111 0.7042060 0.03529745  0.6287315 0.7656058
MOTIVES-boredom          0.6799806 0.6747773 0.04579328  0.5782904 0.7544747
MOTIVES-exciting         0.6780549 0.6724577 0.04843200  0.5626863 0.7581771
SATISFACTION-satisf      0.8379200 0.8370855 0.01873005  0.7971987 0.8714605
SATISFACTION-expecta     0.7968131 0.7949301 0.02327413  0.7454291 0.8380411
SATISFACTION-recommend 0.8504628 0.8499264 0.01665941  0.8116400 0.8786438

$paths
                          Original Mean.Boot  Std.Error   perc.025  perc.975
ACTIVITY -> SATISFACTION 0.1109361 0.1192742 0.04083167 0.04680154 0.1898248
MOTIVES -> SATISFACTION  0.2779547 0.2817236 0.03298507 0.21635329 0.3450580

$rsq
                Original Mean.Boot Std.Error   perc.025  perc.975
SATISFACTION 0.09929988 0.1066307 0.0202159 0.06793349 0.1505098
```

```
$total.efs
                           Original Mean.Boot  Std.Error   perc.025  perc.975
ACTIVITY -> MOTIVES       0.0000000 0.0000000 0.00000000 0.00000000 0.0000000
ACTIVITY -> SATISFACTION 0.1109361 0.1192742 0.04083167 0.04680154 0.1898248
MOTIVES -> SATISFACTION  0.2779547 0.2817236 0.03298507 0.21635329 0.3450580
```

For more details on the package, we suggest to refer to Sanchez (2013)[41].

The cSEM package

The cSEM package (Rademaker and Schuberth, 2020) is the most recent addition to the collection of R packages for PLS-SEM. Even if it is still in an early development stage, it includes a wide range of modern composite-based methodologies such as PLSc-SEM, GSCA and others[42], and we consider it the most promising for the future.

Compared to plspm, the cSEM package has the nice feature of allowing the specification of the model using the lavaan syntax (see Table 2.3 for a brief summary or run the command ?lavaan::model.syntax for a more detailed presentation). More specifically, the steps required by the cSEM package are:

1. specify a model using the lavaan syntax,
2. fit the model using the csem() function,
3. apply some postestimation functions to the object returned by csem().

The csem() function includes different arguments, some of which regard methods we are not covering in this book. Therefore, we list below only those that are relevant for our discussion, that is:

- .data, a data frame or matrix containing the observed values for the manifest variables; it can also be a list of data frames or matrices, in which case estimation is repeated for each dataset,
- .model, a length-one character vector with the model specification using lavaan syntax,
- .approach_2ndorder, a length-one character vector indicating the approach to use for dealing with second-order constructs; possible values are "2stage" (default) or "mixed",
- .disattenuate, a length-one logical vector indicating whether to apply disattenuation to yield consistent loadings and path estimates (i.e., whether to apply PLSc-SEM or not); default is TRUE,

[41]The book is freely available at https://www.gastonsanchez.com/PLS_Path_Modeling_with_R.pdf.

[42]For a thorough introduction we suggest to read the package's vignettes, which is also available at https://m-e-rademaker.github.io/cSEM/.

- `.conv_criterion`, a length-one character vector specifying the criterion to use for convergence; possible values are `"diff_absolute"` (default), `"diff_squared"` or `"diff_relative"`,
- `.PLS_weight_scheme_inner`, a length-one character vector indicating the inner weighting scheme to use; possible values are `"centroid"`, `"factorial"` or `"path"` (default),
- `.iter_max`, the maximum number of iterations allowed (default is 100),
- `.tolerance`, the tolerance criterion for convergence (default is `1e-05`),
- `.resample_method`, a length-one character vector specifying the resampling method to use; either `"none"`, `"jackknife"` or `"bootstrap"`,
- `.R`, the number of bootstrap replications.

The `csem()` function returns an object with three class attributes depending on the type of data and/or the model specified. The first class attribute is always `cSEMResults`. The second one can either be `cSEMResults_default`, `cSEMResults_multi` or `cSEMResults_2ndorder`, with the first indicating the basic model estimation, the second when a list of datasets is provided in input, and the third if the model contains higher order constructs. Finally, the returned object also includes a third class attribute, `cSEMResults_resampled`, in case resampling has been requested. As you can see, the returned object is not only fairly complex but also very flexible thus allowing for further extensions in the future. Each object of class `cSEMResults_default` is a list with two elements:

- `Estimates`, a list containing all the estimated quantities,
- `Information`, a list containing the characteristics of the model.

In all cases, the object returned by `csem()` can be inspected through a range of postestimation functions such as `summarize()`, `infer()` and `assess()` (check the package documentation for the full list). More specifically:

- `summarize()` provides a complete summary of the estimated parameters,
- `infer()` computes common inferential quantities such as standard errors, confidence intervals, test statistics and p-values using resampling,
- `assess()` prints the values of the indexes for assessing the quality of the fitted model.

The following code fits the tourists satisfaction example using the `csem()` interface and summarizes the corresponding results[43]:

[43] We note that currently `cSEM` is not able to handle missing values so that you have to remove them beforehand. The package release we used is 0.3.0.9000.

```
if (!require(cSEM, quietly = TRUE)) {
  install.packages("devtools")
  library(devtools)
  install_github("M-E-Rademaker/cSEM")
}
library(cSEM)

tour_mod <- "
  # measurement model
  ACTIVITY <~ entertain + visittown + nature + fishing
  MOTIVES =~ energy + getaway + boredom + exciting
  SATISFACTION =~ recommend + satisf + expecta
  # structural model
  SATISFACTION ~ ACTIVITY + MOTIVES
"
tour_res <- csem(.data = tour_data_nomiss,
  .model = tour_mod, .PLS_weight_scheme_inner = "path",
  .disattenuate = FALSE, .tolerance = 1e-07)

summarize(tour_res)
assess(tour_res)
```

```
________________________________________________________________________________
----------------------------------- Overview -----------------------------------

General information:
------------------------
Estimation status                  = Ok
Number of observations             = 882
Weight estimator                   = PLS-PM
Inner weighting scheme             = path
Type of indicator correlation      = Pearson
Path model estimator               = OLS
Second-order approach              = NA
Type of path model                 = Linear
Disattenuated                      = No

Construct details:
------------------
Name          Modeled as     Order         Mode

ACTIVITY      Composite      First order   modeB
MOTIVES       Common factor  First order   modeA
SATISFACTION  Common factor  First order   modeA

----------------------------------- Estimates ----------------------------------

Estimated path coefficients:
============================
  Path                          Estimate  Std. error   t-stat.   p-value
  SATISFACTION ~ ACTIVITY         0.1109          NA        NA        NA
  SATISFACTION ~ MOTIVES          0.2780          NA        NA        NA

Estimated loadings:
===================
  Loading                         Estimate  Std. error   t-stat.   p-value
  ACTIVITY =~ entertain             0.2479          NA        NA        NA
  ACTIVITY =~ visittown             0.1448          NA        NA        NA
  ACTIVITY =~ nature                0.7188          NA        NA        NA
```

```
  ACTIVITY =~ fishing              0.6799       NA        NA        NA
  MOTIVES =~ energy                0.7338       NA        NA        NA
  MOTIVES =~ getaway               0.7071       NA        NA        NA
  MOTIVES =~ boredom               0.6799       NA        NA        NA
  MOTIVES =~ exciting              0.6779       NA        NA        NA
  SATISFACTION =~ recommend        0.8505       NA        NA        NA
  SATISFACTION =~ satisf           0.8379       NA        NA        NA
  SATISFACTION =~ expecta          0.7968       NA        NA        NA

Estimated weights:
==================
  Weights                        Estimate  Std. error   t-stat.   p-value
  ACTIVITY <~ entertain            0.3916       NA        NA        NA
  ACTIVITY <~ visittown            0.0140       NA        NA        NA
  ACTIVITY <~ nature               0.7047       NA        NA        NA
  ACTIVITY <~ fishing              0.5800       NA        NA        NA
  MOTIVES <~ energy                0.4821       NA        NA        NA
  MOTIVES <~ getaway               0.3324       NA        NA        NA
  MOTIVES <~ boredom               0.2594       NA        NA        NA
  MOTIVES <~ exciting              0.3464       NA        NA        NA
  SATISFACTION <~ recommend        0.4138       NA        NA        NA
  SATISFACTION <~ satisf           0.3831       NA        NA        NA
  SATISFACTION <~ expecta          0.4104       NA        NA        NA

Estimated construct correlations:
=================================
  Correlation               Estimate  Std. error   t-stat.   p-value
  ACTIVITY ~~ MOTIVES         0.1579       NA        NA        NA

Estimated indicator correlations:
=================================
  Correlation                     Estimate  Std. error   t-stat.   p-value
  entertain ~~ visittown           -0.0193       NA        NA        NA
  entertain ~~ nature              -0.1917       NA        NA        NA
  entertain ~~ fishing             -0.0143       NA        NA        NA
  visittown ~~ nature               0.1803       NA        NA        NA
  visittown ~~ fishing              0.0193       NA        NA        NA
  nature ~~ fishing                 0.1493       NA        NA        NA

------------------------------------ Effects -----------------------------------

Estimated total effects:
========================
  Total effect                    Estimate  Std. error   t-stat.   p-value
  SATISFACTION ~ ACTIVITY           0.1109       NA        NA        NA
  SATISFACTION ~ MOTIVES            0.2780       NA        NA        NA
________________________________________________________________________________

________________________________________________________________________________

Construct             AVE           R2           R2_adj
MOTIVES             0.4901          NA            NA
SATISFACTION        0.6868        0.0993        0.0973

-------------- Common (internal consistency) reliability estimates -------------

Construct       Cronbachs_alpha   Joereskogs_rho   Dijkstra-Henselers_rho_A
MOTIVES             0.6658            0.7934               1.0000
SATISFACTION        0.7716            0.8679               1.0000

----------- Alternative (internal consistency) reliability estimates -----------

Construct           RhoC          RhoC_mm     RhoC_weighted
MOTIVES            0.7934         0.9804         1.0000
SATISFACTION       0.8679         0.9998         1.0000

Construct       RhoC_weighted_mm       RhoT        RhoT_weighted
```

```
MOTIVES           0.7914          0.6658          0.6707
SATISFACTION      0.8676          0.7716          0.7720

--------------------------- Distance and fit measures -------------------------

[ -- output partly skipped -- ]

----------------------- Variance inflation factors (VIFs) ----------------------

  Dependent construct: 'SATISFACTION'

Independent construct    VIF value
ACTIVITY                  1.0256
MOTIVES                   1.0256

-------------------------- Effect sizes (Cohen's f^2) --------------------------

  Dependent construct: 'SATISFACTION'

Independent construct       f^2
ACTIVITY                  0.0133
MOTIVES                   0.0836

------------------------------ Validity assessment -----------------------------

[ -- output partly skipped -- ]

Fornell-Larcker matrix

              MOTIVES SATISFACTION
MOTIVES      0.4900849    0.0873043
SATISFACTION 0.0873043    0.6867689


------------------------------------ Effects -----------------------------------

Estimated total effects:
========================
  Total effect               Estimate  Std. error   t-stat.   p-value
  SATISFACTION ~ ACTIVITY      0.1109          NA        NA        NA
  SATISFACTION ~ MOTIVES       0.2780          NA        NA        NA

Estimated indirect effects:
===========================
  Indirect effect    Estimate  Std. error   t-stat.   p-value
  NA                       NA          NA        NA        NA

[ -- output partly skipped -- ]
```

We note that the output of the `summarize()` function contains a lot of NAs. This is due to the fact that we did not explicitly require to use the bootstrap to perform inference. To do that, we need to include the `.resample_method = "bootstrap"` argument as in the next example:

```
1 tour_boot <- csem(.data = tour_data_nomiss,
2   .model = tour_mod, .PLS_weight_scheme_inner = "path",
3   .disattenuate = TRUE, .tolerance = 1e-07,
4   .resample_method = "bootstrap", .R = 1000,
5   .seed = 1406)
6 summarize(tour_boot, .ci = "CI_percentile")
```

```
________________________________________________________________________________
---------------------------------- Overview ----------------------------------

General information:
------------------------
Estimation status                  = Ok
Number of observations             = 882
Weight estimator                   = PLS-PM
Inner weighting scheme             = path
Type of indicator correlation      = Pearson
Path model estimator               = OLS
Second-order approach              = NA
Type of path model                 = Linear
Disattenuated                      = Yes (PLSc)

Resample information:
---------------------
Resample methode                   = bootstrap
Number of resamples                = 500
Number of admissible results       = 499
Approach to handle inadmissibles = drop
Sign change option                 = none
Random seed                        = 1406

Construct details:
------------------
Name           Modeled as     Order         Mode

ACTIVITY       Composite      First order   modeB
MOTIVES        Common factor  First order   modeA
SATISFACTION   Common factor  First order   modeA

---------------------------------- Estimates ----------------------------------

Estimated path coefficients:
============================
                                                                    CI_percentile
  Path                        Estimate  Std. error   t-stat.  p-value         95%
  SATISFACTION ~ ACTIVITY       0.1012      0.0390    2.5939   0.0095 [ 0.0338; 0.1829 ]
  SATISFACTION ~ MOTIVES        0.3902      0.0466    8.3705   0.0000 [ 0.3068; 0.4818 ]

Estimated loadings:
===================
                                                                          CI_percentile
  Loading                           Estimate  Std. error   t-stat.  p-value         95%
  ACTIVITY =~ entertain               0.2479      0.1945    1.2745   0.2025 [-0.1262; 0.5920 ]
  ACTIVITY =~ visittown               0.1448      0.2191    0.6607   0.5088 [-0.2996; 0.5935 ]
  ACTIVITY =~ nature                  0.7188      0.1392    5.1643   0.0000 [ 0.3661; 0.8927 ]
  ACTIVITY =~ fishing                 0.6799      0.1509    4.5046   0.0000 [ 0.3012; 0.8890 ]
  MOTIVES =~ energy                   0.7465      0.0694   10.7620   0.0000 [ 0.6020; 0.8911 ]
  MOTIVES =~ getaway                  0.5148      0.0718    7.1735   0.0000 [ 0.3857; 0.6556 ]
  MOTIVES =~ boredom                  0.4016      0.0806    4.9817   0.0000 [ 0.2218; 0.5446 ]
  MOTIVES =~ exciting                 0.5364      0.0743    7.2194   0.0000 [ 0.3806; 0.6778 ]
  SATISFACTION =~ recommend           0.7474      0.0546   13.6930   0.0000 [ 0.6424; 0.8578 ]
  SATISFACTION =~ satisf              0.6920      0.0585   11.8382   0.0000 [ 0.5640; 0.8001 ]
  SATISFACTION =~ expecta             0.7413      0.0631   11.7563   0.0000 [ 0.6182; 0.8610 ]

Estimated weights:
==================
                                                                          CI_percentile
  Weights                           Estimate  Std. error   t-stat.  p-value         95%
  ACTIVITY <~ entertain               0.3916      0.1897    2.0640   0.0390 [ 0.0053; 0.7263 ]
  ACTIVITY <~ visittown               0.0140      0.2221    0.0632   0.9496 [-0.4331; 0.4814 ]
  ACTIVITY <~ nature                  0.7047      0.1502    4.6919   0.0000 [ 0.3334; 0.9110 ]
  ACTIVITY <~ fishing                 0.5800      0.1666    3.4817   0.0005 [ 0.2005; 0.8214 ]
  MOTIVES <~ energy                   0.4821      0.0546    8.8327   0.0000 [ 0.3802; 0.5921 ]
  MOTIVES <~ getaway                  0.3324      0.0449    7.4010   0.0000 [ 0.2499; 0.4221 ]
  MOTIVES <~ boredom                  0.2594      0.0444    5.8404   0.0000 [ 0.1571; 0.3363 ]
  MOTIVES <~ exciting                 0.3464      0.0441    7.8578   0.0000 [ 0.2575; 0.4278 ]
  SATISFACTION <~ recommend           0.4138      0.0287   14.4310   0.0000 [ 0.3579; 0.4737 ]
  SATISFACTION <~ satisf              0.3831      0.0306   12.5044   0.0000 [ 0.3222; 0.4441 ]
  SATISFACTION <~ expecta             0.4104      0.0358   11.4651   0.0000 [ 0.3356; 0.4826 ]

Estimated construct correlations:
```

```
=================================
                                                                  CI_percentile
  Correlation            Estimate  Std. error   t-stat.   p-value       95%
  ACTIVITY ~~ MOTIVES      0.1923     0.0570    3.3722    0.0007 [ 0.0736; 0.2985 ]

Estimated indicator correlations:
=================================
                                                                      CI_percentile
  Correlation               Estimate  Std. error   t-stat.   p-value       95%
  entertain ~~ visittown     -0.0193     0.0329    -0.5856    0.5582 [-0.0861; 0.0443 ]
  entertain ~~ nature        -0.1917     0.0330    -5.8029    0.0000 [-0.2544;-0.1248 ]
  entertain ~~ fishing       -0.0143     0.0351    -0.4060    0.6847 [-0.0820; 0.0533 ]
  visittown ~~ nature         0.1803     0.0344     5.2448    0.0000 [ 0.1146; 0.2458 ]
  visittown ~~ fishing        0.0193     0.0321     0.6023    0.5469 [-0.0393; 0.0789 ]
  nature ~~ fishing           0.1493     0.0300     4.9774    0.0000 [ 0.0880; 0.2053 ]

----------------------------------- Effects ------------------------------------

Estimated total effects:
========================
                                                                       CI_percentile
  Total effect               Estimate  Std. error   t-stat.   p-value       95%
  SATISFACTION ~ ACTIVITY     0.1012     0.0390     2.5939    0.0095 [ 0.0338; 0.1829 ]
  SATISFACTION ~ MOTIVES      0.3902     0.0466     8.3705    0.0000 [ 0.3068; 0.4818 ]
________________________________________________________________________________
```

Note that in the previous code chunk, we provided the random number generator seed directly inside the `csem()` function through the `.seed` argument and not through the global `R` seed.

Finally, to get the consistent parameter estimates (i.e., to use PLSc-SEM), we need to set the `.disattenuate` argument to `TRUE`.

Appendix: Technical Details

A formal definition of PLS-SEM

We assume that we collected data on n units for P variables $x_{i1}, x_{i2}, \ldots, x_{iP}$, with $i = 1, \ldots, n$. For convenience, we partition the P observed variables in Q blocks

$$\boldsymbol{X} = [\boldsymbol{X}_1, \ldots, \boldsymbol{X}_q, \ldots, \boldsymbol{X}_Q],$$

where the generic block $\boldsymbol{X}_q$ refers to set of indicators used to measure the qth latent variable in the model. As we described in the chapter, a construct can be measured according to either a reflective or a formative model. In a reflective block it is assumed that each manifest variable is related to the corresponding latent variable by a simple linear regression model, that is

$$x_{ipq} = \lambda_{pq0} + \lambda_{pq}\xi_{iq} + \varepsilon_{ipq}, \tag{3.29}$$

where λ_{pq0} is a location parameter (i.e., the intercept), λ_{pq} is the so called **outer loading**, or simply **loading**, associated with the pth manifest in the qth block, and ε_{ipq} is the error term that also include the measurement error. As in standard OLS regression analysis, it is assumed that the error term has zero mean and it satisfies the so called *predictor specification* assumption, that is

$$E\left(x_{ipq}|\xi_{iq}\right) = \lambda_{pq0} + \lambda_{pq}\xi_{iq}. \tag{3.30}$$

This assumption guarantees that the error term is uncorrelated with the predictor, that is the latent variable ξ_{iq}. The measurement model for a formative block assumes that the latent variable is a linear combination of the corresponding indicators, that is

$$\xi_{iq} = \sum_{p=1}^{P_q} \pi_{pq}x_{ipq} + \delta_{iq}, \tag{3.31}$$

where P_q denotes the number of indicators in the qth block, π_{pq} is the coefficient linking each manifest variable x_{ipq} to the latent measure it determines, and δ_{iq} is the error term representing the part of information about ξ_{iq} which is not accounted for by the indicators in the block. We will refer to the π_{pq}s as the **indicator weights** or **outer coefficients**. As for reflective constructs, we still assume predictor specification, that is

$$E\left(\xi_{iq}|\boldsymbol{X}_q\right) = \sum_{p=1}^{P_q} \pi_{pq}x_{ipq}. \tag{3.32}$$

The structural model specifies the relationships between exogenous and endogenous latent variables. In PLS-SEM it is assumed that the jth endogenous construct, with $j = 1,\ldots,J$, is related to a set of M_j predictors (sometimes also called predecessors) by the following multiple linear regression model

$$\xi_{ij} = \beta_{0j} + \sum_{m:\xi_m\rightarrow\xi_j}^{M_j} \beta_{mj}\xi_{im} + \zeta_{ij}, \tag{3.33}$$

where β_{mj} is the coefficient linking the mth exogenous latent variable to the jth endogenous one and ζ_{ij} is the error term. The coefficients in the structural model are commonly known as **path** or **structural coefficients**. Also the structural equations are assumed to satisfy the predictor specification hypothesis, that is

$$E\left(\xi_{ij}|\{\boldsymbol{\xi}_m\}\right) = \beta_{0j} + \sum_{m:\xi_m\rightarrow\xi_j}^{M_j} \beta_{mj}\xi_{im}. \tag{3.34}$$

No matter how a construct is measured, either reflectively or formatively, the distinctive feature of PLS-SEM is that upon convergence of the algorithm, latent variables scores $\hat{\boldsymbol{\xi}}_q$ corresponding to the qth latent variable $\boldsymbol{\xi}_q$ are computed using the following *weight relation*

$$\hat{\xi}_{iq} = \sum_{p=1}^{P_q} w_{pq}x_{ipq}, \tag{3.35}$$

where the coefficients w_{pq} are known as the **outer weights**. This equation characterizes PLS-SEM as a *component-based* approach to structural equation modelling, in contrast to *factor-based* CB-SEM.

We conclude noting that the outer weights must not be confused with the outer coefficients defined in equation (3.31). Even if the weight relation (3.35) resembles the formative measurement model, they are two distinct statements. In particular, the weight relation represents the rule used in PLS-SEM to compute the latent scores as a linear combination of the corresponding indicators, but it does not specify the "causal direction" between a construct and its manifest variables. Nonetheless, these directions determine how the w_{pq} weights are estimated (see Section 3.3).

More Details on the Consistent PLS-SEM Approach

We provide here more technical details for the consistent PLS-SEM (PLSc-SEM) approach described in the chapter. The presentation here is based on Dijkstra and Henseler (2015a).

The traditional PLS-SEM estimation in the first step of PLSc-SEM produces estimates of the (outer) weights for all latent variables (see Section 3.3.2, Step 3). For the generic qth latent variable $\boldsymbol{\xi}_q$ we denote the corresponding estimated weights as $\widehat{\boldsymbol{w}}_q$. Then, the consistent reliability coefficient estimator for the qth latent variable $\rho_A(\boldsymbol{\xi}_q)$ introduced by Dijkstra and Henseler (2015a) is defined as

$$\rho_A(\boldsymbol{\xi}_q) = \left(\widehat{\boldsymbol{w}}_q^\top \widehat{\boldsymbol{w}}_q\right)^2 \cdot \frac{\widehat{\boldsymbol{w}}_q^\top \left(\boldsymbol{S} - \mathrm{diag}(\boldsymbol{S})\right) \widehat{\boldsymbol{w}}_q}{\widehat{\boldsymbol{w}}_q^\top \left(\widehat{\boldsymbol{w}}_q \widehat{\boldsymbol{w}}_q^\top - \mathrm{diag}(\widehat{\boldsymbol{w}}_q \widehat{\boldsymbol{w}}_q^\top)\right) \widehat{\boldsymbol{w}}_q}, \tag{3.36}$$

where $\boldsymbol{S}$ is the sample covariance matrix of the indicators representing the observed measures for the qth latent variable, while $\mathrm{diag}(\boldsymbol{A})$ is a matrix operator that generates a diagonal matrix using the diagonal elements of $\boldsymbol{A}$.

Step 3 of the PLSc-SEM algorithm applies a correction to the correlations between the latent variables returned by the standard PLS-SEM approach. This correction (Nunnally and Bernstein, 1994, p. 241) is defined as

$$\mathrm{Cor}(\boldsymbol{\xi}_k, \boldsymbol{\xi}_q) = \frac{\mathrm{Cor}(\boldsymbol{\xi}_k^*, \boldsymbol{\xi}_q^*)}{\sqrt{\rho_A(\boldsymbol{\xi}_k^*)\, \rho_A(\boldsymbol{\xi}_q^*)}}, \tag{3.37}$$

where $\boldsymbol{\xi}_q^*$ denotes the scores for the qth latent variable that is returned by the traditional PLS-SEM algorithm.

Then, consistent path coefficients are computed as we illustrated in equation (3.26) but using the correlations as in (3.37). Finally, loadings are estimated consistently as

$$\widehat{\boldsymbol{\lambda}}_q = \widehat{\boldsymbol{w}}_q \cdot \frac{\sqrt{\rho_A(\boldsymbol{\xi}_q)}}{\widehat{\boldsymbol{w}}_q^\top \widehat{\boldsymbol{w}}_q}. \tag{3.38}$$

4

PLS Structural Equation Modelling: Assessment and Interpretation

In this chapter we will describe how to practically assess and interpret the results of a PLS-SEM model. To do so, we also make use of the output obtained from the estimation of a comprehensive PLS-SEM model. We first go through the measurement part and then continue with an evaluation of the structural part of the model. That is, criteria to evaluate the measurement and structural models are explained through an application.

4.1 Introduction

As it is common in every statistical analysis, following the estimation of a model you should always assess the goodness of the results obtained. In PLS-SEM a quite large set of indexes is used that only marginally overlap those available in the CB-SEM approach. In the next sections we present only the most popular tools for making these assessments, but others are also available in the literature (for a recent account see Hair et al., 2017).

One of the most frequently criticized limitations of PLS-SEM is that it does not optimize any predefined global fit criterion, like CB-SEM does. Indeed, as we have described in detail in the previous chapter, the estimates returned by PLS-SEM are the results of an iterative procedure that originally was shown to represent the solution of a so called *fixed-point problem* (Wold, 1965; see also Lyttkens, 1973 for a review). Practically, this implies that there is no way to statistically compare two alternative PLS-SEM models and assess the significance of the comparison as it is typically done in CB-SEM with likelihood ratio tests. On one hand, the lack of a global fit measure has motivated a lot of research that allowed to generate a deeper understanding of the method itself (Amato et al., 2005; Hanafi, 2007; Kramer, 2007; Esposito Vinzi et al., 2010; Tenenhaus and Tenenhaus, 2011; Esposito Vinzi and Russolillo, 2013). On the other hand, given the emphasis of PLS-SEM on predictions, we think that the issue of identifying the fit measure that is optimized by the PLS-SEM is not fundamental since many other tools to assess its quality are available, as we describe hereafter.

4.2 Assessing the Measurement Part

A full PLS-SEM model may contain different types of measurement models which in tandem make up what we may refer to as the measurement part of a model. An examination of the measurement part should take place prior to interpreting the results from the structural part. The reason for this sequential process is simply because we only can rely on results concerning the links between the different constructs (i.e., hypothesis testing making up the structural part) in the model when these constructs exhibit good psychometric properties. The properties that are sought depend on the type of measurement models used, with reflective and formative being the two most commonly used types.

4.2.1 Reflective measurement models

As far as reflective measurement models are concerned, one should examine unidimensionality of the construct, indicator reliability, construct reliability (i.e., composite reliability) and construct validity (i.e., convergent and discriminant validity).

4.2.1.1 Unidimensionality

Dimensionality is about finding out the number of constructs that may be reflected by a set of items. In theory, we can obtain as many constructs as items, suggesting that each of our items reflects a different construct. However, as it is for principal component analysis whose purpose is data reduction, we commonly search for a fewer number of constructs that could explain most of the variance in the items. Further, each construct should itself be unidimensional, suggesting that the items involved should primarily reflect one aspect of a phenomenon. Mathematically, this means that the construct extracted captures more variance in the items than any other possible constructs.

We examine eigenvalues from a principal component analysis (see Section 2.2) associated with each construct to find out whether the construct is unidimensional or multidimensional. An eigenvalue (θ) represents the amount of variance a construct explains in a set of items. It is generally recommended that constructs with eigenvalues above 1 are retained. The reason for this is that a construct should capture at least as much variance as an item's contribution to the correlation matrix. More specifically, if you hypothesize that a certain set of items should reflect a single construct, you would expect only one component with an eigenvalue above 1 (and preferably much larger than one) to demonstrate the construct's unidimensionality.

Figure 4.1 illustrates the idea of dimensionality of items. On the left side of the figure, we see a construct that is unidimensional in that all the four items are primarily explained by one single construct (i.e., the only construct with $\theta > 1$). On the right side, we observe exactly the same items being explained by two separate constructs (i.e., two constructs with $\theta > 1$), illustrating the multidimensional nature

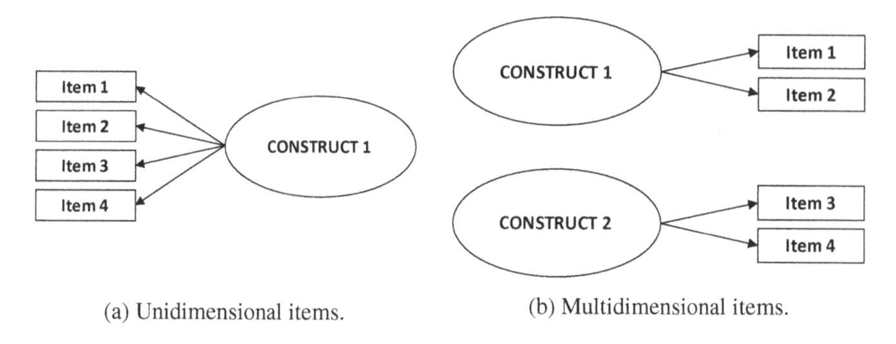

(a) Unidimensional items. (b) Multidimensional items.

FIGURE 4.1: Dimensionality of items.

of these items. In the case of multidimensionality, the measurement model should be respecified in order to meet the requirement of unidimensionality.

4.2.1.2 Construct reliability

While principal component analysis helps us in assessing the dimensionality, reliability analysis examines the internal consistency (i.e., homogeneity) of a construct, a condition which is necessary (as changes in one item mean changes in others) in a sound measurement model. The most popular index used to check the reliability of a construct is the **Cronbach's alpha**, which provides an estimate for the reliability of a construct based on the item intercorrelations.

Cronbach's alpha does however assume τ-equivalence (i.e., all items are equally important in expressing the constructs), which may not always be the case in reality. Thus, Chin (1998b) suggests that **Dillon-Goldstein's rho** (DG rho)[1] is indeed a better measurement of reliability than Cronbach's alpha. Indeed, DG rho does not assume τ-equivalence since it is based on the results from the measurement model (i.e., loadings) rather than the intercorrelations among the items in the raw data.

All the common reliability coefficients including DG rho vary between 0 and 1. The higher the coefficient, the more reliable the construct. However, regardless of the coefficient used, the reliability of a construct should, as a rule of thumb, be larger than 0.7 but below 0.93. The reason why we do not want to have a reliability coefficient above 0.93 is due to the fact that the items in such cases are redundant, meaning that they simply contain information (i.e., they represent answers to similar questions) about the same phenomenon.

4.2.1.3 Construct validity

A construct can be claimed to be valid when both convergent and discriminant validity are demonstrated. **Convergent validity** is the extent to which a set of items reflecting the same construct are positively correlated. The higher the correlation

[1]For the technical definition of Dillon-Goldstein's rho as well as the other criteria, see the technical appendix at the end of the chapter.

between the items, the more variance the items have in common. Since the items compose a construct together, high correlation between each item and the construct would imply high correlation among items, suggesting accordingly a high convergence of the items on a common construct. Thus, to establish convergent validity we seek high positive correlations between the items and the construct.

As we know from principal component/factor analysis, loadings are equivalent to correlations between a set of items and a common construct. We further know from bivariate regression analysis that squared correlation indicates the amount of variance an independent variable accounts for in the dependent variable. Applied to our case here, the squared loading would then indicate the share of variance of an item captured by a construct. As it is common in the latent variable domain to expect that a construct should at least capture half of the variance in each of its associated items, we would usually opt for factor loadings[2] not less than 0.7, yielding a shared variance of about 50 percent (i.e., $0.7^2 = 0.49 \approx 0.50$). This quantity reflects what is known as **item/indicator reliability**. It should also be added that each individual factor loading should be statistically significant depending on the chosen significance level.

Since each squared loading tells us how much variance a construct captures in an item, taking the average of all the squared correlations between the items and the construct would tell us how much variance on average the construct captures in its associated items. This quantity is also referred to as **average variance extracted** (AVE), indicating the **communality** (COM) of a construct. As we want a construct to capture at least half of each of items' variance, we would accordingly expect the construct to also capture on average half of the variance in all of its associated items. In other words, communality/AVE of a construct should at least be 0.5. Figure 4.2 illustrates this idea clearly in that at least 50 percent of the variance in item1 (0.701^2), item2 (0.802^2), item3 (0.864^2) and item4 (0.789^2) is captured by the construct. Further, the construct captures on average nearly 63 percent variance from its items.

Discriminant validity is about the distinctiveness of constructs, showing the extent to which a construct captures variance of its associated items relative to that of items associated with other constructs in the measurement model. The higher the correlation between a construct and its items as compared to its correlation with the other items in the model, the more distinct the construct is. Average variance extracted (average of squared loadings/correlations) by a construct can be considered an omnibus indication of the correlation between a construct and its items. Further, squared correlation between two different constructs indicates how much variance each construct captures in/shares with each other's items.

As such, to be able to establish discriminant validity for these two constructs, we should expect each of the construct's AVE to be larger than the squared correlation between them. This is known as the **Fornell-Larcker criterion** (Fornell and Larcker, 1981). The same idea applies equally to a model containing several constructs, in that

[2]The fact that in PLS-SEM the measurement error, unlike in covariance-based SEM, is not taken into account, stresses the importance of having factor loadings not less than 0.7. In fact, a data analyst should try to opt for even higher values than 0.7. Only in some rare occasions (e.g., an exploratory/pilot study) one could settle down with factor loadings as low as 0.6.

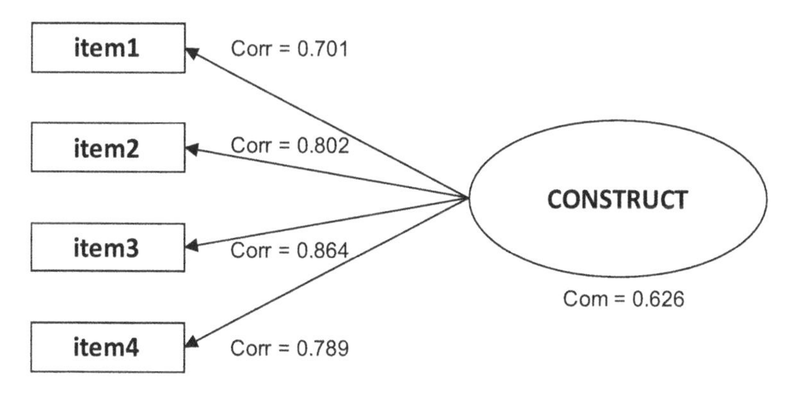

FIGURE 4.2: An example showing convergent validity. Com indicates a communality.

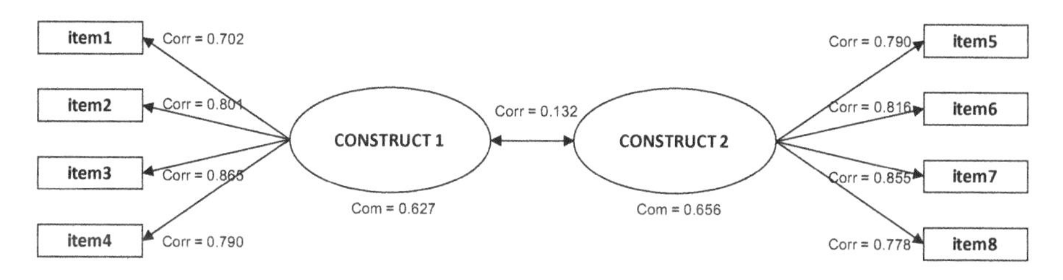

FIGURE 4.3: An example showing discriminant validity. Com indicates a communality.

we this time should expect a higher AVE of a construct than its squared correlation with any other construct in the model. Figure 4.3 illustrates the idea of discriminant validity with an example showing that the AVEs/COMs of both CONSTRUCT 1 (0.627) and CONSTRUCT 2 (0.656) are indeed larger than their squared correlation (0.132^2), demonstrating clearly that these two constructs are distinct enough (i.e., discriminant validity is achieved).

Although we in some cases cannot show empirical evidence of discriminant validity for some constructs in a sample data, a strong theoretical reasoning supported by previous studies may still justify for the inclusion of these constructs as distinct phenomena in the measurement model.

4.2.2 Higher order reflective measurement models

As far as the examination of higher order reflective measurement models is concerned, a multi-stage procedure must be followed. First, the measures in the lowest order are subjected to the same criteria as those that apply to reflective measures explained above (alternatively, see Table 4.1). Then, the measures in the higher order

TABLE 4.1: Summary of the criteria for assessing measurement models.

Reflective	**Formative**
1. There should be only one eigenvalue above 1 associated with a construct (unidimensionality)	1. Items should measure what they are supposed to represent (content validity)
2. Construct reliability coefficient (DG rho) should be larger than 0.7 (homogeneity)	2. Variance inflation values should be less than 2.5 (absence of multicollinearity)
3. Standardized loadings should be above 0.7 (item reliability)	3. Statistically significant weights
4. Average variances extracted (AVE) should be above 0.5 (convergent validity)	
5. AVE should be greater than squared correlations (discriminant validity)	

model should be assessed based on the level of construct reliability and average variance extracted as well as loadings. In other words, construct reliability and average variance extracted values of second-order constructs should respectively be larger than 0.7 and 0.5 as well. Further, the loadings of the lower order constructs on the higher order constructs (i.e., correlations between constructs) should be above 0.6–0.7 and statistically significant (see Figure 4.4). This procedure continues in the same manner regardless how complex a higher order measurement model may be.

4.2.3 Formative measurement models

Due to the nature (i.e., items predicting a construct) of the relationship between the items and constructs, formative measurement models require different assessment criteria than those concern reflective measurement models. There are various criteria suggested by different scholars for evaluating formative measurement models. However, we focus on the most common three criteria here (see Table 4.1). These include an examination of the content validity, multicollinearity between the items forming a construct as well as the statistical significant effect of the items on the construct.

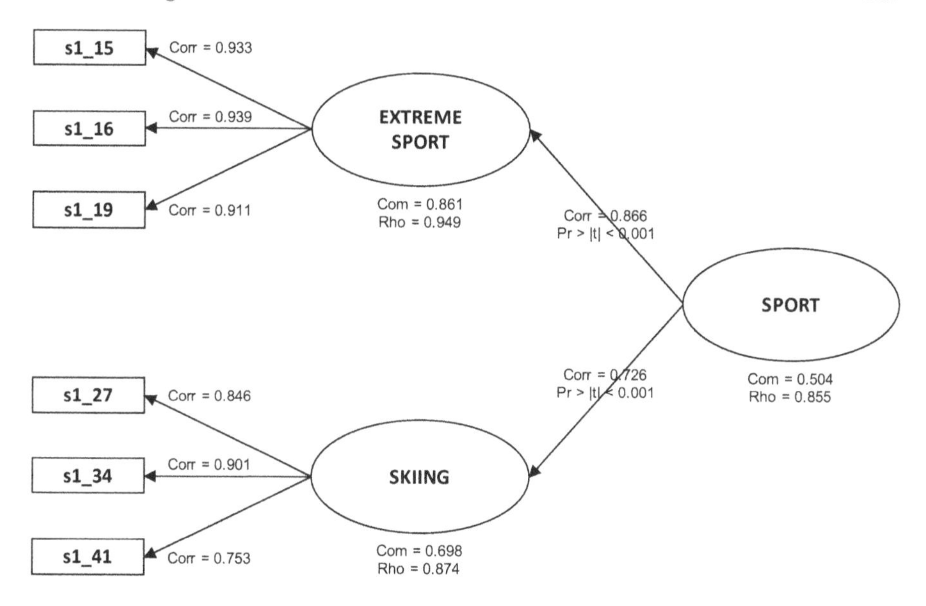

FIGURE 4.4: Evaluation of a second-order reflective measurement model. Com indicates a communality while Rho refers to Dillon-Goldstein's reliability coefficient.

4.2.3.1 Content validity

Content validity is about items measuring what they are claimed to define. Consequently, we should assess each item's operationalization and relevance in relation to the content of the construct. As each item contributes to defining an aspect of the construct, it is important to make sure that all the possible aspects of a construct are operationalized and measured through items. An example that readily can illustrate this issue can be a set of items measuring one's interest in "football", "basketball", "volleyball" and "handball" activity which all together are supposed to represent one's general interest in ball-based sports activity. Leaving out any of these four activities in the measurement model will miss out on an important aspect of the construct of interest. A thorough examination of the content validity may also significantly contribute to distinguishing between formative and reflective measures.

4.2.3.2 Multicollinearity

While high correlation among items is required in reflective measurement models, the reverse is the case for formative measurement models in that items are treated as predictors of a construct (outcome). This relationship represents, in other words, any typical statistical model tested using linear regression analysis. One of the main assumptions in linear regression analysis (see Section A.2.6 for a review) is indeed the absence of (severe) multicollinearity among predictors, an assumption which consequently applies to formative measurements as well. The reason why we want to avoid multicollinearity is the fact that standard error of the coefficients/weights on

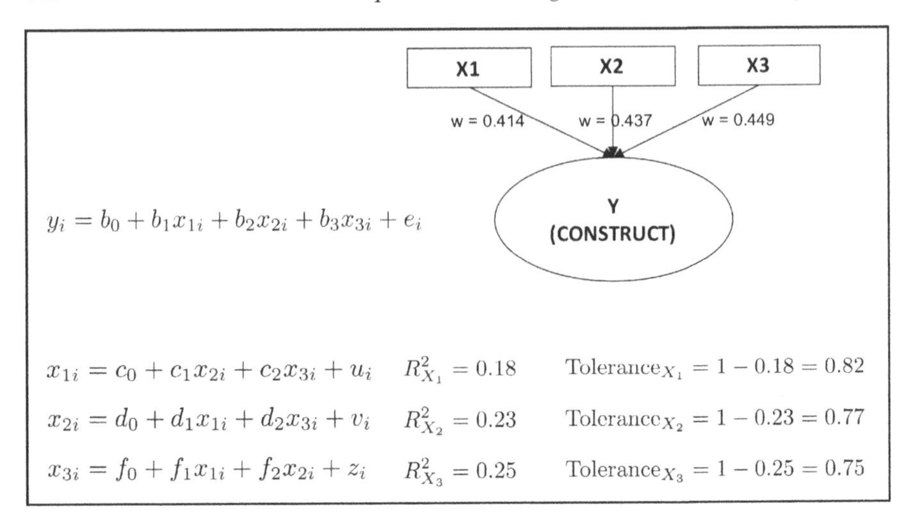

FIGURE 4.5: Formative measures with their tolerance values extracted.

each item will otherwise be biased upwards, leading thus to invalid test statistics which again may result in non-significant coefficients/weights.

Tolerance value is one diagnostic that can be used to examine possible multicollinearity existent among the items. The tolerance value shows the amount of variance in an item that cannot be explained by the remaining items in the equation. Literally speaking, we ask ourselves how much variance in an item, can we tolerate, being explained by the remaining items. There is a general consensus in the literature that we can tolerate a maximum of 60 percent, leaving us with 40 percent unexplained. This 40 percent is usually referred to as the tolerance value. In other words, tolerance values lower than 0.40 may be a sign of multicollinearity. The higher the tolerance value, the less severe the multicollinearity, indicating a weaker covariation among the items.

To be able to obtain tolerance values for each item, we can regress each item on the rest of the items forming the construct, and then extract the R-squared value for each equation. Subtracting each R-squared value from 1 will give us the tolerance value. In Figure 4.5 we provide an example showing this procedure. Incidentally, we observe here that none of the tolerance values is less than 0.4, indication of no multicollinearity among $X1$, $X2$ and $X3$ forming Y (construct).

We can indeed shorten the above procedure by regressing the component/factor score of Y (estimated in the measurement model) on $X1$, $X2$ and $X3$. The tolerance values obtained for these three items from this regression analysis are equivalent to those above. Using this parsimonious approach provide us directly with **variance inflation factor** (VIF) values. Squared root of VIF value for an item tells us how many times (e.g., $\sqrt{2.5} = 1.58$) the standard error of a coefficient has increased as a result of its correlation with the remaining items in the equation. There is an inverse relation between tolerance and VIF values. A tolerance value of 0.4 corresponds directly to a VIF value of 2.5 (i.e., 1/0.4). As such, tolerance values above 0.4 and VIF values less than 2.5 are signs of no serious multicollinearity among the items. As it follows from the reasoning above, in this context we are sceptical about using VIF values up to 10 as acceptable as it is commonly done in linear regression. A VIF value of 10 corresponds to a tolerance value of 0.1, meaning that you simply accept that 90 percent of an item's variance is explained by the rest of the items in a model. This, in our opinion, is a too liberal cut-off value to use in PLS-SEM.

What to do with the items affected by severe multicollinearity? One pragmatic and easy solution is to leave out one of the items from the model. Theoretical argumentation should incidentally be used deciding which of the two highly correlated items shall be left out. Another remedy is to transform (e.g., summated score, mean score, etc.) the highly correlated items and re-estimate the measurement model. A third solution is to try to increase the sample size, if possible and convenient, as large sample sizes tend to reduce the variance of coefficients.

4.2.3.3 Weights

While we examine loadings of reflective measures, we assess weights of formative measures. Weights in PLS-SEM do work in an identical manner as coefficients do in linear regression analysis. As such, we should examine statistical significance and magnitudes of items' weights to be able to judge how well the items are doing forming a construct (i.e., how good the formative measurement model is). The size of a weight reflects an item's contribution to the construct. For instance, in Figure 4.5, all the three weights are nearly of same size, implying that they contribute equally to the *Y* construct.

Unlike for loadings though, there are no definite rules for deciding how large weights should be. In theory, items with even the smallest weights should still be kept in the model given that they cover an aspect of its construct. The magnitudes of weights can incidentally be used to compare relative contribution of different items. What we additionally can look at is the statistical significance of these weights. Ideally, we should also expect the weights to be statistically significant. However, in some occasions we even may consider keeping non-significant weights in the formative measurement model also given strong theoretical reasoning.

4.3 Assessing the Structural Part

Having established a psychometrically sound (i.e., reliable and valid) measurement model, we can go ahead and examine the structural model (hypothesized causal relations between variables). The assessment of a structural part in PLS-SEM is indeed similar to that commonly practised when examining a statistical model tested using linear regression. In other words, we first should evaluate the share of a dependent variable's variance explained (R-squared) by one or a combination of several independent variables. Second, the sign, significance and size (so called "3S") of path coefficients should be considered. Finally, the goodness-of-fit of the full model should be assessed (see Table 4.2).

TABLE 4.2: Summary of the criteria for assessing structural models.

1. The higher the R-squared value, the better the model fits the data. R-squared values of 0.19, 0.33 and 0.67 represent respectively small, moderate and large effects.
2. The sign of path coefficients should be examined in relation to hypothesis statements and model specification issues. β values of 0.05, 0.1 and 0.25 represent respectively a small, medium and large effect.
3. Use standard errors from bootstrapping procedure. P-values less than 0.05 indicate statistically significant relationship.
4. Examine f^2 values as defined in equation (3.28) to determine the effect size of each variable. f^2 values of 0.02, 0.15 and 0.35 represent respectively a small, medium and large effect[3].
5. Examine the issue of multicollinearity. Variance inflation values should be less than 2.5.
6. The relative GoF value should be above 0.90.

4.3.1 R-squared

The R-squared is a popular goodness-of-fit metric in linear regression that quantifies the percentage of variance in the dependent variable explained by a set of independent variables (for a review see Section A.2.2). However, there are no definite thresholds as to how big R-squared should be. The evaluation of R-squared values should be based on a substantial consideration rather than a pure statistical quantity. In some fields (e.g., health) within a particular research area, an R-squared of 0.05 can be considered satisfactory while the same value may not be viewed enough in another field (e.g., marketing). Different fields and research questions may perceive R-squared differently. Further, low R-squared values may regardless be of importance if there is not much previous research done on a particular research topic. One other factor to take into account is indeed the number of predictors included in the model. Generally, a model with few predictors resulting in large R-squared can be

[3]The f^2 index is not currently provided by the `plssem` command.

asserted to do a better prediction job. A common sense thing to do is relate the evaluation of R-squared to the tradition in a particular research field.

The evaluation of R-squared should also take into account the target dependent variable(s). As a full PLS-SEM model is generally a complex one including several dependent variables, we may expect better prediction of some of these as compared to the others in the model. Nevertheless, Falk and Miller (1992) suggest that the variance explained for each dependent variable should, as a rule of thumb, be greater than 10 percent regardless. Further, as for the target dependent variable, we can follow the general guidelines in Chin (1998b): R-squared values of 0.19, 0.33 and 0.67 represent respectively small, moderate and large effects.

4.3.2 Goodness-of-fit

As R-squared alone cannot be used to evaluate the quality of the whole structural model consisting of several equations, an additional global criterion of goodness-of-fit (GoF) has been proposed by Tenenhaus et al. (2004), whose intent is to account for the model performance at both the measurement and the structural model (all of the R-squared values are taken into consideration) with a focus on overall prediction performance of the model (see Chin, 2010). As such, the GoF index is obtained as the geometric mean of the average communality index and the average of R-squared value (Tenenhaus et al., 2004). This GoF measure is usually referred to as the **absolute GoF**. A normalized version, the so called **relative GoF**, has also been proposed (Esposito Vinzi et al., 2010). The relative GoF is bounded between 0 and 1 and values equal to or higher than 0.90 indicate a good model.

However, note that goodness-of-fit in the context of PLS-SEM shall not be considered as equivalent to goodness-of-fit measures (e.g., chi-square statistic or root mean square error of approximation) commonly used to evaluate the quality of a model in the domain of covariance-based structural equation modelling[4]. Goodness-of-fit here refers simply to the predictive ability of a full PLS-SEM model. As long as a researcher's purpose is to develop a complex model to explain as much variance in the dependent variables as possible, the GoF index is readily a suitable criterion to judge how well the model is doing.

4.3.3 Path coefficients

Path coefficients are estimates that help us to assess the hypothesized relationships in the structural model. This assessment is done through an examination of the sign, significance and size of path coefficients. These path coefficients are commonly presented in a standardized form which is equivalent to standardized betas in linear regression. Standardized path coefficients are obtained from estimation of equations in which all the variables are transformed into a same metric (i.e., mean is 0 and standard deviation is 1). This transformation expresses the observational values' distance

---

[4]Some scholars have made a case against using GoF for model validation (see Henseler and Sarstedt, 2013).

from the mean in terms of the number of standard deviations. This is done to account for the differences in the range and variances in the variables. As such, having all the variables expressed on the same metric allows us to compare the coefficients of the variables directly.

Standardized coefficients generally range between -1 and 1, telling us about the strength of the relationship between an independent (X) and a dependent variable (Y). The closer a path coefficient is to ± 1, the stronger the relationship (positive/negative). And naturally, the closer a path coefficient is to 0, the weaker the relationship. Since standardized coefficients are not raw score coefficients, their interpretation will also be different. A standardized coefficient shows the change (in standard deviation units) in the expected value of Y for one standard deviation increase in X. In a complex model with several independent variables, the standardized coefficients can be used to judge the relative importance of these variables regardless of their original measurement scales. Briefly, a variable with a standardized coefficient higher than that of another variable in the model can be said to have a stronger effect on the dependent variable. Standardized coefficients may give us a rough idea of how important a variable is in terms of its contribution to explaining the variance in the dependent variable. Using the benchmark provided by Keith (2016), standardized coefficients of 0.05, 0.1 and 0.25 correspond respectively to small, moderate and large effects.

Another property of path coefficients to be studied is their sign showing the direction of the relationship. As in regression analysis, minus sign indicates a negative relationship (mean Y decreases as X increases) while positive sign indicates a positive relationship (mean Y increases as X increases). The sign of path coefficients helps us to assess whether or not the direction of a relationship between two variables that we hypothesized in our model is supported. The examination of coefficient signs may also indeed help discovering possible misspecification of a model. That is, surprisingly unexpected signs may in some cases be an indication of multicollinearity, exclusion of a relevant variable or even non-linearity. Thus, researchers should look at the coefficient signs in relation to model specification as well. Nevertheless, unexpected signs may also simply occur as a function of multivariate modelling.

Speaking of multicollinearity, this is an issue that also should be considered in the evaluation of a structural model. The explanation and criterion that we provided in Section 4.2.3.2 on formative measurement models, apply directly to assessing multicollinearity between variables in the structural model as well when the model is estimated using OLS.

A third aspect of path coefficients to examine is the statistical significance. Statistical significance is usually reported in form of a p-value. A p-value is a function of the test statistic, and the test statistic is represented by the ratio of a path coefficient and its standard error. In PLS-SEM context, the standard error is obtained using the bootstrapping procedure that we treated in Section 2.1. The most common test statistic level used as a cut-off value to decide whether or not a path coefficient is statistically significant is 1.96, corresponding to a significance level α of (approximately) 0.05. That is, a test statistic (in absolute value) above 1.96 or accordingly a p-value less than $\alpha = 0.05$ indicates a statistically significant path coefficient. Technically, this is to say that the path coefficient is statistically significantly different from zero at a level of 5%. Two other commonly used cut-off test statistic values are 2.57 (α of 0.01) and 1.65 (α of 0.1).

A test describing the relationship between two variables is usually set up as follows:

$H_0 : \beta_1 = 0$ (there is no relation), where H_0 is the null hypothesis

$H_1 : \beta_1 \neq 0$ (there is a relation), where H_1 is the alternative hypothesis

Assuming an α value of 0.05, a p-value less than 0.05 for a path coefficient provides evidence for a non-zero coefficient suggesting that there is a statistically significant relation between two variables. In this case, we conclude that we reject H_0. On the other hand, a p-value above 0.05 provides no evidence for non-zero coefficient indicating that there is not a statistically significant relation between two variables. Then, we conclude that we fail to reject H_0.

Furthermore, using the standard error obtained from the bootstrap procedure, we can construct a confidence interval for every path coefficient using $(b \pm 1.96 \times SE_b)$ assuming a confidence level of (approximately) 0.95, where SE_b represents the estimated standard error of b. As in linear regression analysis, if the confidence interval does not include zero we reject H_0, whereas if the confidence interval includes zero we fail to reject H_0. This is an alternative approach to significance testing that in some cases may be useful to adopt.

4.4 Assessing a PLS-SEM Model: A Full Example

We are in this section going to estimate and evaluate a comprehensive PLS-SEM model using the `plssem` package in Stata. Before starting to evaluate the measurement and structural models in line with the criteria that we have treated so far, we first need to explain and exhibit how to set up a model using `plssem` in Stata.

4.4.1 Setting up the model using `plssem`

A detailed introduction of the `plssem` package in Stata is provided in Chapter 3. Benefiting from this introduction, in this section we will provide a further application involving the dataset we already used in the previous chapter for introducing the basics of PLS-SEM. A description of the data is available in Section 3.2.3. This new application goes beyond the introductory examples we discussed so far. In particular, we will set up and estimate the more complicated but realistic model reproduced in Figure 4.6. As the path diagram shows, here we have got four different measurement models. The first measurement model includes a second-order reflective construct MOTIVES expressed by two first-order reflective constructs, namely Escape and Novelty which are expressed by two indicators each. The second latent variable, SATISFACTION, is measured reflectively by two indicators. The third measurement model is a formative one, ACTIVITY, formed by four different indicators. The fourth and final measurement model represents a single-item construct, RECOMMENDATION, associated only with one indicator. The data for this example are available in file `ch3_MotivesActivity.dta`. To let the output be more easily interpreted, we first rename the observed variables as follows:

```
rename spm1_6 energy
rename spm1_7 getaway
rename spm1_8 boredom
rename spm1_9 exciting
rename spm3_2 entertain
rename spm3_6 visittown
rename spm3_8 nature
rename spm3_12 fishing
rename spm15_3 recommend
rename spm15_7 satisf
rename spm15_8 expecta
```

Let us now set up the entire model in a cumulative fashion. That is, we will first specify one component of the model and then add the others one by one. We can start with the specification of the second-order construct.

```
plssem (Escape > energy getaway) ///
       (Novelty > boredom exciting) ///
       (MOTIVES > energy getaway boredom exciting), ///
       structural(Escape MOTIVES, ///
                  Novelty MOTIVES)
```

As seen in the syntax above, we first specify the two first-order constructs (Escape and Novelty) with two indicators attached to each. We then specify the second-order construct MOTIVES by connecting it to the indicators of both Escape and Novelty. Finally, in the structural part of the `plssem` command, we regress each of the first-order constructs on the second-order construct itself. This specification is also in line with the graphical representation in Figure 4.6 assuming reflective measurement model (arrows from construct to its indicators). As we already know, this way of specifying a second-order construct is referred to as repeated indicator approach.

We next extend the model by including the formative measurement construct (ACTIVITY). We proceed exactly in the same manner as above. However, this time we use the < operator (instead of >) to define the relationship between the indicators (entertain, visittown, nature, fishing) and the construct (ACTIVITY). As you see in Figure 4.6, the direction of the arrows between the items and construct is now reversed in that the construct gets now predicted by the items in a formative model. The resulting syntax is shown below[5].

[5]We repeat that running these lines of code will produce no result. The reason is that the construct ACTIVITY must first be connected to one other construct in the structural part of the model. This is done in the final syntax statement that we present later in this section.

```
1  plssem (Escape > energy getaway) ///
2         (Novelty > boredom exciting) ///
3         (MOTIVES > energy getaway boredom exciting) ///
4         (ACTIVITY < entertain visittown nature fishing), ///
5         structural(Escape MOTIVES, ///
6                    Novelty MOTIVES)
```

As for the construct with a single item, we model the constructs with single items in the exact manner that we follow in modelling formative relations (i.e., using <) just above. The reason for this is simply because we assume that the construct is made up of the single item alone. Constructs with single indicator will appear in an oval-shaped figure with an arrow pointing at them from the item, as shown in Figure 4.6. In the syntax below, you will observe that any single-indicator construct still must be specified in the measurement part

```
1  plssem (Escape > energy getaway) ///
2         (Novelty > boredom exciting) ///
3         (MOTIVES > energy getaway boredom exciting) ///
4         (ACTIVITY < entertain visittown nature fishing) ///
5         (RECOMMENDATION < recommend), ///
6         structural(Escape MOTIVES, ///
7                    Novelty MOTIVES)
```

Our final measurement model represents the construct SATISFACTION expressed by two indicators in a reflective way. This addition does then complete the specification of the overall measurement part of our example study.

```
1  plssem (Escape > energy getaway) ///
2         (Novelty > boredom exciting) ///
3         (MOTIVES > energy getaway boredom exciting) ///
4         (ACTIVITY < entertain visittown nature fishing) ///
5         (RECOMMENDATION < recommend) ///
6         (SATISFACTION > satisf expecta), ///
7         structural(Escape MOTIVES, ///
8                    Novelty MOTIVES)
```

Having drawn and set up the measurement model, the next step is to connect the construct/indicator variables to each other depending on the study's hypothesis. This is readily done by typing first the dependent variable followed by its predictors in the structural section of the syntax:

```
1  plssem (Escape > energy getaway) ///
2         (Novelty > boredom exciting) ///
3         (MOTIVES > energy getaway boredom exciting) ///
4         (ACTIVITY < entertain visittown nature fishing) ///
5         (RECOMMENDATION < recommend) ///
```

```
6          (SATISFACTION > satisf expecta), ///
7          structural(Escape MOTIVES, ///
8                     Novelty MOTIVES, ///
9                     ACTIVITY MOTIVES, ///
10                    SATISFACTION ACTIVITY MOTIVES, ///
11                    RECOMMENDATION SATISFACTION ACTIVITY)
```

Once all the variables are connected to each other, the structural model of the study is also established (see Figure 4.6) and ready to be estimated.

4.4.2 Estimation using `plssem` in Stata

To be able to estimate our model in Figure 4.6, including the measurement and structural model, we first need to choose among various options and settings for the estimation procedure. Treatment of the items (manifest variables) is the first matter to make a decision about. The most common choice in the domain of structural equation modelling is to standardize the manifest variable (i.e., mean of 0, and variance of 1). Two possible reasons for standardizing the items before estimation are (1) to make the items comparable if they are on different scales and (2) to take into account severely unequal variances of the items. `plssem` standardizes the manifest variables by default.

Another choice one should make concerns the type of weighting scheme to be used for the inner (i.e., structural) model estimation. As we described in the previous chapter, the most common schemes used are the centroid scheme, the factorial scheme and the path scheme. The weights they produce are a function of the linear correlation between the latent/observed variables, yielding similar results. However, the path scheme is generally the one recommended as it is the only one that takes into account the causal order in the structural model (see Henseler, 2010). Furthermore, the use of centroid scheme is definitely discouraged for the estimation of models including higher order constructs (see Hair et al., 2017). As such, the path scheme is the default choice in `plssem`.

Further, to be able to obtain the bootstrapped standard error estimates and subsequently the p-values of different parameters (loadings, weights, path coefficients etc.), we need to explicitly choose the `boot` option (see Section 3.5). Thus, you need to specify the number of bootstrap replications to use. You can further set a seed number for the bootstrap calculation to be able to reproduce the results:

```
1  plssem (Escape > energy getaway) ///
2         (Novelty > boredom exciting) ///
3         (MOTIVES > energy getaway boredom exciting) ///
4         (ACTIVITY < entertain visittown nature fishing) ///
5         (RECOMMENDATION < recommend) ///
6         (SATISFACTION > satisf expecta), ///
7         structural(Escape MOTIVES, ///
8                    Novelty MOTIVES, ///
9                    ACTIVITY MOTIVES, ///
```

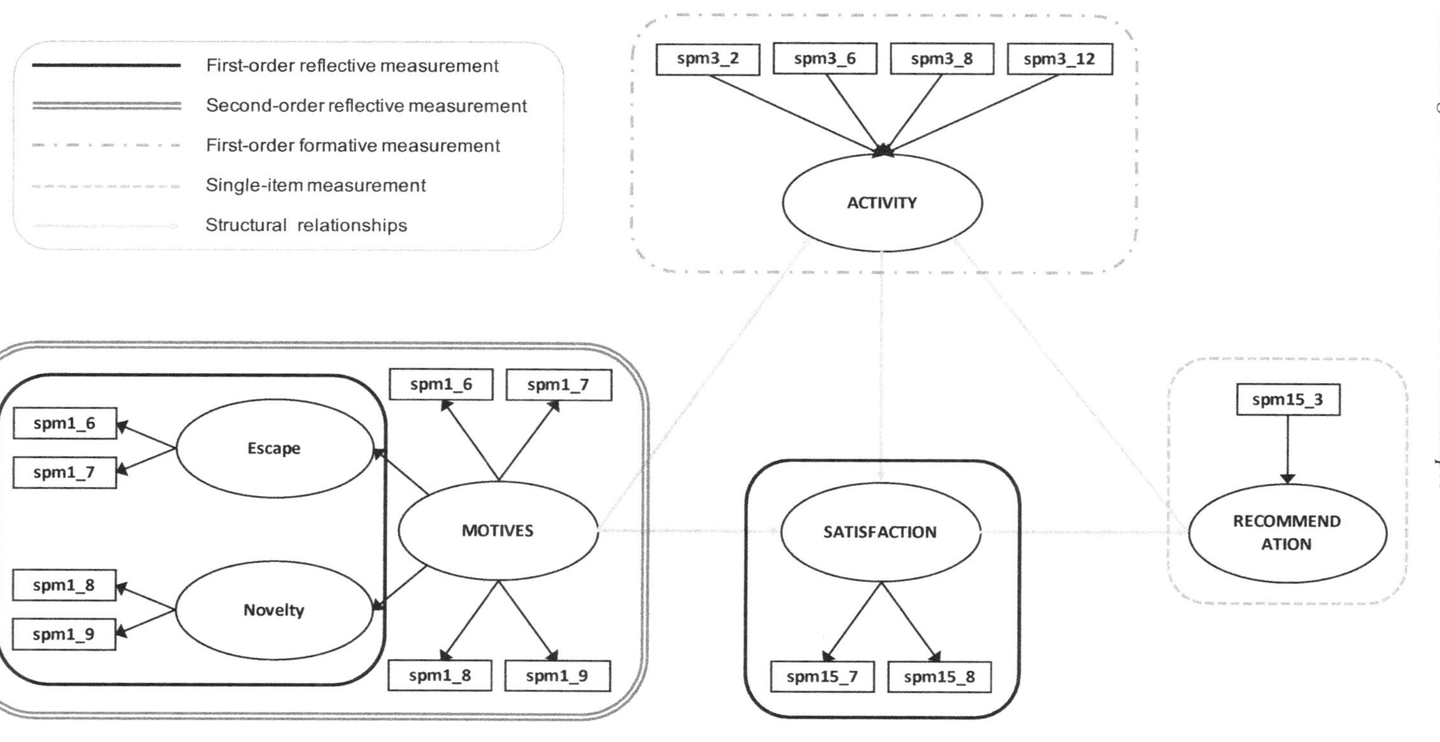

FIGURE 4.6: Measurement and structural parts of a full PLS-SEM model that includes both first- and second-order constructs.

```
10                    SATISFACTION ACTIVITY MOTIVES, ///
11                    RECOMMENDATION SATISFACTION ACTIVITY) ///
12                    boot(1000) seed(123456)
```

Treatment of missing values is one (optional) final issue to give thought to prior to estimating a PLS-SEM model. In `plssem`, there are two approaches available for handling missing data. One considers the traditional mean substitution approach, which estimates missing data by using the mean (hence only for quantitative variables), and the so called nearest neighbours approach (see Section 3.6). The default option of `plssem` is listwise deletion of the missing values in that it removes the observations with a missing value on any of the indicators in the model. Let us here for the sake of demonstration use the mean imputation option. Finally, we require also reporting the measurement model's parameters p-values with the `loadpval` option:

```
1  plssem (Escape > energy getaway) ///
2         (Novelty > boredom exciting) ///
3         (MOTIVES > energy getaway boredom exciting) ///
4         (ACTIVITY < entertain visittown nature fishing) ///
5         (RECOMMENDATION < recommend) ///
6         (SATISFACTION > satisf expecta), ///
7         structural(Escape MOTIVES, ///
8                    Novelty MOTIVES, ///
9                    ACTIVITY MOTIVES, ///
10                   SATISFACTION ACTIVITY MOTIVES, ///
11                   RECOMMENDATION SATISFACTION ACTIVITY) ///
12                   boot(1000) seed(123456) ///
13                   missing(mean) loadpval
```

We estimate the model depicted in Figure 4.6 with the settings and options exhibited in the above syntax. We chose to standardize the indicators, use path scheme, perform bootstrap procedure with 1000 replications with the seed number 123456, and finally choose the `mean` option for imputing the missing values. The results from this estimation are provided in Figure 4.7.

4.4.3 Evaluation of the example study model

In line with our suggestion, we first examine the measurement part of the model and make sure that we have psychometrically sound measurement models prior to assessing the results from the structural model.

4.4.3.1 Measurement part

The measurement part of the example study model consists of three first-order reflective measurement models (Escape, Novelty and SATISFACTION), one second-order reflective measurement model (MOTIVES), one first-order formative measurement model (ACTIVITY) and a single-item measurement model.

```
Partial least squares SEM                          Number of obs          =      1000
                                                   Average R-squared      =      0.38586
                                                   Average communality    =      0.64920
Weighting scheme: path                             Absolute GoF           =      0.50050
Tolerance: 1.00e-07                                Relative GoF           =      0.88253
Initialization: indsum                             Average redundancy     =      0.35900

Measurement model - Standardized loadings
-------------------------------------------------------------------------------------------------------
             | Reflective:  Reflective:  Reflective:   Formative:     Formative:     Reflective:
             |     Escape       Novelty      MOTIVES     ACTIVITY  RECOMMENDA~N   SATISFACTION
-------------+-----------------------------------------------------------------------------------------
      energy |      0.809
     getaway |      0.860
     boredom |                   0.887
    exciting |                   0.877
      energy |                                0.627
     getaway |                                0.722
     boredom |                                0.774
    exciting |                                0.745
   entertain |                                             0.721
   visittown |                                             0.303
      nature |                                             0.262
     fishing |                                             0.558
   recommend |                                                            1.000
      satisf |                                                                           0.885
     expecta |                                                                           0.832
-------------+-----------------------------------------------------------------------------------------
    Cronbach |      0.567        0.715        0.686                       1.000          0.646
          DG |      0.821        0.875        0.810                       1.000          0.849
       rho_A |      0.574        0.715        0.692        1.000          1.000          0.659
-------------------------------------------------------------------------------------------------------

Measurement model - Standardized loading p-values (Bootstrap)
-------------------------------------------------------------------------------------------------------
             | Reflective:  Reflective:  Reflective:   Formative:     Formative:     Reflective:
             |     Escape       Novelty      MOTIVES     ACTIVITY  RECOMMENDA~N   SATISFACTION
-------------+-----------------------------------------------------------------------------------------
      energy |      0.000
     getaway |      0.000
     boredom |                   0.000
    exciting |                   0.000
      energy |                                0.000
     getaway |                                0.000
     boredom |                                0.000
    exciting |                                0.000
   entertain |                                             0.000
   visittown |                                             0.023
      nature |                                             0.093
     fishing |                                             0.000
   recommend |                                                            0.000
      satisf |                                                                           0.000
     expecta |                                                                           0.000
-------------------------------------------------------------------------------------------------------

Discriminant validity - Squared interfactor correlation vs. Average variance extracted (AVE)
-------------------------------------------------------------------------------------------------------
             |     Escape      Novelty      MOTIVES     ACTIVITY  RECOMMENDA~N  SATISFACTION
-------------+-----------------------------------------------------------------------------------------
      Escape |      1.000        0.160        0.657        0.030        0.046         0.043
     Novelty |      0.160        1.000        0.741        0.050        0.012         0.037
     MOTIVES |      0.657        0.741        1.000        0.057        0.036         0.057
    ACTIVITY |      0.030        0.050        0.057        1.000        0.016         0.007
RECOMMENDA~N |      0.046        0.012        0.036        0.016        1.000         0.411
SATISFACTION |      0.043        0.037        0.057        0.007        0.411         1.000
-------------+-----------------------------------------------------------------------------------------
         AVE |      0.697        0.778        0.517                                   0.737
-------------------------------------------------------------------------------------------------------

Structural model - Standardized path coefficients (Bootstrap)
-------------------------------------------------------------------------------------
    Variable |     Escape      Novelty     ACTIVITY  RECOMMENDA~N  SATISFACTION
-------------+-----------------------------------------------------------------------
     MOTIVES |      0.811        0.861        0.239                       0.232
             |    (0.000)      (0.000)      (0.000)                     (0.000)
    ACTIVITY |                                            0.075         0.026
             |                                          (0.002)       (0.544)
SATISFACTION |                                            0.635
             |                                          (0.000)
-------------+-----------------------------------------------------------------------
        r2_a |      0.657        0.741        0.056       0.415         0.055
-------------------------------------------------------------------------------------
 p-values in parentheses
```

FIGURE 4.7: Results for the estimation of the `plssem` model reported in Figure 4.6.

First-order reflective measurement models

We start with the examination of the unidimensionality of the three first-order reflective constructs[6]. To do so, we perform a factor analysis with principal component extraction method using the `factor` command in Stata as shown below.

```
factor energy getaway, pcf
factor boredom exciting, pcf
factor satisf expecta, pcf
```

The results, not reported here, show that the two indicators reflected by the construct Escape are associated with two eigenvalues one of which is 1.4 indicating that 70% (1.4/2) of the variance in these two indicators is captured by Escape. As for Novelty indicators the first eigenvalue is 1.6 showing that 80% (1.6/2) of the variance in the two indicators is accounted for by Novelty. Finally, Satisfaction items are associated with two eigenvalues one of which is 1.5, suggesting that 75% (1.5/2) of the indicators is explained by Satisfaction. The fact each of these three constructs captures considerable amount of variance (at least 70 percent) in their respective items strengthens the unidimensionality further. In other words, we are not losing a lot of information (variance) contained in the items.

We continue by examining the construct reliability of our reflective constructs. As shown in Figure 4.7, the construct reliability coefficients (DG rho) for Escape, Novelty and Satisfaction are all clearly above the suggested threshold of 0.7. This confirms the homogeneity of the three constructs.

Next, we need to assess the reliability of our indicators expressing their respective constructs. As can be observed in Figure 4.7, all of the standardized loadings associated with the three reflective constructs are above the acceptable level of 0.7 and statistically significant at 0.01 using the bootstrap standard errors. This confirms that enough amount (at least 50 percent) variance of each indicator is contained in its construct.

Next, we go ahead and examine the communality of a construct measured in terms of average variance extracted (AVE). As seen in Figure 4.7, the average extracted variances for all three of the reflective constructs are above the suggested figure of 0.5. This confirms that each of our constructs captures enough (at least 50 percent) of the variance in its associated items.

Finally, all of the AVEs are clearly above the squared correlations between any two first-order reflective constructs in the model. This demonstrates discriminant validity of the three constructs, suggesting that they are distinct enough (shares more variance with their own items than with the other constructs' items) from each other.

[6]Although it generally makes more sense to check for unidimensionality when there are more than two items per construct, we still want to show you how you go about examining unidimensionality with our reflective constructs expressed by only two items each. The same idea applies to constructs with several items.

Second-order reflective measurement model

We continue our measurement model assessment by checking the construct reliability and communality (AVE) as well as indicator (first-order constructs) reliability of the second-order reflective construct MOTIVES. As seen in Figure 4.6, MOTIVES is expressed by two first-order constructs, namely Escape and Novelty. As shown in Figure 4.7, both the construct reliability (DG rho) and communality (AVE) are above the recommended level of respectively 0.7 and 0.5. Further, in the structural part of the results in Figure 4.7, the loadings[7] of the first-order constructs (i.e., 0.811 and 0.861) on the second-order constructs are above 0.7 and statistically significant using the bootstrap standard errors.

Formative measurement model

The first matter in the evaluation of a formative construct is to examine the content validity of the construct. This is an issue that should be dealt with through theoretical/contextual reasoning rather than statistical tests. In our model shown in Figure 4.6, ACTIVITY is a formative construct. To be able to measure people's level of activity people are asked to indicate whether they participated in four common activities on their vacation in the example study region. It is assumed that there is not necessarily high correlation among these four activity items. As such, it makes sense that responses to these four activities can indeed measure a person's activity level.

The second feature of the formative construct to consider is the statistical significance of the indicators. As seen in Figure 4.7, apart from the third item (i.e., nature) all of the items' weights are statistically significant at 0.05 using bootstrap standard errors. Despite the non-significant weight on the third item, it is included in the model as it theoretically makes sense to have this item included in the formation of an overall activity construct. Leaving out this item from the model may lead to loss of one important aspect of the whole activity measure.

Finally, as suggested earlier, we examine possible multicollinearity among the four items[8]. We first obtain the estimated latent scores for the construct ACTIVITY. These scores will be observed in the variable list right after the estimation. Then we regress this latent score on the four indicators. None of the VIF values is above 2.5, suggesting that multicollinearity does not cause any serious problems for the stability of our estimates and test statistics. This procedure is easily done in Stata as follows (output not reported):

```
1  quietly regress ACTIVITY entertain visittown nature fishing
2  estat vif
```

[7]We refer to these parameters as loadings instead of path coefficients because they regard the relationships between second-order and first-order constructs.

[8]We are aware of the limitations of computing VIFs for binary predictors. However, we follow this practice here only for illustration purposes.

Single-item measurement model

When it comes to single item constructs, they are modelled in the same manner as formative constructs in that the single item alone forms a construct regardless of the conceptual nature (reflective/formative) of the item. As such all of the single item's variance is included in its corresponding construct (i.e., communality equal to 1) which is the case for our single item construct RECOMMENDATION (see Figure 4.6).

4.4.3.2 Structural part

The structural part of our model in Figure 4.6 contains five different hypotheses to be tested. We now examine whether or not these hypotheses are supported, and we assess the predictive power of the overall model in terms of the amount of variance explained in the dependent variables.

Hypotheses and path coefficients

We first examine the path coefficients with respect to the hypotheses set up in the model. As shown in Figure 4.7, all of our hypotheses assume direct relations. It is generally a good idea to list these hypotheses as follows as guidance for you and readers of your work:

H1: Motives are positively related to activity

H2: Motives are positively related to satisfaction

H3: Activities are positively related to satisfaction

H4: Activities are positively related to recommendation

H5: Satisfaction is positively related to recommendation

As far as direct effects are concerned, to start with, we observe in Figure 4.7 that MOTIVES are significantly and positively related to ACTIVITY suggesting that increased motivation leads to increased activity participation. Hypothesis 1 is consequently supported. The magnitude of the relation between Motives and Activity can be said to be moderate (0.239).

The results show further that MOTIVES are also significantly related to SATISFACTION suggesting that hypothesis 2 is supported. Incidentally, the size of the effect (0.232) is moderate. Hypothesis 3 gets no support from our empirical data in that ACTIVITY is found not to be significantly and positively related to SATISFACTION. When it comes to hypothesis 4, we see that it too is supported as ACTIVITY is significantly and positively associated with RECOMMENDATION. The strength of this relation can be claimed to be weak though (0.075). The final relation, between SATISFACTION and RECOMMENDATION, is also shown to be significant and positive, a finding that provides certainly support for hypothesis 5. The magnitude of this relation (0.635) can be considered large.

```
Structural model - Multicollinearity check (VIFs)
----------------------------------------------------------------------------------
     Variable |     Escape       Novelty      ACTIVITY   RECOMMEN~N   SATISFAC~N
--------------+-------------------------------------------------------------------
      MOTIVES |      1.000         1.000         1.000                     1.060
     ACTIVITY |                                               1.007        1.060
 SATISFACTION |                                               1.007
----------------------------------------------------------------------------------
```

FIGURE 4.8: Multicollinearity assessment for the structural part of the PLS-SEM model shown in Figure 4.6.

To sum up the examination of the path coefficients of our model, we see that four out five of our hypotheses find support in the data. This result could indeed be seen as an indication that the researcher did a good job building a relevant and useful theoretical model.

Predictive power

The examination of a structural model should also include an assessment of the predictive power of the individual models as well as the whole model. When it comes to individual models, we observed in Figure 4.7 that about 6 percent of the variance in ACTIVITY, nearly 6 percent of the variance in SATISFACTION, and almost 42 percent of the variance in RECOMMENDATION is explained by their respective models. The fact that 42 percent of our target dependent variable is explained indicates a large effect and suggests that the model fits the data well. In plain English, this means that with this model we can effectively predict people's willingness to recommend the destination further to friends and relatives. The unexplained variance (58 percent) may be due to several factors (other relevant variables that are for different reasons not in the model, heterogeneity in the data, etc.).

To be able to judge the overall prediction performance of the whole model, we look at the relative GoF value, which is not that far from the 0.9 threshold. Given that all parameter estimates are in line with our expectations (i.e., sign and significance), we consider 0.88 satisfactory in this example. Furthermore, according to Wetzels et al. (2009) absolute GoF values above 0.36 can be considered acceptable, which is the case here (0.5).

Multicollinearity

As we estimate the model using OLS rule, we need to check for multicollinearity that may cause instability in the estimates. To do so, we simply type in `estat vif` after our `plssem` estimation, which will provide us with the output[9] reported in Figure 4.8. As seen, all the VIF values are below the suggested threshold of 2.5 suggesting that no severe multicollinearity exists in any of the equations of our model.

[9]Note that `plssem` does not allow to run `estat vif` when the model has been estimated using bootstrap. Therefore, to get the output in Figure 4.8, you need to remove the `boot(1000)` option from the code above.

4.5 Summary

A full PLS-SEM model consists of a measurement and a structural part. The former is about the relationship between items and constructs while the latter deals with the links among the constructs. A psychometrically sound measurement is a prerequisite for interpreting the results from the structural part. A measurement model can mainly be reflective or formative. To evaluate reflective measurement models we need to examine construct and item reliabilities as well as convergent and discriminant validities. As for the assessment of formative measurement models the most important condition is the absence of severe multicollinearity among the items. The structural part can be as simple as a simultaneous linear regression model or a more advanced mediated/moderated regression models. For the evaluation of the structural model we examine path coefficients (sign, size and significance) as well as predictive power of individual dependent variables and the overall model through R-squared and GoF index.

Appendix: R Commands

We show here how to fit the model presented in Section 4.4 using the `cSEM` package introduced in Chapter 3.

We first remind that the `cSEM` package is not able to deal with missing values. Therefore, either we omit all the missing data or we impute them beforehand. We decide to use mean imputation. Among the many `R` packages available for imputing missing values, here we use `mice` (van Buuren and Groothuis-Oudshoorn, 2011):

```
tour_data <- read.csv(file.path(path_data,
   "ch3_MotivesActivity.csv"))

# mean imputation
if (!require(mice, quietly = TRUE)) {
  install.packages("mice")
}
library(mice)

tour_imp <- mice(tour_data, method = "mean")
tour_comp <- complete(tour_imp)
```

The path diagram for the tourists satisfaction example is shown in Figure 4.6, which shows that the model includes the second-order reflective construct MOTIVES. The approach we used in the chapter to deal with such hierarchical constructs is the repeated indicators approach (see Section 3.10). The `cSEM` package does not implement this approach directly. Nevertheless, we can still use it by properly specifying the model and appending the repeated indicators to the dataset. These indicators need to be renamed because the `csem()` function does not allow for one indicator to be attached to multiple constructs:

```
tour_comp$energy_tmp <- tour_comp$energy
tour_comp$getaway_tmp <- tour_comp$getaway
tour_comp$boredom_tmp <- tour_comp$boredom
tour_comp$exciting_tmp <- tour_comp$exciting
```

The code below defines and fits the corresponding model using 1000 bootstrap replications:

```
library(cSEM)

# repeated indicators appraoch
tour_mod <- "
  # measurement model
  Escape =~ energy + getaway
  Novelty =~ boredom + exciting
  ACTIVITY <~ entertain + visittown + nature + fishing
  SATISFACTION =~ satisf + expecta
  RECOMMENDATION <~ recommend

  # 2nd order construct
  MOTIVES =~ energy_tmp + getaway_tmp + boredom_tmp + exciting_tmp

  # structural model
  Escape ~ MOTIVES
  Novelty ~ MOTIVES
  ACTIVITY ~ MOTIVES
  SATISFACTION ~ ACTIVITY + MOTIVES
  RECOMMENDATION ~ SATISFACTION + ACTIVITY
"

tour_boot <- csem(.data = tour_comp,
  .model = tour_mod, .PLS_weight_scheme_inner = "path",
  .disattenuate = FALSE, .tolerance = 1e-07,
  .resample_method = "bootstrap", .R = 1000,
  .seed = 1406)
```

The results are given by:

```
summarize(tour_boot, .ci = "CI_percentile")
assess(tour_boot,
  .quality_criterion = c("ave",
    "cronbachs_alpha", "cronbachs_alpha_weighted",
    "effects", "reliability", "r2",
```

```
6   "r2_adj", "fl_criterion"))
7 calculateGoF(tour_boot)
```

```
________________________________________________________________________________
--------------------------------- Overview ---------------------------------

General information:
------------------------
Estimation status                  = Ok
Number of observations             = 1000
Weight estimator                   = PLS-PM
Inner weighting scheme             = path
Type of indicator correlation      = Pearson
Path model estimator               = OLS
Second-order approach              = NA
Type of path model                 = Linear
Disattenuated                      = No

Resample information:
---------------------
Resample methode                   = bootstrap
Number of resamples                = 1000
Number of admissible results       = 1000
Approach to handle inadmissibles   = drop
Sign change option                 = none
Random seed                        = 1406

Construct details:
------------------
Name                Modeled as     Order         Mode

MOTIVES             Common factor  First order   modeA
Escape              Common factor  First order   modeA
Novelty             Common factor  First order   modeA
ACTIVITY            Composite      First order   modeB
SATISFACTION        Common factor  First order   modeA
RECOMMENDATION      Composite      First order   modeB

--------------------------------- Estimates ---------------------------------

Estimated path coefficients:
============================
                                                                      CI_percentile
  Path                           Estimate  Std. error   t-stat.   p-value          95%
  Escape ~ MOTIVES                 0.8107      0.0135   59.9586    0.0000 [ 0.7816; 0.8348 ]
  Novelty ~ MOTIVES                0.8611      0.0089   96.4351    0.0000 [ 0.8430; 0.8768 ]
  ACTIVITY ~ MOTIVES               0.2387      0.0355    6.7341    0.0000 [ 0.1650; 0.3044 ]
  SATISFACTION ~ MOTIVES           0.2318      0.0362    6.4098    0.0000 [ 0.1652; 0.3013 ]
  SATISFACTION ~ ACTIVITY          0.0257      0.0421    0.6101    0.5418 [-0.0570; 0.1077 ]
  RECOMMENDATION ~ ACTIVITY        0.0746      0.0244    3.0623    0.0022 [ 0.0263; 0.1221 ]
  RECOMMENDATION ~ SATISFACTION    0.6349      0.0252   25.2301    0.0000 [ 0.5856; 0.6798 ]

Estimated loadings:
===================
                                                                      CI_percentile
  Loading                        Estimate  Std. error   t-stat.   p-value          95%
  MOTIVES =~ energy_tmp            0.6275      0.0294   21.3650    0.0000 [ 0.5674; 0.6797 ]
  MOTIVES =~ getaway_tmp           0.7217      0.0190   37.9566    0.0000 [ 0.6816; 0.7550 ]
  MOTIVES =~ boredom_tmp           0.7737      0.0152   50.7676    0.0000 [ 0.7437; 0.8011 ]
  MOTIVES =~ exciting_tmp          0.7448      0.0178   41.7409    0.0000 [ 0.7072; 0.7785 ]
  Escape =~ energy                 0.8092      0.0184   44.0801    0.0000 [ 0.7673; 0.8393 ]
  Escape =~ getaway                0.8597      0.0096   89.8755    0.0000 [ 0.8406; 0.8770 ]
  Novelty =~ boredom               0.8867      0.0074  119.7272    0.0000 [ 0.8722; 0.9003 ]
  Novelty =~ exciting              0.8772      0.0088   99.8278    0.0000 [ 0.8591; 0.8936 ]
  ACTIVITY =~ entertain            0.7212      0.1200    6.0117    0.0000 [ 0.4261; 0.8892 ]
  ACTIVITY =~ visittown            0.3027      0.1296    2.3364    0.0195 [ 0.0224; 0.5416 ]
  ACTIVITY =~ nature               0.2618      0.1615    1.6205    0.1051 [-0.0656; 0.5618 ]
  ACTIVITY =~ fishing              0.5580      0.1243    4.4911    0.0000 [ 0.2886; 0.7614 ]
  SATISFACTION =~ satisf           0.8849      0.0089   99.4754    0.0000 [ 0.8652; 0.9001 ]
  SATISFACTION =~ expecta          0.8316      0.0153   54.2646    0.0000 [ 0.7995; 0.8582 ]
  RECOMMENDATION =~ recommend      1.0000      0.0000       Inf    0.0000 [ 1.0000; 1.0000 ]

Estimated weights:
==================
                                                                      CI_percentile
  Weights                        Estimate  Std. error   t-stat.   p-value          95%
  MOTIVES <~ energy_tmp            0.3116      0.0109   28.6885    0.0000 [ 0.2883; 0.3314 ]
  MOTIVES <~ getaway_tmp           0.3522      0.0096   36.5451    0.0000 [ 0.3334; 0.3720 ]
```

```
  MOTIVES <~ boredom_tmp              0.3662      0.0104   35.3694    0.0000 [ 0.3471; 0.3883 ]
  MOTIVES <~ exciting_tmp             0.3584      0.0086   41.7477    0.0000 [ 0.3425; 0.3763 ]
  Escape <~ energy                    0.5562      0.0149   37.3640    0.0000 [ 0.5254; 0.5843 ]
  Escape <~ getaway                   0.6397      0.0193   33.1639    0.0000 [ 0.6050; 0.6830 ]
  Novelty <~ boredom                  0.5776      0.0099   58.3882    0.0000 [ 0.5594; 0.5981 ]
  Novelty <~ exciting                 0.5561      0.0089   62.2940    0.0000 [ 0.5396; 0.5752 ]
  ACTIVITY <~ entertain               0.7769      0.1069    7.2644    0.0000 [ 0.5186; 0.9156 ]
  ACTIVITY <~ visittown               0.2330      0.1331    1.7504    0.0800 [-0.0507; 0.4683 ]
  ACTIVITY <~ nature                  0.2774      0.1549    1.7903    0.0734 [-0.0326; 0.5619 ]
  ACTIVITY <~ fishing                 0.5315      0.1228    4.3285    0.0000 [ 0.2724; 0.7469 ]
  SATISFACTION <~ satisf              0.6319      0.0181   34.8657    0.0000 [ 0.5971; 0.6663 ]
  SATISFACTION <~ expecta             0.5301      0.0162   32.8048    0.0000 [ 0.5006; 0.5634 ]
  RECOMMENDATION <~ recommend         1.0000      0.0000       Inf    0.0000 [ 1.0000; 1.0000 ]

Estimated indicator correlations:
=================================
                                                                         CI_percentile
  Correlation                Estimate  Std. error   t-stat.   p-value         95%
  entertain ~~ visittown       0.0088      0.0319    0.2775    0.7814 [-0.0523; 0.0693 ]
  entertain ~~ nature         -0.1758      0.0319   -5.5093    0.0000 [-0.2384;-0.1163 ]
  entertain ~~ fishing        -0.0168      0.0306   -0.5478    0.5838 [-0.0740; 0.0453 ]
  visittown ~~ nature          0.2101      0.0314    6.6838    0.0000 [ 0.1448; 0.2661 ]
  visittown ~~ fishing         0.0085      0.0313    0.2728    0.7850 [-0.0537; 0.0689 ]
  nature ~~ fishing            0.1355      0.0296    4.5733    0.0000 [ 0.0699; 0.1885 ]

------------------------------------ Effects ------------------------------------

Estimated total effects:
========================
                                                                                 CI_percentile
  Total effect                        Estimate  Std. error   t-stat.   p-value         95%
  Escape ~ MOTIVES                      0.8107      0.0135   59.9586    0.0000 [ 0.7816; 0.8348 ]
  Novelty ~ MOTIVES                     0.8611      0.0089   96.4351    0.0000 [ 0.8430; 0.8768 ]
  ACTIVITY ~ MOTIVES                    0.2387      0.0355    6.7341    0.0000 [ 0.1650; 0.3044 ]
  SATISFACTION ~ MOTIVES                0.2380      0.0345    6.8952    0.0000 [ 0.1734; 0.3055 ]
  SATISFACTION ~ ACTIVITY               0.0257      0.0421    0.6101    0.5418 [-0.0570; 0.1077 ]
  RECOMMENDATION ~ MOTIVES              0.1689      0.0237    7.1386    0.0000 [ 0.1256; 0.2162 ]
  RECOMMENDATION ~ ACTIVITY             0.0909      0.0371    2.4501    0.0143 [ 0.0155; 0.1580 ]
  RECOMMENDATION ~ SATISFACTION         0.6349      0.0252   25.2301    0.0000 [ 0.5856; 0.6798 ]

Estimated indirect effects:
===========================
                                                                            CI_percentile
  Indirect effect                Estimate  Std. error   t-stat.   p-value         95%
  SATISFACTION ~ MOTIVES           0.0061      0.0096    0.6367    0.5243 [-0.0141; 0.0246 ]
  RECOMMENDATION ~ MOTIVES         0.1689      0.0237    7.1386    0.0000 [ 0.1256; 0.2162 ]
  RECOMMENDATION ~ ACTIVITY        0.0163      0.0267    0.6097    0.5421 [-0.0383; 0.0690 ]
_________________________________________________________________________________

_________________________________________________________________________________

Construct               AVE            R2           R2_adj
MOTIVES              0.5170            NA               NA
Escape               0.6969        0.6572           0.6569
Novelty              0.7779        0.7415           0.7412
SATISFACTION         0.7373        0.0572           0.0554
ACTIVITY                 NA        0.0570           0.0561
RECOMMENDATION           NA        0.4164           0.4152

-------------- Common (internal consistency) reliability estimates -------------

Construct     Cronbachs_alpha    Joereskogs_rho    Dijkstra-Henselers_rho_A
MOTIVES              0.6857            0.8098                      1.0000
Escape               0.5667            0.8212                      1.0000
Novelty              0.7145            0.8751                      1.0000
SATISFACTION         0.6460            0.8487                      1.0000

----------------------------- Validity assessment ------------------------------

Fornell-Larcker matrix

                MOTIVES     Escape    Novelty SATISFACTION
MOTIVES      0.51698288 0.65721532 0.74149264   0.05662834
Escape       0.65721532 0.69687345 0.16034385   0.04304957
Novelty      0.74149264 0.16034385 0.77787604   0.03674087
SATISFACTION 0.05662834 0.04304957 0.03674087   0.73728206


------------------------------------ Effects ------------------------------------
```

```
Estimated total effects:
========================
  Total effect                         Estimate  Std. error   t-stat.   p-value
  Escape ~ MOTIVES                       0.8107      0.0135   59.9586    0.0000
  Novelty ~ MOTIVES                      0.8611      0.0089   96.4351    0.0000
  ACTIVITY ~ MOTIVES                     0.2387      0.0355    6.7341    0.0000
  SATISFACTION ~ MOTIVES                 0.2380      0.0345    6.8952    0.0000
  SATISFACTION ~ ACTIVITY                0.0257      0.0421    0.6101    0.5418
  RECOMMENDATION ~ MOTIVES               0.1689      0.0237    7.1386    0.0000
  RECOMMENDATION ~ ACTIVITY              0.0909      0.0371    2.4501    0.0143
  RECOMMENDATION ~ SATISFACTION          0.6349      0.0252   25.2301    0.0000

Estimated indirect effects:
===========================
  Indirect effect                  Estimate  Std. error   t-stat.   p-value
  SATISFACTION ~ MOTIVES             0.0061      0.0096    0.6367    0.5243
  RECOMMENDATION ~ MOTIVES           0.1689      0.0237    7.1386    0.0000
  RECOMMENDATION ~ ACTIVITY          0.0163      0.0267    0.6097    0.5421

Variance accounted for (VAF):
=============================
  Effects                          Estimate  Std. error   t-stat.   p-value
  SATISFACTION ~ MOTIVES             0.0258          NA        NA        NA
  RECOMMENDATION ~ MOTIVES           1.0000          NA        NA        NA
  RECOMMENDATION ~ ACTIVITY          0.1793          NA        NA        NA
________________________________________________________________________________
[1] 0.4671579
```

All the results match with those provided by Stata apart from the goodness-of-fit measure which is 0.4672 with `cSEM` and 0.5005 with Stata. This difference is due to the fact that Stata's `plssem`, differently from `cSEM`, disregards the formative and single-item constructs in computing the index.

We conclude this appendix by highlighting that the `csem()` function includes the `.approach_2ndorder` argument for dealing directly with second-order constructs, which can take value either `"2stage"` (default) or `"mixed"`. These are different from the repeated indicators approach and refer to the methods described in Agarwal and Karahanna (2000) and Ringle et al. (2012) respectively. The code reported below fits the same model but using the `"2stage"` approach (the output is not reported):

```
# two-stage approach
tour_mod2 <- "
  # measurement model
  Escape =~ energy + getaway
  Novelty =~ boredom + exciting
  ACTIVITY <~ entertain + visittown + nature + fishing
  RECOMMENDATION <~ recommend
  SATISFACTION =~ satisf + expecta

  # 2nd order construct
  MOTIVES =~ Escape + Novelty

  # structural model
  Escape ~ MOTIVES
  Novelty ~ MOTIVES
  ACTIVITY ~ MOTIVES
  SATISFACTION ~ ACTIVITY + MOTIVES
  RECOMMENDATION ~ SATISFACTION + ACTIVITY
"
```

```
18 tour_boot2 <- csem(.data = tour_comp,
19    .model = tour_mod2, .PLS_weight_scheme_inner = "path",
20    .disattenuate = FALSE, .tolerance = 1e-07,
21    .approach_2ndorder = "2stage",
22    .resample_method = "bootstrap", .R = 1000,
23    .seed = 1406)
24 summarize(tour_boot2, .ci = "CI_percentile")
25 assess(tour_boot2)
```

Note in particular that the model's specification is different from that we used with the repeated indicators approach.

Appendix: Technical Details

Tools for assessing the measurement part of a PLS-SEM model

We provide here the technical details for the criteria introduced in the chapter to assess the measurement part of a PLS-SEM model. We use the same notation as in Chapter 3.

For a reflective block, you should first check the unidimensionality, which can be examined using the following tools:

- *principal component analysis* (PCA), according to which a block can be considered unidimensional if the first eigenvalue of the correlation matrix of the block's indicators is larger than 1 and the remaining ones are instead smaller than 1, or at least far from the first one (see the appendix at page 2.6 for more technical details on PCA). One suggestion that is found in the literature is to build the first principal component to be positively correlated with all, or at least the majority of, the indicators (Tenenhaus et al., 2005).

- *Cronbach's alpha*, which is the most popular measure of internal consistency for a set of items. If the P_q indicators for the generic qth block are on their original scale, the Cronbach's alpha is defined as

$$\alpha_q = \frac{\sum_{p \neq p'} \text{Cov}(\boldsymbol{x}_{pq}, \boldsymbol{x}_{p'q})}{\text{Var}\left(\sum_{p=1}^{P_q} \boldsymbol{x}_{pq}\right)} \cdot \frac{P_q}{P_q - 1}. \tag{4.1}$$

In case the items are standardized, the formula above simplifies to

$$\alpha_q = \frac{\sum_{p \neq p'} \text{Cor}(\boldsymbol{x}_{pq}, \boldsymbol{x}_{p'q})}{P_q + \sum_{p \neq p'} \text{Cor}(\boldsymbol{x}_{pq}, \boldsymbol{x}_{p'q})} \cdot \frac{P_q}{P_q - 1}. \tag{4.2}$$

The Cronbach's alpha is a number in between 0 and 1, with values closer to 1 indicating a stronger consistency between the items. Conventionally, a block is considered unidimensional if its Cronbach's alpha is larger than 0.7.

- *Dillon-Goldstein's rho*, which for the generic qth reflective block is defined as

$$\rho_q = \frac{\left(\sum_{p=1}^{P_q} \lambda_{pq}\right)^2 \cdot \mathrm{Var}(\boldsymbol{\xi}_q)}{\left(\sum_{p=1}^{P_q} \lambda_{pq}\right)^2 \cdot \mathrm{Var}(\boldsymbol{\xi}_q) + \sum_{p=1}^{P_q} \mathrm{Var}(\boldsymbol{\varepsilon}_{pq})}, \tag{4.3}$$

where $\mathrm{Var}(\boldsymbol{\varepsilon}_{pq})$ indicates the variance of the error term. Since both the indicators and the latent variables are usually assumed to be standardized, the variance $\mathrm{Var}(\boldsymbol{\xi}_q)$ of the qth construct is equal to 1. Moreover, an estimate of $\mathrm{Var}(\boldsymbol{\varepsilon}_{pq})$ is represented by $1 - \sum_{p=1}^{P_q} \lambda_{pq}^2$, so that equation (4.3) is estimated with

$$\widehat{\rho}_q = \frac{\left(\sum_{p=1}^{P_q} \widehat{\lambda}_{pq}\right)^2}{\left(\sum_{p=1}^{P_q} \widehat{\lambda}_{pq}\right)^2 + \sum_{p=1}^{P_q} (1 - \widehat{\lambda}_{pq}^2)}. \tag{4.4}$$

As for the Cronbach's alpha, the closer $\widehat{\rho}_q$ is to 1, the more reliable are the block's indicators as a unidimensional measure of the construct. Typically, a block can be considered unidimensional if the Dillon-Goldstein's $\widehat{\rho}_q$ is at least 0.7.

For what regards convergent and discriminant validity, the most popular criteria used in PLS-SEM employ the *average variance extracted* (AVE) (Fornell and Larcker, 1981), which provides a measure of the average amount of variance of the indicators in the block that is accounted for by the corresponding construct (a similar concept as the communalities in factor analysis). The AVE for the generic qth reflective block with unstandardized indicators is defined as

$$AVE_q = \frac{\sum_{p=1}^{P_q} \lambda_{pq}^2}{\sum_{p=1}^{P_q} \mathrm{Var}(\boldsymbol{x}_{pq})}, \tag{4.5}$$

which reduces to

$$AVE_q = \frac{1}{P_q} \sum_{p=1}^{P_q} \lambda_{pq}^2 \tag{4.6}$$

in case of standardized indicators. Since the standardized loading λ_{pq} corresponds to the correlation between the pth indicator and the qth latent variable, expression (4.6) is also referred to as the **communality** of the qth reflective block, that is

$$\mathrm{Com}_q = \frac{1}{P_q} \sum_{p=1}^{P_q} \mathrm{Cor}\left(\boldsymbol{x}_{pq}, \widehat{\boldsymbol{\xi}_q}\right)^2 \tag{4.7}$$

The assessment of formative blocks uses criteria that we already introduced in other chapters (see in particular Appendix A).

A measure of goodness for the whole measurement model is given by the **average communality**, that is defined by

$$\overline{\mathrm{Com}} = \frac{1}{P} \sum_{q=1}^{Q} P_q \cdot \mathrm{Com}_q = \frac{1}{P} \sum_{q=1}^{Q} \sum_{p=1}^{P_q} \mathrm{Cor}\left(\boldsymbol{x}_{pq}, \widehat{\boldsymbol{\xi}_q}\right)^2, \tag{4.8}$$

which provides a weighted average of the block-specific communalities with weights given by the number of indicators in each block[10].

Tools for assessing the structural part of a PLS-SEM model

As we already described, the goodness of the structural model depends on the proportion of variance of an endogenous latent variable that is explained by the corresponding exogenous latent predictors, an information that is provided by the R-squared index. However, the quality of the measurement part for the endogenous variables as measured by the communalities must also be taken into account. Therefore, for each endogenous latent variable we can define a new measure called **redundancy** that provides the amount of variance for the manifest variables in the block that is explained by the corresponding exogenous latent predictors:

$$\text{Red}_j = \text{Com}_j \times R_j^2. \tag{4.9}$$

The redundancy indexes are available only for reflective blocks since the computation requires computing the communalities. The goodness of the whole structural model can be assessed with the **average redundancy** defined as

$$\overline{\text{Red}} = \frac{1}{J}\sum_{j=1}^{J} \text{Red}_j. \tag{4.10}$$

Finally, the **absolute goodness-of-fit (GoF) index** measures the overall quality of a PLS-SEM model and it is defined as the geometric mean of the average communality, which provides an assessment of the measurement model quality, and the average R-squared, $\bar{R}^2 = \sum_{j=1}^{J} R_j^2/J$, which instead provides a measure of the structural model goodness:

$$\text{GoF}_{\text{abs}} = \sqrt{\overline{\text{Com}} \times \bar{R}^2}. \tag{4.11}$$

A normalized version of the GoF index, the **relative GoF**, that is guaranteed to be bounded between 0 and 1, has been introduced by Tenenhaus et al. (2004) and it is defined by dividing each term in (4.11) by a corresponding upper bound. More specifically, regarding the first term (i.e., the average communality), assuming all the variables are standardized, an upper bound for the sum of the squared correlations between the indicators in the qth block and the corresponding latent variable is given by the first eigenvalue for the PCA on the qth block indicators $\lambda_1^{(q)}$, that is

$$\sum_{p=1}^{P_q} \text{Cor}\left(\boldsymbol{x}_{pq}, \widehat{\boldsymbol{\xi}}_q\right)^2 \leq \lambda_1^{(q)}, \tag{4.12}$$

[10] Some authors define the average communality excluding the blocks composed of a single indicator because in that case the communality is always equal to 1, which will artificially inflate the results (Esposito Vinzi and Russolillo, 2013).

so that we can define the quantity

$$T_1 = \frac{1}{P} \sum_{q=1}^{Q} \frac{\sum_{p=1}^{P_q} \text{Cor}\left(\boldsymbol{x}_{pq}, \widehat{\boldsymbol{\xi}_q}\right)^2}{\lambda_1^{(q)}}. \tag{4.13}$$

For what regards the second term in (4.11) (i.e., the average R-squared), a possible majorization is provided by the first canonical correlation of the canonical analysis (see for example Mardia et al., 1979, Chapter 10) between the manifest variables associated with the jth endogenous latent variable $\boldsymbol{\xi}_j$ and the manifest variables associated with the latent variables predicting $\boldsymbol{\xi}_j$. Denoting by ρ_j the first canonical correlation, we get that the normalized second term is given by

$$T_2 = \frac{1}{J} \sum_{j=1}^{J} \frac{R_j^2}{\rho_j^2}. \tag{4.14}$$

Then, the relative GoF is defined as the geometric mean of T_1 and T_2, that is

$$\text{GoF}_{\text{rel}} = \sqrt{\frac{1}{P} \sum_{q=1}^{Q} \frac{\sum_{p=1}^{P_q} \text{Cor}\left(\boldsymbol{x}_{pq}, \widehat{\boldsymbol{\xi}_q}\right)^2}{\lambda_1^{(q)}} \times \frac{1}{J} \sum_{j=1}^{J} \frac{R_j^2}{\rho_j^2}}. \tag{4.15}$$

Other procedures for assessing the predictive power of a PLS-SEM model (e.g., *blindfolding*) have also been introduced in the literature (see for example Esposito Vinzi and Russolillo, 2013, and the references mentioned therein).

Part II

Advanced Methods

5

Mediation Analysis With PLS-SEM

In this chapter we explain what mediation analysis is as well as presenting two approaches to testing mediational hypotheses. The first approach is that of Baron and Kenny modified by Iacobucci and colleagues, and the second one is that of Zhao and his colleagues. Before introducing them, we think that from a pedagogical point of view it is better to first explain the original approach consisting of a series of separate regression analyses. Following these explanations, we show how in practice the reader can perform different types of mediational analyses using the postestimation command `estat mediate` after the `plssem` command in Stata. The applications we show include mediation models with one observed mediator, one latent mediator as well as multiple latent mediators.

5.1 Introduction

Statistical mediation analysis is, in a nutshell, about quantifying the indirect effect of an independent variable (X) on the dependent variable (Y) through a third variable called the mediator (M). In other words, mediation analysis unearths the mechanism, be it emotional, cognitive, biological or otherwise by which X influences Y (Hayes, 2013). As an example, a mediation analysis can provide evidence for the fact that parents' educational level does influence their children's educational level, which then will be linked with their annual salary (or other occupational outcomes for that matter). That is, mediation analysis does help the researcher analyse complex statistical models including more than only direct relationships. Examining indirect effects (in addition to direct effects) has thus been an increasingly popular approach adopted by scholars in the social sciences. One major factor that has facilitated this adoption is the readily available statistical packages developed for this purpose.

5.2 Baron and Kenny's Approach to Mediation Analysis

Social scientists typically adopt Baron and Kenny (1986) approach (from here on referred to as BK approach) explained also recently elsewhere by Kenny (2016) to

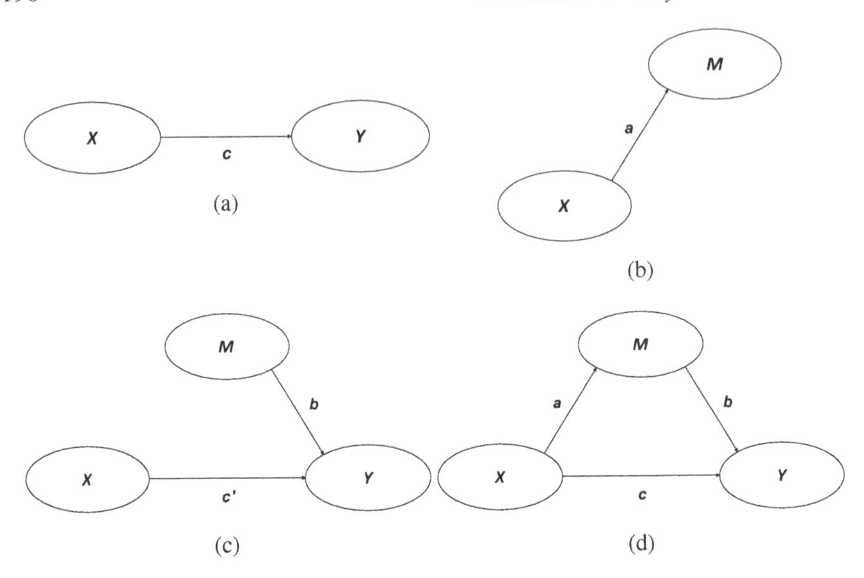

FIGURE 5.1: Diagrammatic representation of statistical mediation analysis.

conduct a mediation analysis. The BK approach consists of four distinct steps to be followed in establishing complete mediation. These steps are explained below and accordingly shown diagrammatically in Figure 5.1:

Step 1. Regress Y on X to estimate path c, which must be statistically significant implying that there is an effect to be mediated (see Figure 5.1a):

$$Y = \beta_0 + cX + \varepsilon \tag{5.1}$$

Step 2. Regress M on X to estimate path a, which must be statistically significant providing evidence of a relationship between the independent and mediator variable (see Figure 5.1b):

$$M = \gamma_0 + aX + \delta \tag{5.2}$$

Step 3. Regress Y on M (by controlling for X) to estimate path b, which must be statistically significant. X is controlled for as Y and M may be correlated because X causes both (see Figure 5.1c). This estimation provides us with path c' as well.

$$Y = \theta_0 + bM + c'X + \nu \tag{5.3}$$

Step 4. Path c' must be zero, a situation indicating that the magnitude of path c' is reduced to zero after controlling for the mediator. Note that the BK approach does not favour deciding whether path c' is equal to zero in terms of statistical significance alone. The reason is that trivially small coefficients can be statistically significant with large sample sizes and very large coefficients can be non-significant with small sample sizes (Kenny, 2016).

If all the four steps above are met, then one can claim that M completely mediates the relationship between X and Y. However, if the first three steps are met but the step 4 is not met, one can assert that M partially mediates the relationship.

Partial mediation here implies a reduction in path c, which according to Baron and Kenny (1986, p. 1176) is more realistic to encounter. The question that rises then is how big the reduction $(c - c')$ should be to claim that there exists a partial mediation (Jose, 2013). The BK approach is thus commonly followed by the Sobel's test (Sobel, 1987) which assesses the statistical significance of this reduction[1]. Since testing the reduction $(c - c')$ is equivalent to testing the mediated path $a \cdot b$, the Sobel's test is based on the following test statistic (Iacobucci et al., 2007)

$$z = \frac{a \cdot b}{\sqrt{b^2 s_a^2 + a^2 s_b^2}}, \tag{5.4}$$

where a and s_a (standard error of a) come from step 2, while b and s_b (standard error of b) come from step 3 of the BK approach described above[2]. If $|z| > 1.96$, then the mediation ($c - c'$ or $a \cdot b$) is statistically significant at the 0.05 level.

5.2.1 Modifying the Baron-Kenny approach

The original BK approach suggests that one estimates the first three steps separately using regression. However, Iacobucci et al. (2007) have demonstrated with a series of Monte Carlo simulations that using the regression technique suffers from a serious drawback (even in the simplest mediation model including X, M and Y) when compared with the structural equation modelling (SEM) technique. Their simulations show that the regression technique consistently produces larger standard errors for the path coefficients than does the SEM technique (see Figure 5.2) as a result of the fact that the latter estimates all the model parameters simultaneously[3].

A further advantage of the SEM technique is that it inherently can facilitate mediation analysis including multi-item scales. The conclusion is then that the SEM technique should be the standard framework for conducting mediation analysis. Consequently, Iacobucci et al. (2007, p. 153), by modifying the BK approach, propose the following series of steps for conducting mediation analysis via structural equation modelling:

[1] It goes without saying, if path c' is zero then there is no need for the Sobel's test.

[2] A slightly different way of computing the standard error is through the delta method. The delta method uses $\sqrt{b^2 s_a^2 + a^2 s_b^2 + s_a^2 s_b^2}$ as the denominator in equation (5.4).

[3] We remark that the simulations in Iacobucci et al. (2007) were using covariance-based SEM.

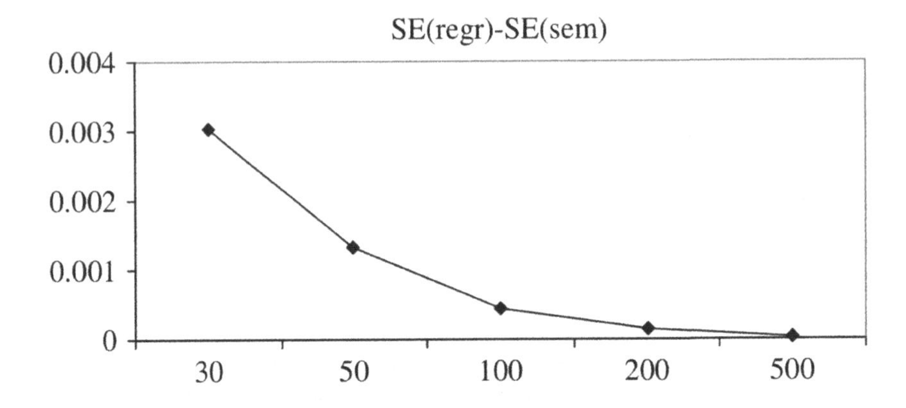

FIGURE 5.2: Comparing standard errors from the regression and SEM techniques (from Iacobucci et al., 2007, p. 144).

Step 1. Fit one model (Figure 5.1d) via SEM so the direct and indirect paths are fit simultaneously so as to estimate either effect while partialling out, or statistically controlling for, the other.

1. If either one or both are not significant there is no mediation and the researcher should stop.
2. Some mediation is indicated when both $X \rightarrow M$ and $M \rightarrow Y$ coefficients are significant and the researcher goes to the next step.

Step 2. Compute the Sobel's z to test explicitly the relative sizes of the indirect (mediated) versus direct paths. Conclusions hold as follows:

1. If the z is significant and the direct path $X \rightarrow Y$ is not, the mediation is complete.
2. If both the z and the direct path $X \rightarrow Y$ are significant, the mediation is partial.
3. If the z is not significant but the direct path $X \rightarrow Y$ is, the mediation is partial in the presence of a direct effect.
4. If neither the z nor the direct path $X \rightarrow Y$ are significant, the mediation is partial in the absence of a direct effect.

Step 3. The researcher can report the results categorically as "no", "partial" or "full" mediation.

5.2.2 Alternative to the Baron-Kenny approach

Zhao et al. (2010) do agree with Iacobucci et al. (2007) that the SEM technique is an optimal framework for conducting mediation analysis. They however go a step further and suggest the BK approach (i.e., three regression estimations and the Sobel's test) be replaced with only one test: the bootstrap test of the indirect effect $a \cdot b$ (see Figure 5.1d). They argue that to establish mediation, all that matter is that the indirect effect is statistically significant based on the bootstrap test. Based on this reasoning, the following steps are recommended by the authors for testing mediation hypotheses:

Step 1. If neither the bootstrap test of the indirect effect $(a \cdot b)$ nor the $X \rightarrow Y$ coefficient (c) is significant, then there is no-effect non-mediation (i.e., no mediation).

Step 2. If the bootstrap test of the indirect effect $(a \cdot b)$ is not significant but $X \rightarrow Y$ coefficient (c) is significant, then there is direct-only non-mediation (i.e., no mediation).

Step 3. If the bootstrap test of the indirect effect $(a \cdot b)$ is significant and $X \rightarrow Y$ coefficient (c) is not significant, then there is indirect-only mediation (i.e., full mediation).

Step 4. If both the bootstrap test of the indirect effect $(a \cdot b)$ and $X \rightarrow Y$ coefficient (c) are significant and their coefficients point in same direction, then there is complementary mediation (i.e., partial mediation).

Step 5. If both the bootstrap test of the indirect effect $(a \cdot b)$ and $X \rightarrow Y$ coefficient (c) are significant and their coefficients point in opposite direction, then there is competitive mediation (i.e., partial mediation).

The reason why Zhao et al. (2010) categorically suggest the use of the bootstrap test of indirect effects is due to the fact that the Sobel's test has low power because it by default uses a normal approximation presuming a symmetric distribution when the sampling distribution of $a \cdot b$ is known to be highly skewed (Kenny, 2016). This is still the case even when a and b per se are normally distributed (Jose, 2013), as illustrated in Figure 5.3.

As we already know (see Section 2.1), bootstrapping is a technique for generating an empirical sampling distribution of a statistic (which in our case is the mediated/indirect effect). We remind that this distribution comes about by computing and collecting the indirect effects from each of B (e.g., 1000) resamples[4] drawn with replacement from the original sample data (see Figure 5.4). From this bootstrap distribution, the standard error and accordingly a confidence interval are obtained, which can be used to test the statistical significance of the indirect effect. As for any regression coefficient, the rule is that if the confidence interval of the indirect effect does not include the value of zero, one can conclude that the indirect effect is statistically significant.

[4]Each sample drawn from the original sample must be the same size as that of the original sample, that is n.

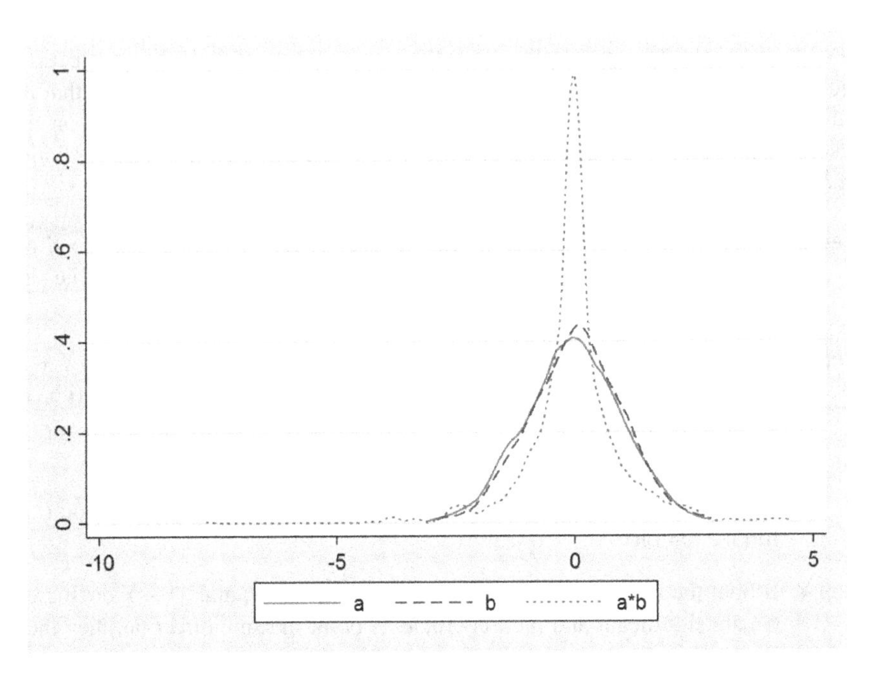

FIGURE 5.3: Two normally distributed variables (a and b) and the corresponding non-normal distribution of their product $(a \cdot b)$.

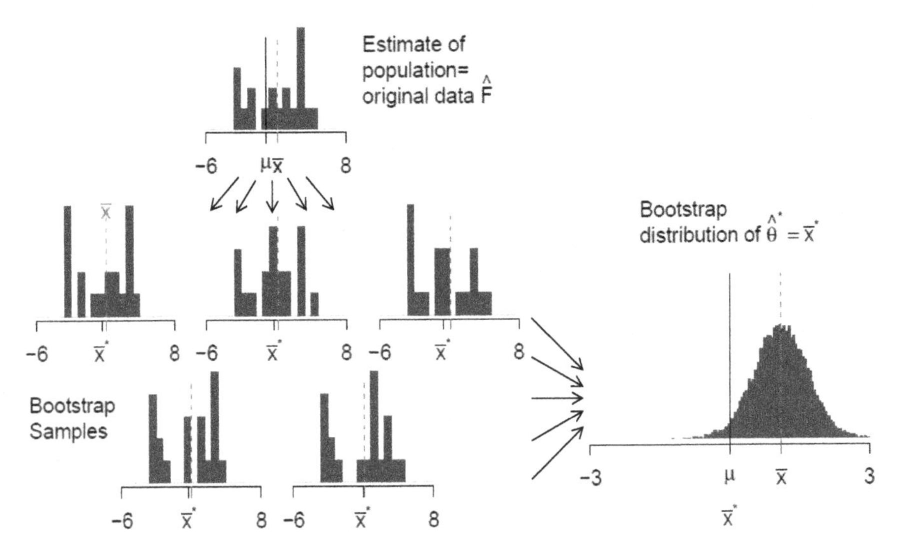

FIGURE 5.4: The process of generating the bootstrap distribution (see Hesterberg, 2015).

5.2.3 Effect size of the mediation

One way of determining the effect size of an indirect effect is to examine the completely standardized coefficient. According to Kenny (2016), a small effect size would be 0.01, a medium effect would be 0.09 and a large effect would be 0.25. He further notes that if X is a dichotomous variable, the effect size would resemble Cohen's $d$and thus a small effect size would be 0.02, a medium effect would be 0.15 and a large effect would be 0.40 (Cohen, 1988) .

A second way of gauging the effect size of an indirect effect is to take the **ratio of the indirect effect to the total effect** (RIT) given by

$$RIT = \frac{a \cdot b}{(a \cdot b) + c}, \tag{5.5}$$

Say for instance that the figure obtained from equation (5.5) is about 0.40. We can then interpret this as that a mediated effect explains 40% of the total effect of the independent variable (X) on a dependent variable (Y) or that 40% of the effect of the independent variable (X) on (Y) is mediated by the mediator variable (M) (MacKinnon, 2008). In cases in which the mediated effect and direct effect have opposite signs, a remedying option is to take the absolute values of the quantities to go into equation (5.5) (Alwin and Hauser, 1975).

A third and final measure to evaluate the effect size of an indirect effect is to take the **ratio of the indirect effect to the direct effect** (RID) as shown in the following formula:

$$RID = \frac{a \cdot b}{c}, \tag{5.6}$$

Suppose that the number resulting from equation (5.6) is about 2. The researcher can then interpret this as that the mediated effect is about 2 times as large as the direct effect (MacKinnon, 2008).

5.3 Examples in Stata

In this section we are going to estimate different (simple to complex) types of mediation models using the `plssem` package in Stata and then based on these estimated models we are going to test the assumed mediational hypotheses with the help of the postestimation command `estat mediate`. Since we already know how the `plssem` package works (see Chapter 3), we describe here how the `estat mediate` command as well as its options work. We start by presenting the full syntax of the command. Note that specifications included in the brackets are optional while the rest is obligatory:

```
1 estat mediate indep(varname) med(varname) dep(varname)
2   [breps(#) seed(#) zlc rit rid bca level(#) digits(#)]
```

where

- `indep(varname)` specifies the name of the independent variable.
- `med(varname)` specifies the name of the mediating variable.
- `dep(varname)` specifies the name of the dependent variable.
- `breps(#)` specifies the number of bootstrap replications to be performed. The default is 50.
- `seed(#)` sets the seed number for the bootstrap estimation to ensure replicability.
- `zlc` tests the mediational hypotheses based on the approach by Zhao et al. (2010) in addition to the tests based on the adjusted Baron and Kenny approach.
- `rit` provides the ratio of the indirect effect to the total effect.
- `rid` provides the ratio of the indirect effect to the direct effect.
- `bca` provides the bias-corrected accelerated (BCa) bootstrap confidence intervals. The default is the percentile confidence intervals. We warn you that this option usually makes the computation much longer because it requires using the jackknife method (see Section 2.6).
- `level(#)` indicates the confidence level to use.
- `digits(#)` indicates the number of digits to show in the output.

We are now ready to estimate our example mediational models. We will start with the simplest form of a mediational model including only observed variables with one mediator and no covariates.

5.3.1 Example 1: A single observed mediator variable

In this example we are going to estimate a model with three observed variables measuring people's age (`age`), job tenure (`tenure`) and hourly wage (`wage`), which are included in the dataset called `wageed`. Our mediational hypothesis here is that `tenure` (*M*) will mediate the relationship between `age` (*X*) and `wage` (*Y*). Diagrammatically, this model is equivalent to the one illustrated in Figure 5.1d. After loading the data, we estimate the whole mediation model using the `plssem` command:

```
1 use http://www.stata-press.com/data/r15/wageed.dta, clear

2 plssem (Age > age) ///
3        (Tenure > tenure) ///
4        (Wage > wage), ///
```

```
Structural model - Standardized path coefficients
---------------------------------------------
     Variable |      Tenure          Wage
--------------+------------------------------
          Age |       0.632         0.255
              |     (0.000)       (0.000)
       Tenure |                     0.322
              |                   (0.000)
--------------+------------------------------
         r2_a |       0.399         0.272
---------------------------------------------
 p-values in parentheses
```

FIGURE 5.5: Structural model estimates for the `plssem` model discussed in Section 5.3.1.

```
5          structural(Tenure Age, ///
6                     Wage Tenure Age)
```

The `plssem` command will provide us with the output (some parts are omitted) reported in Figure 5.5, from which we can observe the path coefficients on a $(X \rightarrow M)$, b $(M \rightarrow Y)$ and c $(X \rightarrow Y)$. Notice that our model above corresponds to a latent model with three constructs expressed by a single indicator each. As learnt in Chapter 4, we specify a latent variable for a single indicator in `plssem` command in the same way as we would do if we had more than one indicator. The results from the structural part of our `plssem` model estimation above includes two separate equations showing the direct relationships, which can be expressed as:

$$\widehat{\texttt{Tenure}} = 0.632 \cdot \texttt{Age}$$
$$\widehat{\texttt{Wage}} = 0.255 \cdot \texttt{Age} + 0.322\,\texttt{Tenure},$$

with all path coefficients being highly statistically significant.

Based on these estimates, we can now use the `estat mediate` command to test the mediational hypothesis automatically without having to make all the estimations ourselves saving us the time and more importantly providing accurate calculations. The command is shown below and the corresponding output is reported in Figure 5.6:

```
1 estat mediate, indep(Age) med(Tenure) dep(Wage) ///
2   seed(12345) breps(1000) zlc rit rid
```

Let us now go through the output produced by `estat mediate`. First, as the most salient result we observe that the indirect effect of `age` (via `tenure`) on `wage` is about 0.203, which could be considered moderate effect. Further, all the three procedures (Sobel, delta method and bootstrapping) for testing the significance of this indirect effect show that the indirect effect is statistically significant. Next, following

```
Significance testing of (standardized) indirect effect
-------------------------------------------------------------------------------------
 Statistics              |              Sobel              Delta          Bootstrap
-------------------------+-----------------------------------------------------------
 Indirect effect         |              0.203              0.203              0.203
 Standard error          |              0.010              0.010              0.010
 Z statistic             |             21.298             21.298             20.836
 P-value                 |              0.000              0.000              0.000
 Confidence interval     |     (0.185, 0.222)     (0.185, 0.222)     (0.185, 0.223)
-------------------------------------------------------------------------------------
 confidence level: 95%
 bootstrap replications: 1000

 Baron & Kenny approach to testing mediation
 STEP 1 - Tenure:Age (X -> M) with b = 0.632 and p = 0.000
 STEP 2 - Wage:Tenure (M -> Y) with b = 0.322 and p = 0.000
 STEP 3 - Wage:Age (X -> Y) with b = 0.255 and p = 0.000
          As STEP 1, STEP 2 and STEP 3 as well as the Sobel's test above
          are significant the mediation is partial

 Zhao, Lynch & Chen's approach to testing mediation
 STEP 1 - Wage:Age (X -> Y) with b = 0.255 and p = 0.000
          As the bootstrap test above is significant, STEP 1 is
          significant and their coefficients point in same direction,
          you have complementary mediation (partial mediation)

 RIT  =   (Indirect effect / Total effect)
          (0.203 / 0.458) = 0.444
          Meaning that about 44.4% of the effect of Age
          on Wage is mediated by Tenure

 RID  =   (Indirect effect / Direct effect)
          (0.203 / 0.255) = 0.799
          That is, the mediated effect is about 0.799 times as
          large as the direct effect of Age on Wage
```

FIGURE 5.6: Test of the mediation effect example discussed in Section 5.3.1. Note that in the notation `Tenure:Age`, the variable reported before : represents the dependent variable, whereas the variable indicated after : represents the independent variable.

the steps suggested by BK and ZLC (see Section 5.2) we can conclude that there is a partial mediation or alternatively put, `tenure` partially mediates the effect of `age` on `wage`. From this output we can also see that about 44 percent of the effect of `age` on `wage` is mediated by `tenure`. In other words, we could also state that the mediated effect is about 0.8 times (quite close to 1 meaning as much) as the direct effect of `age` on `wage`.

5.3.2 Example 2: A single latent mediator variable

In our second example, we are still going to have three variables in our model however this time these variables will be latent ones each represented by more than one indicator, which indeed is a more typical example of models estimated using the structural equation modelling technique. The observed variables to be used are found in the data available in the `ch5_envbehav.dta` file.

Based on the relevant literature, we assume that PERSONAL NORMS will mediate the relationship between ENVIRONMENTAL CONCERN and ENVIRINMEN-

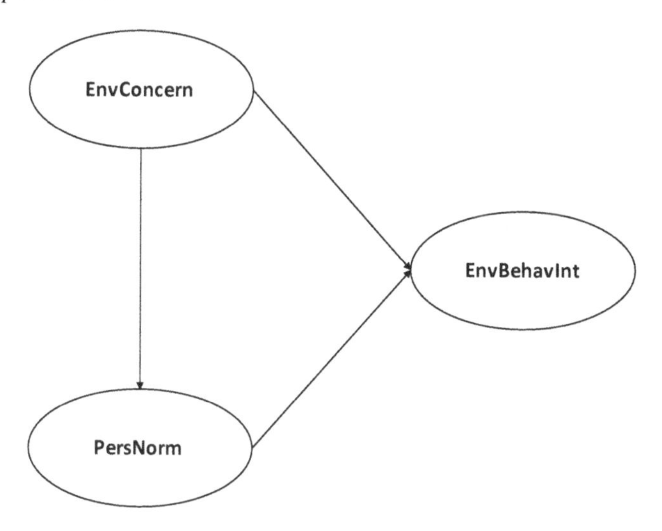

FIGURE 5.7: Structural model for the PLS-SEM model described in Section 5.3.2.

TAL BEHAVIOUR INTENTION. Put it differently, ENVIRONMENTAL CONCERN will influence ENVIRONEMNTAL BEHAVIOUR INTENTION indirectly (via PERSONAL NORMS) as well as directly as depicted in Figure 5.7. Since we are here concerned primarily with mediational model, we show only the structural part of the full PLS-SEM model.

For the purposes of testing our mediational hypothesis we operationalize ENVIRONMENTAL CONCERN (`EnvConcern`), ENVIRONMENTAL BEHAVIOUR INTENTION (`EnvBehavInt`) and PERSONAL NORMS (`PersNorm`) using an ordinal scale asking the respondents to indicate on five-point scale (from 1 = totally disagree to 5 = totally agree) to what extent they agree with the ten statements listed in Table 5.1.

As we did with the previous example, we have to estimate the mediational model with the `plssem` command prior to asking for the details of the mediational hypothesis test using `estat mediate`. The output of the PLS-SEM analysis is reported in Figure 5.8.

```
1  use ch5_envbehav, clear

2  plssem (EnvConcern > sp1e sp1m sp1o) ///
3         (PersNorm > sp3a sp3b sp3c) ///
4         (EnvBehavInt > sp2a sp2b sp2c sp2d), ///
5         structural(PersNorm EnvConcern, ///
6                    EnvBehavInt PersNorm EnvConcern)
```

Since our mediational model in the current example includes latent variables, we need to examine the psychometric properties of the model based on the same criteria that we presented in Chapter 4. This is also the reason why we provide the complete

```
Iteration 1: outer weights rel. diff. = 2.57e-01
Iteration 2: outer weights rel. diff. = 1.69e-02
Iteration 3: outer weights rel. diff. = 8.62e-04
Iteration 4: outer weights rel. diff. = 6.47e-05
Iteration 5: outer weights rel. diff. = 3.64e-06
Iteration 6: outer weights rel. diff. = 2.84e-07
Iteration 7: outer weights rel. diff. = 1.62e-08

Partial least squares SEM                       Number of obs        =       947
                                                Average R-squared    =   0.16744
                                                Average communality  =   0.56054
Weighting scheme: path                          Absolute GoF         =   0.30636
Tolerance: 1.00e-07                             Relative GoF         =   0.96308
Initialization: indsum                          Average redundancy   =   0.09361

Measurement model - Standardized loadings
-----------------------------------------------------------
              | Reflective:   Reflective:   Reflective:
              |  EnvConcern      PersNorm   EnvBehavInt
--------------+--------------------------------------------
         sp1e |      0.721
         sp1m |      0.710
         sp1o |      0.809
         sp3a |                    0.828
         sp3b |                    0.724
         sp3c |                    0.715
         sp2a |                                  0.765
         sp2b |                                  0.743
         sp2c |                                  0.753
         sp2d |                                  0.708
--------------+--------------------------------------------
     Cronbach |      0.608         0.634         0.736
           DG |      0.792         0.801         0.831
        rho_A |      0.621         0.660         0.752
-----------------------------------------------------------

Discriminant validity - Squared interfactor correlation vs. Average variance extracted (AVE)
-----------------------------------------------------------
              |  EnvConcern      PersNorm   EnvBehavInt
--------------+--------------------------------------------
   EnvConcern |      1.000         0.116         0.112
     PersNorm |      0.116         1.000         0.177
  EnvBehavInt |      0.112         0.177         1.000
--------------+--------------------------------------------
          AVE |      0.560         0.574         0.551
-----------------------------------------------------------

Structural model - Standardized path coefficients
-------------------------------------------
     Variable |    PersNorm   EnvBehavInt
--------------+----------------------------
   EnvConcern |       0.341         0.217
              |     (0.000)       (0.000)
     PersNorm |                     0.347
              |                   (0.000)
--------------+----------------------------
         r2_a |       0.115         0.217
-------------------------------------------
 p-values in parentheses
```

FIGURE 5.8: Estimates for the model discussed in Section 5.3.2.

TABLE 5.1: Content of the latent and manifest variables for the example in Section 5.3.2.

Latent variable	Manifest variable	Content
`EnvConcern`	`sp1e`	Humans are severely abusing the environment
	`sp1m`	The balance of nature is very delicate and easily upset
	`sp1o`	If things continue on their present course, we will soon experience a major ecological catastrophe
`EnvBehavInt`	`sp2a`	I would be willing to sign a petition to support an environmental cause
	`sp2b`	I would consider joining a group or club which is concerned with the environment
	`sp2c`	I would be willing to pay more taxes to support greater governmental control of pollution
	`sp2d`	I would be willing to pay more each month for electricity if it meant cleaner air
`PersNorm`	`sp3a`	Feel moral obligation to buy environmentally friendly products for my household
	`sp3b`	Feel moral obligation to recycle household waste
	`sp3c`	Feel moral obligation to pay attention to advertisements about products which are safe for the environment

results of the `plssem` results. Without explaining the detailed assessment of the measurement model here, we could confirm that our measurement model looks satisfactory. The results from the structural part of the model estimated above contains two separate equations showing the direct relationships (between the two dependent variables and their corresponding independent variable/s), that is

$$\begin{aligned}\widehat{\texttt{PersNorm}} &= 0.341 \cdot \texttt{EnvConcern} \\ \widehat{\texttt{EnvBehavInt}} &= 0.217 \cdot \texttt{EnvConcern} + 0.347 \cdot \texttt{PersNorm},\end{aligned}$$

with all coefficients being highly statistically significant.

Finally, we can now use the `estat mediate` command to test our initial mediational hypothesis and get the results reported in Figure 5.9.

```
Significance testing of (standardized) indirect effect
------------------------------------------------------------------------------------
 Statistics              |          Sobel             Delta            Bootstrap
-------------------------+----------------------------------------------------------
 Indirect effect         |          0.118             0.118              0.118
 Standard error          |          0.015             0.015              0.016
 Z statistic             |          7.945             7.945              7.432
 P-value                 |          0.000             0.000              0.000
 Confidence interval     |   (0.089, 0.147)    (0.089, 0.147)     (0.088, 0.151)
------------------------------------------------------------------------------------
 confidence level: 95%
 bootstrap replications: 1000

 Baron & Kenny approach to testing mediation
 STEP 1 - PersNorm:EnvConcern (X -> M) with b = 0.341 and p = 0.000
 STEP 2 - EnvBehavInt:PersNorm (M -> Y) with b = 0.347 and p = 0.000
 STEP 3 - EnvBehavInt:EnvConcern (X -> Y) with b = 0.217 and p = 0.000
          As STEP 1, STEP 2 and STEP 3 as well as the Sobel's test above
          are significant the mediation is partial

 Zhao, Lynch & Chen's approach to testing mediation
 STEP 1 - EnvBehavInt:EnvConcern (X -> Y) with b = 0.217 and p = 0.000
          As the bootstrap test above is significant, STEP 1 is
          significant and their coefficients point in same direction,
          you have complementary mediation (partial mediation)

 RIT  =   (Indirect effect / Total effect)
          (0.118 / 0.335) = 0.352
          Meaning that about 35.2% of the effect of EnvConcern
          on EnvBehavInt is mediated by PersNorm

 RID  =   (Indirect effect / Direct effect)
          (0.118 / 0.217) = 0.544
          That is, the mediated effect is about 0.544 times as
          large as the direct effect of EnvConcern on EnvBehavInt
```

FIGURE 5.9: Test of the mediation effect example discussed in Section 5.3.2.

```
1 estat mediate, indep(EnvConcern) med(PersNorm) ///
2   dep(EnvBehavInt) ///
3   seed(12345) breps(1000) zlc rit rid
```

We can now go through the results and provide the necessary interpretations. First, we observe that the indirect effect of `EnvConcern` (via `PersNorm`) on `EnvBehavInt` is about 0.118. We further see that all the three procedures (Sobel, delta and bootstrap) for testing the significance of this indirect effect confirm that the indirect effect is statistically significant. Finally, according to the steps proposed by BK and ZLC approaches we also see that there is a partial mediation. That is, `PersNorm` partially mediates the effect of `EnvConcern` on `EnvBehavInt`. From the same output it could be observed that about 35 percent of the effect of `EnvConcern` on `EnvBehavInt` is mediated by `PersNorm`. We could alternatively claim that the mediated effect is about 0.5 times (nearly half as much) as the direct effect of `EnvConcern` on `EnvBehavInt`.

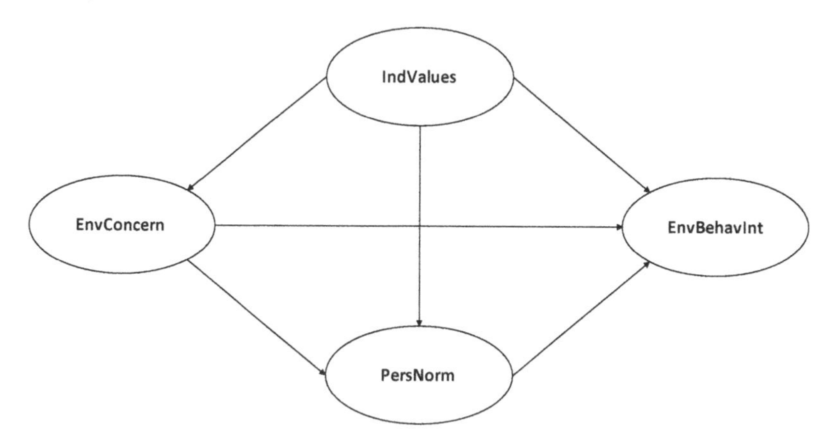

FIGURE 5.10: Structural model for the PLS-SEM model described in Section 5.3.3.

5.3.3 Example 3: Multiple latent mediator variables

In our third example, we are simply going to extend our model from the previous example by adding one more latent variable to it, namely `IndValues` representing people's individualistic personal values. This latent variable is operationalized using an ordinal scale asking the respondents to indicate on five-point scale (from 1 = not important at all to 5 = very important) how important the following three values were to them: sense of accomplishment, self-fulfilment and self-respect. Again, based on the theory reviewed, in addition to the mediational hypothesis in the previous example, we propose two more mediational hypotheses in our new model. Briefly and as depicted in Figure 5.10, we hypothesize that INDIVIDUALISTIC VALUES will influence ENVIRONMENTAL BEHAVIOUR INTENTION indirectly both via ENVIRONMENTAL CONCERN and PERSONAL NORMS.

Now that we have set up our model, we can estimate it using the `plssem` command as follows (the output is reported in Figure 5.11):

```
1  plssem (IndValues > sp8_6 sp8_7 sp8_9) ///
2         (EnvConcern > sp1e sp1m sp1o) ///
3         (PersNorm > sp3a sp3b sp3c) ///
4         (EnvBehavInt > sp2a sp2b sp2c sp2d), ///
5         structural(EnvConcern IndValues, ///
6                    PersNorm IndValues EnvConcern, ///
7                    EnvBehavInt PersNorm EnvConcern IndValues)
```

Here too, we will have to examine the psychometric properties of the measurement model prior to examining the structural model including the two additional mediational hypotheses. Again, without going into the details, we can confirm that the measurement model below is satisfactory. In this case, the structural model includes three separate equations, as there are three dependent variables in the model, which we mathematically express as:

```
Iteration 1: outer weights rel. diff. = 3.86e-01
Iteration 2: outer weights rel. diff. = 2.96e-02
Iteration 3: outer weights rel. diff. = 1.00e-03
Iteration 4: outer weights rel. diff. = 5.81e-05
Iteration 5: outer weights rel. diff. = 3.70e-06
Iteration 6: outer weights rel. diff. = 2.22e-07
Iteration 7: outer weights rel. diff. = 1.50e-08

Partial least squares SEM                   Number of obs         =      925
                                            Average R-squared     =      0.12187
                                            Average communality   =      0.59844
Weighting scheme: path                      Absolute GoF          =      0.27005
Tolerance: 1.00e-07                         Relative GoF          =      0.94544
Initialization: indsum                      Average redundancy    =      0.06821

Measurement model - Standardized loadings
--------------------------------------------------------------------------
             | Reflective:   Reflective:    Reflective:   Reflective:
             |  IndValues    EnvConcern       PersNorm    EnvBehavInt
-------------+------------------------------------------------------------
       sp8_6 |      0.811
       sp8_7 |      0.888
       sp8_9 |      0.850
        sp1e |                   0.714
        sp1m |                   0.716
        sp1o |                   0.810
        sp3a |                                  0.825
        sp3b |                                  0.736
        sp3c |                                  0.713
        sp2a |                                                 0.771
        sp2b |                                                 0.745
        sp2c |                                                 0.746
        sp2d |                                                 0.704
-------------+------------------------------------------------------------
    Cronbach |      0.813        0.607          0.638          0.736
          DG |      0.887        0.792          0.803          0.830
       rho_A |      0.846        0.620          0.663          0.756
--------------------------------------------------------------------------

Discriminant validity - Squared interfactor correlation vs. Average variance extracted (AVE)
--------------------------------------------------------------------------
             |   IndValues    EnvConcern       PersNorm     EnvBehavInt
-------------+------------------------------------------------------------
   IndValues |      1.000        0.025          0.008          0.005
  EnvConcern |      0.025        1.000          0.119          0.119
    PersNorm |      0.008        0.119          1.000          0.174
 EnvBehavInt |      0.005        0.119          0.174          1.000
-------------+------------------------------------------------------------
         AVE |      0.723        0.560          0.577          0.550
--------------------------------------------------------------------------

Structural model - Standardized path coefficients
-------------------------------------------------------------
    Variable |   EnvConcern        PersNorm     EnvBehavInt
-------------+-----------------------------------------------
   IndValues |       0.157          0.036          0.004
             |     (0.000)        (0.252)        (0.880)
  EnvConcern |                      0.340          0.228
             |                    (0.000)        (0.000)
    PersNorm |                                     0.338
             |                                   (0.000)
-------------+-----------------------------------------------
        r2_a |       0.024          0.119          0.218
-------------------------------------------------------------
  p-values in parentheses
```

FIGURE 5.11: Estimates for the model discussed in Section 5.3.3.

```
Significance testing of (standardized) indirect effect
------------------------------------------------------------------------------------------
 Statistics             |           Sobel               Delta              Bootstrap
------------------------+-----------------------------------------------------------------
 Indirect effect        |           0.012               0.012                  0.012
 Standard error         |           0.011               0.011                  0.012
 Z statistic            |           1.140               1.140                  1.046
 P-value                |           0.254               0.254                  0.296
 Confidence interval    |    (-0.009, 0.033)     (-0.009, 0.033)        (-0.009, 0.035)
------------------------------------------------------------------------------------------
 confidence level: 95%
 bootstrap replications: 1000

 Baron & Kenny approach to testing mediation
 STEP 1 - PersNorm:IndValues (X -> M) with b = 0.036 and p = 0.252
 STEP 2 - EnvBehavInt:PersNorm (M -> Y) with b = 0.338 and p = 0.000
          As either STEP 1 or STEP 2 (or both) are not significant,
          there is no mediation

 Zhao, Lynch & Chen's approach to testing mediation
 STEP 1 - EnvBehavInt:IndValues (X -> Y) with b = 0.004 and p = 0.880
          As the bootstrap test above is not significant and STEP 1 is
          not significant you have no effect nonmediation (no mediation)

 RIT  =   (Indirect effect / Total effect)
          (0.012 / 0.017) = 0.732
          Meaning that about 73.2% of the effect of IndValues
          on EnvBehavInt is mediated by PersNorm

 RID  =   (Indirect effect / Direct effect)
          (0.012 / 0.004) = 2.729
          That is, the mediated effect is about 2.729 times as
          large as the direct effect of IndValues on EnvBehavInt
```

FIGURE 5.12: Test of the first mediation effect for the example discussed in Section 5.3.3.

$$
\begin{aligned}
\widehat{\texttt{EnvConcern}} &= 0.157 \cdot \texttt{IndValues} \\
\widehat{\texttt{PersNorm}} &= 0.036 \cdot \texttt{IndValues} + 0.340 \cdot \texttt{EnvConcern} \\
\widehat{\texttt{EnvBehavInt}} &= 0.004 \cdot \texttt{IndValues} + 0.228 \cdot \texttt{EnvConcern} + 0.338 \cdot \texttt{PersNorm},
\end{aligned}
$$

with all coefficients being highly statistically significant apart from those for `IndValues` in the last two equations.

Then, we can use the `estat mediate` command to test one of the mediational hypotheses (i.e., people's individual values affect their personal norms, which then influences their environmental behaviour intention) and get the results reported in Figure 5.12:

```
estat mediate, indep(IndValues) med(PersNorm) ///
  dep(EnvBehavInt) ///
  seed(12345) breps(1000) zlc rit rid
```

When examining the output produced by the `estat mediate` command, we notice that both the BK and ZLC approaches indicate that there is no mediation.

```
Significance testing of (standardized) indirect effect
------------------------------------------------------------------------------------------
 Statistics             |            Sobel                Delta              Bootstrap
------------------------+-----------------------------------------------------------------
 Indirect effect        |            0.036                0.036                  0.036
 Standard error         |            0.009                0.009                  0.010
 Z statistic            |            4.021                4.021                  3.749
 P-value                |            0.000                0.000                  0.000
 Confidence interval    |     (0.018, 0.053)       (0.018, 0.053)         (0.018, 0.055)
------------------------------------------------------------------------------------------
 confidence level: 95%
 bootstrap replications: 1000

 Baron & Kenny approach to testing mediation
 STEP 1 - EnvConcern:IndValues (X -> M) with b = 0.157 and p = 0.000
 STEP 2 - EnvBehavInt:EnvConcern (M -> Y) with b = 0.228 and p = 0.000
 STEP 3 - EnvBehavInt:IndValues (X -> Y) with b = 0.004 and p = 0.880
          As STEP 1, STEP 2 and the Sobel's test above are significant
          and STEP 3 is not significant the mediation is complete

 Zhao, Lynch & Chen's approach to testing mediation
 STEP 1 - EnvBehavInt:IndValues (X -> Y) with b = 0.004 and p = 0.880
          As the bootstrap test above is significant and STEP 1 is not
          significant you have indirect-only mediation (full mediation)

 RIT  =   (Indirect effect / Total effect)
          (0.036 / 0.040) = 0.889
          Meaning that about 88.9% of the effect of IndValues
          on EnvBehavInt is mediated by EnvConcern

 RID  =   (Indirect effect / Direct effect)
          (0.036 / 0.004) = 8.049
          That is, the mediated effect is about 8.049 times as
          large as the direct effect of IndValues on EnvBehavInt
```

FIGURE 5.13: Test of the second mediation effect for the example discussed in Section 5.3.3.

Finally, we can test the second mediational hypothesis using the following command (output in Figure 5.13):

```
1 estat mediate, indep(IndValues) med(EnvConcern) ///
2     dep(EnvBehavInt) ///
3     seed(12345) breps(1000) zlc rit rid
```

From the results, we see that the indirect effect of INDIVIDUALISTIC VALUES on ENVIRONMENTAL BEHAVIOUR INTENTION via ENVIRONMENTAL CONCERN is statistically significant. Further, both BK and ZLC procedures show that ENVIRONMENTAL CONCERN does completely mediate the relationship between INDIVIDUALISTIC VALUES and ENVIRONMENTAL BEHAVIOUR INTENTION. As such, we have support for our second mediational hypothesis.

5.4 Moderated Mediation

Moderated mediation occurs when an indirect effect of an independent variable on a dependent variable (X via M on Y) varies depending on the values of another variable in the model. For instance, we could find out that the indirect effect of environmental concern (via personal norm) on environmentally friendly behaviour will be stronger among highly-educated people than among low-educated people. What we are saying here is that personal norm will mediate a larger share of the effect of environmental concern among highly-educated people. In this example, essentially, moderated mediation will be about testing the difference between the indirect effects for highly-educated and low-educated groups.

As such, we suggest that one uses the multi-group approach (see Section 6.4) to performing moderated mediation. As the first step, the mediation model must be specified and tested in each group (e.g., highly- and low-educated). As we did earlier in this chapter, the measurement model must be examined in each group prior to examining the structural part of the model estimated. As the second step, we would test the mediational hypothesis in each group using the `estat mediate` command as we did just in the previous section. Finally, if the mediational test gives similar/identical results for both groups, then one should obtain bootstrap standard confidence intervals for the difference between the indirect effects for deciding whether this difference is statistically significant or not. This procedure can readily be applied to a moderator including more than two categories using the knowledge provided in the current and next chapter on moderation effects.

When the moderator variable is a continuous one, we suggest to discretize the continuous moderator variable based on a theoretical/practical reasoning or arbitrary values such as one standard deviation below and above the mean. In this case, we would have three groups. We can then use the multi-group approach to testing the moderated mediation as described above. In the current version of `plssem` we do not have yet a procedure for testing the difference between the indirect effects for two or more groups.

5.5 Summary

A mediated effect occurs when an exogenous variable indirectly (via another variable) influences an endogenous variable. We have seen in this chapter that there are two approaches to testing mediational hypotheses. Both of them rightly suggest the use of simultaneous structural equation models as the standard framework for mediation analysis for observed variables, latent variables as well as a combination of observed and latent variables. When latent variables are included in a mediation model, we have also learnt that we still need to examine the measurement part of the model before interpreting the results from the structural part of the model as well as the mediation analysis results.

Appendix: R Commands

None of the R packages for PLS-SEM we presented in the previous chapters currently include any interface for performing a mediation analysis. Therefore, we created a couple of simple functions, called `mediate()` and `print_mediate()`, which replicate the calculations of Stata's `estat mediate` postestimation command. You can find the functions in the `mediate.R` file available in the supplementary material for this chapter. The `mediate()` function performs the computations and it includes the following arguments:

- `frmlist`, a named list of formulas specifying the relationships in the structural model,
- `data`, a data frame containing the sample data,
- `indep`, a length-one character vector providing the name of the independent variable,
- `med`, a length-one character vector providing the name of the mediator variable,
- `dep`, a length-one character vector providing the name of the dependent variable,
- `B`, the number of bootstrap replications,
- `bca`, a length-one logical vector indicating whether to print the bias-corrected accelerated (BCa) bootstrap confidence intervals (`TRUE`) instead of the percentile confidence intervals (`FALSE`),
- `level`, a length-one numeric vector specifying the confidence level to use,
- `fit_plssem`, a `cSEMResults` object containing the fit of a PLS-SEM model performed using the `csem()` function from the `cSEM` package.

Note that the `frmlist` argument must be specified as a list with elements that are named using the names of the endogenous constructs, otherwise the function will not be able to pick the right numbers.

The `print_mediate()` function prints the results and it provides the following arguments:

- `res_mediate`, a list containing the results of the `mediate()` function,
- `zlc`, a length-one logical vector indicating whether to print the Zhao, Lynch and Chen's approach to testing mediation (`TRUE`) or not (`FALSE`),

- `rit`, a length-one logical vector indicating whether to print the ratio of the indirect effect to the total effect (`TRUE`) or not (`FALSE`),
- `rid`, a length-one logical vector indicating whether to print the ratio of the indirect effect to the direct effect (`TRUE`) or not (`FALSE`),
- `digits`, a length-one numeric vector specifying the number of decimal digits to report in the output.

We now use the `mediate()` function for replicating the same examples we presented in Section 5.3, but we clearly skip the comments on the results. The first example involves a single *observed* mediator variable and uses the `wageed` dataset shipped with Stata. After importing the data, the following code computes the PLS-SEM solution using the `cSEM` package and then applies the `mediate()` function (results are reported next):

```
1  if (!require(systemfit, quietly = TRUE)) {
2    install.packages("systemfit")
3  }
4  if (!require(bootstrap, quietly = TRUE)) {
5    install.packages("bootstrap")
6  }
7  if (!require(numDeriv, quietly = TRUE)) {
8    install.packages("numDeriv")
9  }
10 library(cSEM)

11 source(file.path(path_code, "R", "mediate.R"))

12 # Example 1: a single observed mediator variable
13 wage_data <- read.csv(file.path(path_data,
14   "wageed.csv"))

15 wage_mod <- "
16   # measurement model
17   Age =~ age
18   Tenure =~ tenure
19   Wage =~ wage

20   # structural model
21   Tenure ~ Age
22   Wage ~ Tenure + Age
23 "

24 wage_res <- csem(.data = wage_data,
25   .model = wage_mod, .PLS_weight_scheme_inner = "path",
26   .disattenuate = FALSE, .tolerance = 1e-07,
27   .resample_method = "bootstrap", .R = 1000)
```

```
28 wage_hat <- as.data.frame(getConstructScores(wage_res))
29 tenure_frm <- as.formula(Tenure ~ Age)
30 wage_frm <- as.formula(Wage ~ Tenure + Age)
31 frmlist <- list(Tenure = tenure_frm, Wage = wage_frm)
32 dep <- "Wage"
33 med <- "Tenure"
34 indep <- "Age"

35 set.seed(1406)
36 wage_med <- mediate(frmlist, wage_hat, indep, med, dep,
37   B = 1000, fit = wage_res)
38 mediate_print(wage_med, rit = TRUE, rid = TRUE, zlc = TRUE,
39   digits = 4)
```

```
Significance testing of (standardized) indirect effect

                        Sobel       Delta    Bootstrap
Indirect effect        0.2033      0.2033       0.2033
Standard error         0.0095      0.0095       0.0097
Z statistic           21.2976     21.2976      20.9043
P-value                0.0000      0.0000       0.0000
Lower CI               0.1846      0.1846       0.1839
Upper CI               0.2221      0.2221       0.2222
---
confidence level: 95%
bootstrap replications: 1000

Baron & Kenny approach to testing mediation
STEP 1 - Tenure:Age (X -> M) with b = 0.6321 and p = 0
STEP 2 - Wage:Tenure (M -> Y) with b = 0.3217 and p = 1.489e-104
STEP 3 - Wage:Age (X -> Y) with b = 0.2546 and p = 3.13e-75
         As STEP 1, STEP 2 and STEP 3 as well as the Sobel's test above
         are significant the mediation is partial

Zhao, Lynch & Chen's approach to testing mediation
STEP 1 - Wage:Age (X -> Y) with b = 0.2546 and p = 3.13e-75
         As the bootstrap test above is significant, STEP 1 is
         significant and their coefficients point in same direction,
         you have complementary mediation (partial mediation)

RIT  =   (Indirect effect / Total effect)
         (0.2033 / 0.4579) = 0.444
         Meaning that about 44.4% of the effect of Age
         on Wage is mediated by Tenure

RID  =   (Indirect effect / Direct effect)
         (0.2033 / 0.2546) = 0.7987
         That is, the mediated effect is about 0.7987 times as
         large as the direct effect of Age on Wage
```

The second example involves a single *latent* mediator variable and uses the data available in the `ch5_envbehav.dta` file. The structural part of the model is shown in Figure 5.7. In this case, the independent variable is `EnvConcern`, the mediator is `PersNorm` and the dependent one is `EnvBehavInt`. The following code fits the model (output omitted) and then performs the mediation analysis[5]:

```
if (!require(haven, quietly = TRUE)) {
  install.packages("haven")
}

envbehav_data <- read_stata(file =
  file.path(path_data, "ch5_envbehav.dta"))
envbehav_data <- as.data.frame(envbehav_data)      # convert to data.frame
envbehav_data <- envbehav_data[, -c(11, 12, 13)]  # remove unneeded columns
envbehav_data <- na.omit(envbehav_data)            # remove missing values

envbehav_mod <- "
  # measurement model
  EnvConcern =~ sp1e + sp1m + sp1o
  PersNorm =~ sp3a + sp3b + sp3c
  EnvBehavInt =~ sp2a + sp2b + sp2c + sp2d

  # structural model
  PersNorm ~ EnvConcern
  EnvBehavInt ~ PersNorm + EnvConcern
"

envbehav_res <- csem(.data = envbehav_data,
  .model = envbehav_mod, .PLS_weight_scheme_inner = "path",
  .disattenuate = FALSE, .tolerance = 1e-07,
  .resample_method = "bootstrap", .R = 1000)
# summarize(envbehav_res)

envbehav_hat <- as.data.frame(getConstructScores(envbehav_res))
ciccio <- as.formula(PersNorm ~ EnvConcern)
pluto <- as.formula(EnvBehavInt ~ PersNorm + EnvConcern)
frmlist <- list(PersNorm = PersNorm_frm, EnvBehavInt = EnvBehavInt_frm)
dep <- "EnvBehavInt"
med <- "PersNorm"
indep <- "EnvConcern"

set.seed(1406)
envbehav_med <- mediate(frmlist, envbehav_hat, indep, med, dep,
  B = 1000, fit = envbehav_res)
mediate_print(envbehav_med, rit = TRUE, rid = TRUE, zlc = TRUE)
```

```
Significance testing of (standardized) indirect effect

                    Sobel      Delta    Bootstrap
Indirect effect     0.118      0.118        0.118
Standard error      0.015      0.015        0.017
Z statistic         7.945      7.945        6.986
P-value             0.000      0.000        0.000
Lower CI            0.089      0.089        0.089
Upper CI            0.147      0.147        0.155
---
confidence level: 95%
bootstrap replications: 1000

Baron & Kenny approach to testing mediation
```

[5] In this case we use the `haven` package to import the Stata data file because it is more flexible than `foreign`.

```
STEP 1 - PersNorm:EnvConcern (X -> M) with b = 0.341 and p = 6.95e-30
STEP 2 - EnvBehavInt:PersNorm (M -> Y) with b = 0.347 and p = 3.25e-27
STEP 3 - EnvBehavInt:EnvConcern (X -> Y) with b = 0.217 and p = 6.09e-12
         As STEP 1, STEP 2 and STEP 3 as well as the Sobel's test above
         are significant the mediation is partial

Zhao, Lynch & Chen's approach to testing mediation
STEP 1 - EnvBehavInt:EnvConcern (X -> Y) with b = 0.217 and p = 6.09e-12
         As the bootstrap test above is significant, STEP 1 is
         significant and their coefficients point in same direction,
         you have complementary mediation (partial mediation)

RIT  =   (Indirect effect / Total effect)
         (0.118 / 0.335) = 0.352
         Meaning that about 35.2% of the effect of EnvConcern
         on EnvBehavInt is mediated by PersNorm

RID  =   (Indirect effect / Direct effect)
         (0.118 / 0.217) = 0.544
         That is, the mediated effect is about 0.544 times as
         large as the direct effect of EnvConcern on EnvBehavInt
```

Finally, the third example involves multiple latent mediator variables using the same data as in the previous example, but a more complex model (see Figure 5.10 for the corresponding structural part). In this case, we assume that the new construct `IndValues` may influence `EnvBehavInt` indirectly both via the constructs `EnvConcern` and `PersNorm`. The first code chunk below fits the model (output skipped), while the next two perform the mediation analysis separately for the two different pathways:

```
1  envbehav_data <- read_stata(file =
2    file.path(path_data, "ch5_envbehav.dta"))
3  envbehav_data <- as.data.frame(envbehav_data)     # convert to data.frame
4  envbehav_data <- na.omit(envbehav_data)           # remove missing values

5  envbehav_mod <- "
6    # measurement model
7    EnvConcern =~ sp1e + sp1m + sp1o
8    PersNorm =~ sp3a + sp3b + sp3c
9    EnvBehavInt =~ sp2a + sp2b + sp2c + sp2d
10   IndValues =~ sp8_6 + sp8_7 + sp8_9

11   # structural model
12   EnvConcern ~ IndValues
13   PersNorm ~ IndValues + EnvConcern
14   EnvBehavInt ~ PersNorm + EnvConcern + IndValues
15 "

16 envbehav_res <- csem(.data = envbehav_data,
17   .model = envbehav_mod, .PLS_weight_scheme_inner = "path",
18   .disattenuate = FALSE, .tolerance = 1e-07,
19   .resample_method = "bootstrap", .R = 1000)
20 # summarize(envbehav_res)
```

```
1  envbehav_hat <- as.data.frame(getConstructScores(envbehav_res))
2  EnvConcern_frm <- as.formula(EnvConcern ~ IndValues)
3  PersNorm_frm <- as.formula(PersNorm ~ IndValues + EnvConcern)
4  EnvBehavInt_frm <- as.formula(EnvBehavInt ~ PersNorm + EnvConcern + IndValues)
```

```
5  frmlist <- list(EnvConcern = EnvConcern_frm, PersNorm = PersNorm_frm,
6    EnvBehavInt = EnvBehavInt_frm)
7  dep <- "EnvBehavInt"
8  med <- "PersNorm"
9  indep <- "IndValues"
10 set.seed(1406)
11 envbehav_med <- mediate(frmlist, envbehav_hat, indep, med, dep,
12   B = 1000, fit = envbehav_res)
13 mediate_print(envbehav_med, rit = TRUE, rid = TRUE, zlc = TRUE)
```

```
Significance testing of (standardized) indirect effect

                     Sobel       Delta    Bootstrap
Indirect effect      0.012       0.012        0.012
Standard error       0.011       0.011        0.012
Z statistic          1.140       1.140        1.023
P-value              0.254       0.254        0.306
Lower CI            -0.009      -0.009       -0.011
Upper CI             0.033       0.033        0.035
---
confidence level: 95%
bootstrap replications: 1000

Baron & Kenny approach to testing mediation
STEP 1 - PersNorm:IndValues (X -> M) with b = 0.0358 and p = 0.321
STEP 2 - EnvBehavInt:PersNorm (M -> Y) with b = 0.338 and p = 1.82e-26
         As either STEP 1 or STEP 2 (or both) are not significant,
         there is no mediation

Zhao, Lynch & Chen's approach to testing mediation
STEP 1 - EnvBehavInt:IndValues (X -> Y) with b = 0.00444 and p = 0.89
         As the bootstrap test above is not significant and STEP 1 is
         not significant you have no effect nonmediation (no mediation)

RIT  =   (Indirect effect / Total effect)
         (0.012 / 0.017) = 0.732
         Meaning that about 73.2% of the effect of IndValues
         on EnvBehavInt is mediated by PersNorm

RID  =   (Indirect effect / Direct effect)
         (0.012 / 0.004) = 2.729
         That is, the mediated effect is about 2.729 times as
         large as the direct effect of IndValues on EnvBehavInt
```

```
1 dep <- "EnvBehavInt"
2 med <- "EnvConcern"
3 indep <- "IndValues"
4 set.seed(1406)
5 envbehav_med <- mediate(frmlist, envbehav_hat, indep, med, dep,
6   B = 1000, fit = envbehav_res)
7 mediate_print(envbehav_med, rit = TRUE, rid = TRUE, zlc = TRUE)
```

```
Significance testing of (standardized) indirect effect

                     Sobel       Delta    Bootstrap
Indirect effect      0.036       0.036        0.036
Standard error       0.009       0.009        0.009
Z statistic          4.021       4.021        3.839
P-value              0.000       0.000        0.000
Lower CI             0.018       0.018        0.019
```

```
Upper CI            0.053         0.053         0.056
---
confidence level: 95%
bootstrap replications: 1000

Baron & Kenny approach to testing mediation
STEP 1 - EnvConcern:IndValues (X -> M) with b = 0.157 and p = 0.00000126
STEP 2 - EnvBehavInt:EnvConcern (M -> Y) with b = 0.228 and p = 8e-13
STEP 3 - EnvBehavInt:IndValues (X -> Y) with b = 0.00444 and p = 0.89
         As STEP 1, STEP 2 and the Sobel's test above are significant
         and STEP 3 is not significant the mediation is complete

Zhao, Lynch & Chen's approach to testing mediation
STEP 1 - EnvBehavInt:IndValues (X -> Y) with b = 0.00444 and p = 0.89
         As the bootstrap test above is significant and STEP 1 is not
         significant you have indirect-only mediation (full mediation)

RIT  =   (Indirect effect / Total effect)
         (0.036 / 0.04) = 0.889
         Meaning that about 88.9% of the effect of IndValues
         on EnvBehavInt is mediated by EnvConcern

RID  =   (Indirect effect / Direct effect)
         (0.036 / 0.004) = 8.049
         That is, the mediated effect is about 8.049 times as
         large as the direct effect of IndValues on EnvBehavInt
```

6

Moderating/Interaction Effects Using PLS-SEM

In this chapter the reader learns what a moderating/interaction effect is within the framework of two-way linear interactions in PLS-SEM. Three approaches to testing interactions effects in PLS-SEM are accordingly elucidated. These are the product-indicator approach, the two-stage approach and the multi-sample approach, the latter also called multi-group analysis. In the explanation of the multi-sample approach, parametric and permutation tests are also explained. Finally, the chapter concludes with detailed applications of the three approaches.

6.1 Introduction

Social scientists typically specify a statistical model based on the assumption that the effect of an independent (exogenous) variable on a dependent (endogenous) variable is invariant of any other exogenous variable in the model. More precisely, the coefficient of an exogenous variable is assumed to be the same at every level of other exogenous variables in the model. Such linear additive models may however not hold in some situations leading thus at best to less nuanced or even wrong information. Thus, the importance of non-additive statistical models has been continuously stressed in the methodology literature.

In linear additive models, we examine main effects whereas in non-additive models, we are mainly interested in testing moderating or interaction effects (see Figure 6.1). Incidentally, we use the terms moderating effect and interaction effect interchangeably in this chapter. A *moderating effect* is said to occur when a third variable, the *moderator*, affects the relation between an exogenous variable and an endogenous variable. The moderating effect is demonstrated through a significant change in the size or/and direction of the coefficient of the exogenous variable at different values of the moderator variable. The concept of moderating/interaction effect can be best explained through some real life examples.

For instance, an organization psychologist finds out that the effect of authoritarian leadership style on employee effectiveness differs between inexperienced and experienced employees: authoritarian leadership has a significantly larger and positive effect on effectiveness among inexperienced employees than among experienced

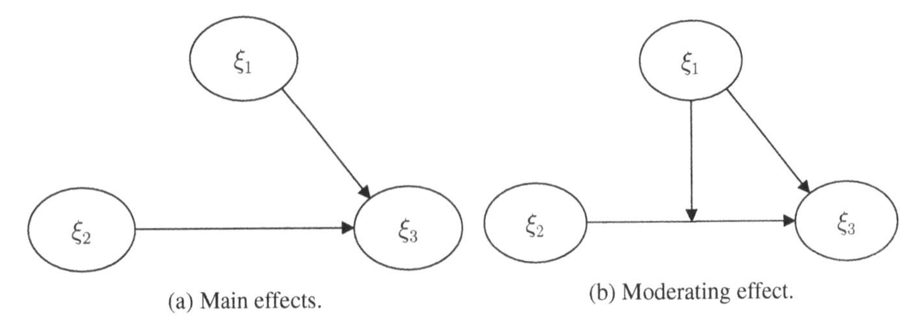

(a) Main effects. (b) Moderating effect.

FIGURE 6.1: Illustration of additive (left panel) and non-additive (right panel) models. ξ_1 and ξ_2 represent exogenous latent variables while ξ_3 indicates an endogenous latent variable.

employees. In this case, authoritarian leadership is the exogenous variable, effectiveness is the endogenous variable and employee status (in/experienced) is the moderator variable. In another example, we may have a political scientist discovering in her research that the effect of attitude towards immigrants on opinions about right-wing politics differs from a low unemployment year (1990) to a high unemployment year (1991). More specifically, attitude towards immigrants has a stronger negative effect on opinions about right-wing politics in the low unemployment year than it has in the high unemployment year. Here, year of employment rate (1990 vs. 1991) is the moderator variable. The researcher can thus claim that year moderates the effect of attitude towards immigrants on opinions about right-wing politics. In a third example, we have a marketing scholar discovering that image of a company moderates the effect of media coverage on willingness to buy a product. In other words, as people have a more favourable image of a company, the effect of media coverage of the company on people's willingness to buy a product from that company increases.

Bear in mind that in the first example, we have a categorical moderator variable (inexperienced/experienced employee). We have a categorical moderator variable in the second example as well. However, the moderator variable in this case includes two time points. In the third example, we have a continuous moderator variable (image). These examples imply that we can readily use both categorical and continuous variables as moderators in PLS-SEM.

In the following sections, we present in detail the three most common approaches used to examine moderating/interactions effects in partial least squares structural equation modelling. These are the product-indicator, the two-stage and the multi-sample approach.

6.2 Product-Indicator Approach

The product-indicator approach used often in PLS-SEM stems from the way interactions are modelled in multiple linear regression analysis[1]. That is, a product term created by multiplying x_1 and x_2 (let's call it x_1x_2) is entered into the regression model together with x_1 and x_2 as $y = \beta_0 + \beta_1 x_1 + \beta_2 x_2 + \beta_3 x_1 x_2 + \varepsilon$. This equation shows a typical interactive model containing two component terms (x_1 and x_2) that are observable. The question that rises is then how one goes about applying this idea to interactions between latent variables.

Chin et al. (2003) have addressed this issue and suggest creating product terms directly between the indicators of the exogenous latent variable and the indicators of the moderator latent variable. As shown in Figure 6.2[2], when the exogenous (ξ_1) and moderator (ξ_2) latent variables are reflected by two (x_{11} and x_{21}) and three indicators (x_{12}, x_{22} and x_{32}) respectively, there will be as many as 6 product indicators which will serve as the reflective indicators of the latent interaction variable (ξ_3). The endogenous variable (ξ_4) is then simply regressed onto the three latent variables in tandem resulting in the following equation

$$\xi_4 = \beta_0 + \beta_1 \xi_1 + \beta_2 \xi_2 + \beta_3 \xi_3 + \zeta, \tag{6.1}$$

where ξ_3 corresponds to the product ($\xi_1 \times \xi_2$).

Here we test whether β_3 is significantly different from 0 ($H_0 : \beta_3 = 0$). The standard error from the bootstrap distribution of β_3 is generated and used to calculate the required p-value. If the p-value is less than 0.05 (or another conventional level), then we can reject the null hypothesis, and thus provide evidence for the presence in the model of an interaction effect ($H_1 : \beta_3 \neq 0$).

We note that both the exogenous, moderator and interaction latent variables must all be reflective constructs to be able to employ the product-indicator approach to test interaction effects in PLS-SEM. Due to pedagogical reasons, in Figure 6.2 we denote ξ_2 as the moderator variable and ξ_1 as the independent variable. However, we can readily reverse this situation and still apply the same product-indicator equation. In other words, the product term ($\xi_1 \times \xi_2$) can equally be interpreted as if either ξ_1 or ξ_2 is the moderator variable. In practice though, the researcher's hypothesis decides what variable is to be treated as the moderator.

To better understand the interpretation of the results from a non-additive latent variable model estimated with an interaction term, we first remind the reader of the way one interprets the coefficients in an additive model (i.e., with no interaction term). An additive version of the model depicted in Figure 6.2 would then look as follows:

$$\xi_4 = \beta_0 + \beta_1 \xi_1 + \beta_2 \xi_2 + \zeta. \tag{6.2}$$

Here we are purely interested in examining the main effects of ξ_1 and ξ_2 on ξ_4.

[1] In Section A.2.5 you can find a review of interaction terms in linear regression.

[2] In this chapter we simplify slightly the notation to allow focusing on the new concepts we illustrate.

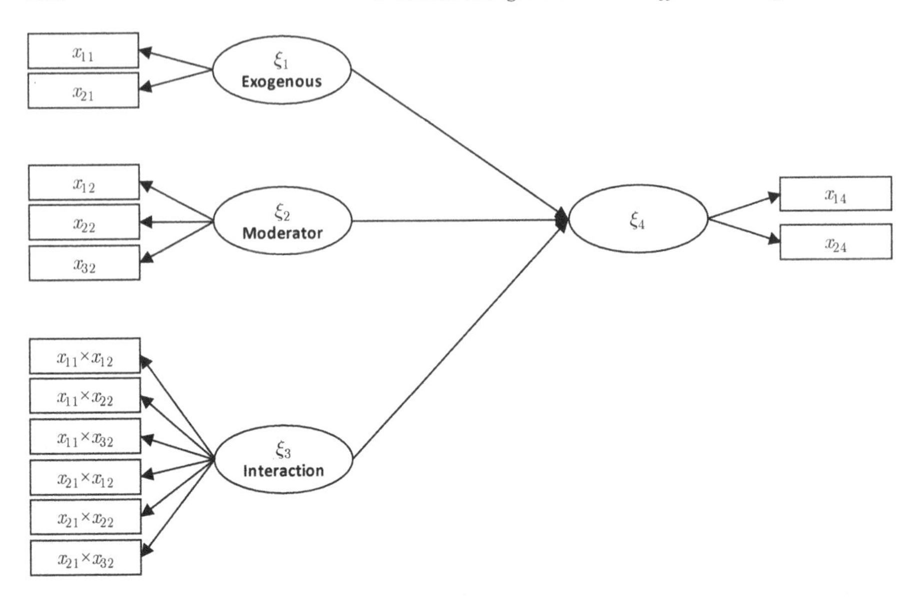

FIGURE 6.2: Depiction of the product-indicator approach in PLS-SEM.

Intuitively put, β_1 reflects the "average" effect of ξ_1 on ξ_4 at all levels of ξ_2. β_1 is subsequently assumed to be constant across ξ_2. More formally, β_1 represents the number of units that mean ξ_4 changes as a result of one unit increase in ξ_1 while holding ξ_2 constant. Accordingly, β_2 represents the number of units the mean of ξ_4 changes as a result of one unit increase in ξ_2 while holding ξ_1 constant.

Let us now extend equation (6.2) by creating an interaction term $(\xi_1 \times \xi_2)$ and including it in the model which then results in the equation below:

$$\xi_4 = \beta_0 + \beta_1 \xi_1 + \beta_2 \xi_2 + \beta_3 (\xi_1 \times \xi_2) + \zeta. \tag{6.3}$$

In the equation above, β_1 and β_2 no longer represent main effects. They now reflect the so called simple main effects. Specifically, β_1 represents the effect of ξ_1 on ξ_4 when ξ_2 is equal to 0, and likewise β_2 represents the effect of ξ_2 on ξ_4 when ξ_1 is equal to 0. More importantly, β_3 represents the change in the slope of ξ_4 on ξ_1 as a result of one unit increase in ξ_2. Had we chosen β_1 as the moderator variable instead, β_3 would represent the change in the slope of ξ_4 on ξ_2 as a result of one unit increase in ξ_1. Mathematically, these two interpretations are equally legitimate.

The model represented by (6.3) illustrates one further important point that, in our opinion, is often neglected or kept implicit in the dissemination of results from interaction analysis in research publications. That is, in the case of interaction between two latent variables (or continuous variables for that matter) the form of the interaction effect is assumed to be linear. However, in some situations the effect of an exogenous variable on an endogenous variable may not necessarily follow a linear pattern.

In our example above, the coefficient β_3 on the interaction term $(\xi_1 \times \xi_2)$ in the product indicator approach is used to test the linear form of interactions. The null hypothesis is that $\beta_3 = 0$, so if β_3 proves to be significantly different from 0, then we can claim that ξ_2 indeed moderates the relation between ξ_1 and ξ_4. As such, a non-significant β_3 is a sign of no linear interaction effect and however it should not be seen as a sign of not any form of interaction at all. Thus, researchers should be cautious about not making hasty conclusions in such cases.

Let us now further elaborate the interpretation of an interactive model using some hypothetical values for the intercept and the coefficients of the predictors and interaction part in (6.3). Let us further assume also that ξ_1, ξ_2 (moderator) and ξ_4 all have a scale ranging from 1 to 5 and all the coefficients in the equation are statistically significant:

$$\begin{aligned}\xi_4 &= \beta_0 + \beta_1\xi_1 + \beta_2\xi_2 + \beta_3(\xi_1 \times \xi_2) + \zeta \\ &= -4 + 2\xi_1 + 3\xi_2 + 1(\xi_1 \times \xi_2) + \zeta.\end{aligned}$$

To obtain the slope of ξ_4 on ξ_1 at different values of ξ_2, we use the pursuing approach:

$$\begin{array}{lll}\xi_4 = -4 + 2\xi_1 + 3 \times (1) + 1(\xi_1 \times 1) + \zeta & \text{when } \xi_2 = 1 \text{ then} & \xi_4 = -1 + 3\xi_1 + \zeta \\ \xi_4 = -4 + 2\xi_1 + 3 \times (2) + 1(\xi_1 \times 2) + \zeta & \text{when } \xi_2 = 2 \text{ then} & \xi_4 = 2 + 4\xi_1 + \zeta \\ \xi_4 = -4 + 2\xi_1 + 3 \times (3) + 1(\xi_1 \times 3) + \zeta & \text{when } \xi_2 = 3 \text{ then} & \xi_4 = 5 + 5\xi_1 + \zeta \\ \xi_4 = -4 + 2\xi_1 + 3 \times (4) + 1(\xi_1 \times 4) + \zeta & \text{when } \xi_2 = 4 \text{ then} & \xi_4 = 8 + 6\xi_1 + \zeta \\ \xi_4 = -4 + 2\xi_1 + 3 \times (5) + 1(\xi_1 \times 5) + \zeta & \text{when } \xi_2 = 5 \text{ then} & \xi_4 = 11 + 7\xi_1 + \zeta\end{array}$$

As seen above, the slope increases by one unit as a result of one unit increase in the moderator variable (ξ_2) exhibiting a linear interaction form.

Until now we have deliberately only worked with raw (i.e., untransformed) data to get a deeper understanding of interactions. We remember from equation (6.3) that a coefficient on the independent variable in an interactive model reflects the slope (effect) when the moderator is zero. Such an interpretation does however not make any sense when the moderator variable does not have 0 in its scale range. One solution to easily come around this situation is indeed standardizing the variables. Like in an interactive model with centred data, the coefficients on the independent variable will still reflect the slope at the value of 0. Since 0 is the mean of a standardized variable, the coefficient will reflect the slope at the mean of the moderator variable. As opposed to raw scales used for the interpretation in the centred solution, the coefficients in the standardized solution will be interpreted in terms of standard deviations. The interpretation of the standardized solution is similar to that in multiple linear regression analysis. That is, one standard deviation increase in X leads to so much standard deviation change in Y. If one wishes to obtain a standardized solution in PLS-SEM, the variables are suggested to be standardized prior to creating the interaction term. The reason is simply that standardized product term (a posteriori) is not equal to product term of standardized indicators (a priori). The latter provides the basis for the correct interpretation in terms of standard deviations.

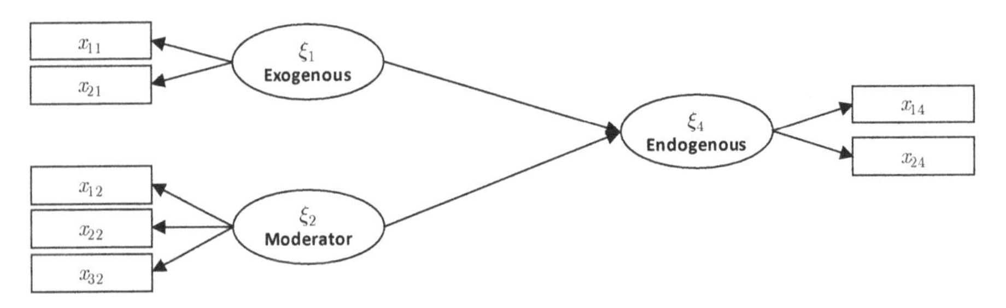

FIGURE 6.3: Two-stage approach: stage one, main effects only.

6.3 Two-Stage Approach

The two-stage approach was originally proposed by Chin et al. (2003) for interactive models with formative constructs (i.e., exogenous and moderator variables). However, as suggested elsewhere (Henseler and Chin, 2010), the two-stage approach may also be employed for testing interaction effects between reflective constructs. Thus, the following explanation of the two-stage approach is equally relevant and valid for interactive models with formative or reflective constructs.

The logic behind the two-stage approach is that we estimate and use latent variable scores as manifest (observed) variables in constructing the different components of an interactive model. More specifically, as the name of the method implies, this approach takes place in two stages. In the first stage, we build up a main effect only (additive) model and estimate the latent variable scores for the components of this model (ξ_1, ξ_2 and ξ_4). The latent variable scores, from the model shown in Figure 6.3, are obtained after ξ_4 is regressed on the exogenous latent variable (ξ_1) and moderator latent variable (ξ_2).

In stage two, we build up the interactive model using the latent variable scores obtained from the first stage. Let us now denote the latent variable scores for the exogenous and moderator variables as ξ_1^{LS}, ξ_2^{LS} and ξ_4^{LS} respectively. As depicted in Figure 6.4, the latent variable scores are used as the manifest variables or indicators reflecting the exogenous (ξ_1) and moderator (ξ_2) variable. The interaction term (ξ_3) is then created as the product of the indicators (i.e., $\xi_1^{LS} \times \xi_2^{LS}$) of the component terms. The latent variable score for the endogenous variable is incidentally used as the indicator of ξ_4 in the interactive model. The endogenous variable is subsequently regressed on the exogenous, moderator and the interaction term in tandem.

The explanation and procedures (including the bootstrap procedure for testing the interaction effect) as to the interpretation of the coefficients in an interactive model provided under the section on product indicator approach is directly transferable to the two-stage approach. In a way, the only difference between product-indicator and two-stage approach is that in the former, constructs are reflected by more than one indicator whereas in the latter the constructs are simply single-indicator quantities. Put it differently, the interpretation of the two-stage interactive models will indeed

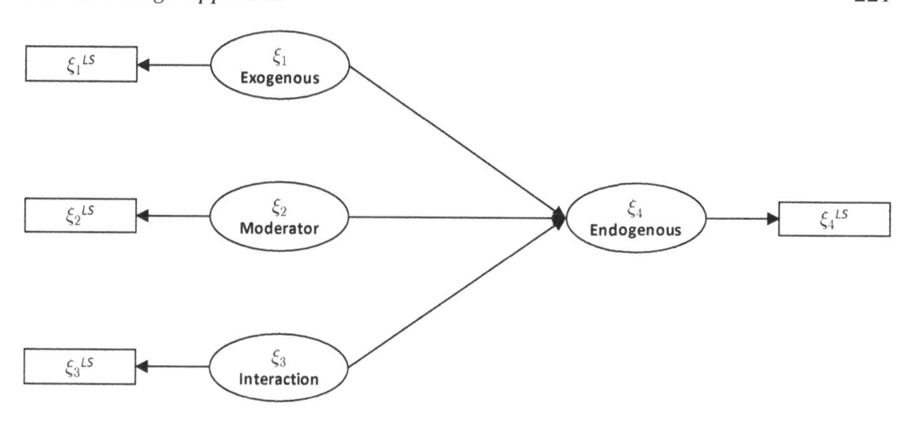

FIGURE 6.4: Two-stage approach: stage two, interactive model.

be same as that used for interpreting interaction effects between two continuous predictors in an ordinary multiple regression analysis.

The two-stage approach as opposed to the product indicator approach is more flexible, and thus it, in our opinion, can readily be used to test models with an interaction between an exogenous latent variable and a dichotomous moderator variable as well. Although we actually can use the same interpretation technique as in the product indicator approach also for interpreting the results from an interactive model with a categorical moderator variable, we will still in the following present and explain how this can be done in practice. To do so, we start with the first step of the two-stage approach, that is the equation below where ξ_1 is the exogenous latent variable and θ_1 represents the dichotomous moderator variable:

$$\xi_4 = \beta_0 + \beta_1 \xi_1 + \beta_2 \theta_1 + \zeta. \tag{6.4}$$

As explained previously, the latent variable scores obtained having estimated the model represented by equation (6.4) are subsequently used to build up the interactive model reflected now by the following equation:

$$\xi_4^{LS} = \beta_0 + \beta_1 \xi_1^{LS} + \beta_2 \theta_1 + \beta_3 \left(\xi_1^{LS} \times \theta_1\right) + \zeta. \tag{6.5}$$

Suppose now that, after the estimation of this model, we get the following hypothetical values for the coefficients:

$$\begin{aligned} \xi_4^{LS} &= \beta_0 + \beta_1 \xi_1^{LS} + \beta_2 \theta_1 + \beta_3 \left(\xi_1^{LS} \times \theta_1\right) + \zeta & (6.6) \\ &= 204 + 1.70 \xi_1^{LS} - 15 \theta_1 - 1.65 \left(\xi_1^{LS} \times \theta_1\right) + \zeta. & (6.7) \end{aligned}$$

The simple main effect of ξ_1 on ξ_4 is 1.70. This is also referred to as the slope for those belonging to the 0 value of the moderator variable (θ_1). The simple main effect θ_1 is -15. This is also referred to the difference between mean scores of those belonging to the 0 and 1 values of the moderator when ξ_1 is zero. The coefficient for the interaction term is -1.65, suggesting that for a unit increase (in this case from 0 to 1), the effect of ξ_1 on ξ_4 (slope) decreases by 1.65 units. In other words, the slope for those belonging to the 0 value of the moderator is 1.65 units higher than that for those belonging to the 1 value of the moderator. Put it less formally, ξ_1 has a larger positive effect on ξ_4 in group 0 as compared to group 1. We can also alternatively arrive at the same results by directly working out the following equations:

- when $\theta_1 = 0$

$$\begin{aligned}\xi_4^{LS} &= 204 + 1.70\xi_1^{LS} - 15 \times 0 - 1.65\left(\xi_1^{LS} \times 0\right) + \zeta \\ &= 204 + 1.70\xi_1^{LS} + \zeta,\end{aligned}$$

  so that the slope for the 0 group is 1.70,

- when $\theta_1 = 1$

$$\begin{aligned}\xi_4^{LS} &= 204 + 1.70\xi_1^{LS} - 15 \times 1 - 1.65\left(\xi_1^{LS} \times 1\right) + \zeta \\ &= 189 + 0.05\xi_1^{LS} + \zeta,\end{aligned}$$

  so that the slope for the 1 group is 0.05.

The difference between the two slopes is 1.65 (i.e., $1.70 - 0.05$), which indeed corresponds to the coefficient of the interaction term in equation (6.6). This point is also graphically depicted in Figure 6.5. The slope of ξ_4 on ξ_1 in group 0 is clearly larger (steeper regression line) than that in group 1, a pattern that is an indication of an interaction effect. If there was not an interaction effect, the two regression lines would be parallel to each other.

Use of the two-stage approach can be further extended to test interaction effects when the moderator variable has more than two categories. Say that the moderator variable has three categories. In this case, we create three dummy variables two of which (say θ_1 and θ_2) along with the exogenous variable they are assumed to interact with are included in the model. In so doing, this time two product terms are created between the two included dummy variables and the exogenous variable. Our equation will then look as the following:

$$\xi_4^{LS} = \beta_0 + \beta_1\xi_1^{LS} + \beta_2\theta_1 + \beta_3\theta_2 + \beta_4\left(\xi_1^{LS} \times \theta_1\right) + \beta_5\left(\xi_1^{LS} \times \theta_2\right) + \zeta. \tag{6.8}$$

As we did earlier, we can plug in different values for the terms in equation (6.8) to obtain the slopes of interest. However, as far as the interaction effects are concerned, we can simply look and interpret the coefficients of the product terms directly. Here, β_4 and β_5 would reflect the slope difference between the dummy group coded as 1 and the dummy group coded as 0 (i.e., the reference category).

Although interactive models including a categorical moderator can be estimated using the two-stage approach, the same can alternatively be done using the multi-sample approach. In the following, we consequently present the multi-sample approach.

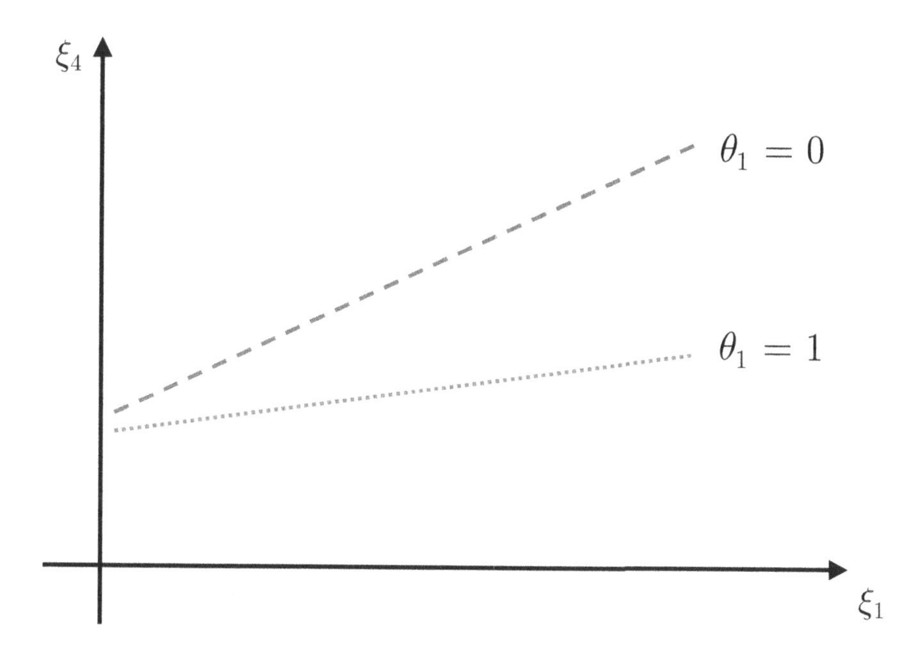

FIGURE 6.5: Interaction between a latent and a dichotomous variable.

6.4 Multi-Sample Approach

The multi-sample approach, as its name implies, is simply about estimating and then comparing the slope/coefficient of interest between two or more groups. This approach is sometimes referred to as multi-group analysis (MGA) in the structural equation modelling literature as well.

The idea behind multi-sample approach can further be illustrated through Figure 6.6. Here we see that when the assumed moderator variable is a categorical one consisting of two categories, each of these two categories represents simply an independent sample. As such, slope(s) of interest ($\xi_1 \rightarrow \xi_4$) is estimated for each category of the moderator variable separately. Subsequently, a statistical test is performed to find out whether the difference (Δ) between the two slopes, say β_{group_1} and β_{group_2}, is statistically different from 0 or not. An interaction can be claimed to be present when Δ is significant and vice versa. In other words, we are testing if the effect of ξ_1 on ξ_4 is statistically different in the two samples.

This idea can readily be extended to a case where we have an assumed categorical moderator variable consisting of K categories (i.e., groups). In this situation, as shown in Figure 6.6, the slope(s) of interest ($\xi_1 \rightarrow \xi_4$) is estimated and compared between the K groups. Again, the differences between the slopes are tested for statistical significance to conclude whether or not an interaction exists.

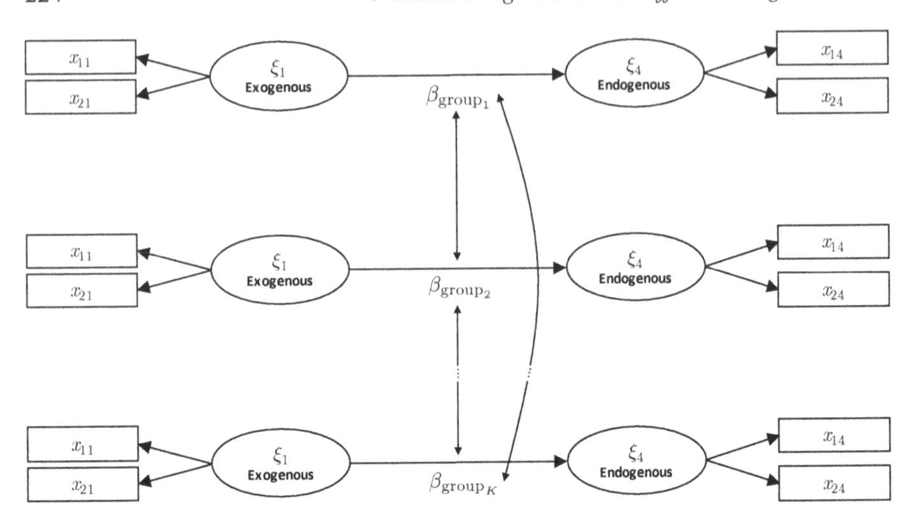

FIGURE 6.6: Multi-sample approach to test interaction effects.

6.4.1 Parametric test

One approach to testing the difference between path coefficients involves applying an independent samples t-test where standard errors are estimated with a bootstrap resampling procedure. This means that we need to obtain the bootstrap standard errors of the estimated path coefficients by resampling B times (usually at least 1000) each group/sample. The standard error estimates are next plugged in the following equation (see for example Hair et al., 2018c):

$$t = \frac{\widehat{\beta}_{\text{group}_1} - \widehat{\beta}_{\text{group}_2}}{\sqrt{\left[\frac{1}{m} + \frac{1}{n}\right] \left[\frac{(m-1)^2}{(m+n-2)} SE^2_{\text{group}_1} + \frac{(n-1)^2}{(m+n-2)} SE^2_{\text{group}_2}\right]}}, \tag{6.9}$$

where the numerator incidentally represents the difference between the (original) slope estimates for the two groups, m and n are the sample sizes and SE_{group_1} and SE_{group_2} are the corresponding standard errors obtained via bootstrapping for the two path coefficient estimates. Under suitable assumptions, the test statistic (6.9) follows a t distribution with $(m+n-2)$ degrees of freedom. The resulting t-test statistic is finally used in a traditional manner to decide whether or not the difference between the slopes of the two samples is statistically different from 0. That is, if the resulting t statistic is larger than the critical value from the t distribution, the difference can be claimed to be statistically different suggesting the existence of an interaction effect.

The testing approach we just described is based on the usual assumptions of equal variance and normality of the data[3]. If the equal variance assumption is not supported

[3] The data referred to here are the path coefficient estimates obtained through the bootstrap procedure.

by the data[4], then the Welch's t-test should be used instead (Welch, 1947). The test statistic in this case is given by:

$$t = \frac{\widehat{\beta}_{\text{group}_1} - \widehat{\beta}_{\text{group}_2}}{\sqrt{\frac{(m-1)}{m} SE^2_{\text{group}_1} + \frac{(n-1)}{n} SE^2_{\text{group}_2}}}.$$

This statistic is asymptotically distributed as a t distribution where the number of degrees of freedom are calculated using the so called Satterthwaite's approximation (Satterthwaite, 1946), that is

$$df = \frac{\left(\frac{(m-1)}{m} SE^2_{\text{group}_1} + \frac{(n-1)}{n} SE^2_{\text{group}_2}\right)^2}{\frac{(m-1)}{m^2} SE^4_{\text{group}_1} + \frac{(n-1)}{n^2} SE^4_{\text{group}_2}} - 2,$$

which is then rounded to the nearest integer.

6.4.2 Permutation test

The permutation test[5] is a distribution-free test (i.e., it doesn't require any distributional assumption) suiting thus well to partial least squares structural equation modelling. The way a permutation test works is that first as done above the difference between the slopes of the two samples (original slope difference) is computed. Then, the two samples are merged making up a single dataset which then gets divided randomly B times (usually at least 500) in two samples again. The slope difference between each of these pair of samples is estimated. All of these slope differences make up an empirical sampling distribution of our test statistic under the null hypothesis that the slope difference is equal to 0. If the corresponding p-value is typically less than 0.05, we reject the null hypothesis that the slope difference is equal to 0.

As a general conclusion regarding the choice between parametric and permutation test, we can suggest that the permutation test should be preferable regardless of small or large sample size if the assumption of normality is severely violated[6]. If not, parametric tests should be the usual choice (see Chin and Dibbern, 2010, page 174).

Before moving on to the applications of the three approaches to examining interactions in the next section, we would like to provide you with some general guidelines as to the choice among these approaches. When you have a reflective moderator construct containing multiple continuous or/and binary indicators and exogenous construct(s) reflected by continuous or/and binary indicators, product indicator approach and two-stage approach can be used interchangeably despite slight differences between them. Further, the two-stage and multi-sample approach can both be used when you have a categorical (two or more than two groups) moderator and reflective exogenous construct(s). Finally, whenever a formative construct is involved either as a moderator or exogenous variable, the two-stage approach is suggested to be used.

[4]To check if your data support the equal variance assumption, you should perform a test on the variances such as the Levene's test (Levene, 1960).

[5]You can find a brief review of permutation tests in the technical appendix of Chapter 7.

[6]Normality here refers to the path coefficient estimates obtained through the bootstrap procedure.

As a closing remark for this section, we also remind that if more than two groups are compared using either the bootstrap or permutation test, some correction for multiple testing must be used to avoid inflating the overall type I error (see for example Hair et al., 2018c, Chapter 4)[7].

6.5 Example Study: Interaction Effects

In this section we apply the three approaches for examining interaction effects we presented in the previous sections using the Stata `plssem` package. In all three of the applications we make use of a real dataset available in the file `ch6_CultureCuriosity.dta`. We already used a portion of these data in Section 3.10, but here we consider a larger set of items. Prior to continuing, we would like to present our moderation/interaction hypothesis as well as the involved variables.

From the relevant literature, we know that both curiosity trait (hereafter CURIOSITY) and cultural upbringing (hereafter CULTURE) have significant effects on people's propensity to travel abroad for holiday (hereafter HOLIDAYING INTEREST). Building up on this knowledge, we further hypothesize that there may be a moderating/interaction effect that should also be included in the equation. That is, we assume that CURIOSITY influences the relationship between CULTURE and HOLIDAYING INTEREST. More specifically, we expect that as CURIOSITY increases, the effect of CULTURE on HOLIDAYING INTEREST decreases. Put informally, CULTURE will have a stronger effect among incurious people than among curious people.

We then operationalize CULTURE, CURIOSITY and HOLIDAYING INTEREST using an ordinal scale asking the respondents to indicate on five-point scale (1 = totally disagree and 5 = totally agree) to what extent they agree with the eight statements listed in Table 6.1. While the operationalization of CURIOSITY takes the original form (i.e., ordinal) for use in the application of product-indicator approach, for the purposes of the remaining two applications (two-stage and multi-sample) we transform CURIOSITY into a categorical moderator consisting of two categories representing incurious and curious samples.

6.5.1 Application of the product-indicator approach

To be able to estimate our interactive model (which will resemble Figure 6.2), we first need to standardize the items of the interacting constructs of CULTURE and CURIOISITY:

[7]Currently, the `plssem` package does not include any correction for multiple testing.

TABLE 6.1: The example study's constructs and indicators.

Latent variable	Manifest variable	Content
CULTURE	V1A	I have grown up in a family that has always gone on vacations at holiday times
	V1B	My parents have had an impact on my interest in going on vacations at holiday times
	V1C	There is an expectation from my friends/colleagues that one should go on vacation at holiday times
CURIOSITY	V2A	I like to discover new places to go to
	V2E	I like learning about things that are unfamiliar/foreign to me
	V2F	I become fascinated when learning new information
HOLIDAYING INTEREST	V3A	I like going on vacations
	V3B	I like travelling abroad for holidaying

```
foreach var of varlist V1A V1B V1C V2A V2E V2F {
  egen `var'_std = std(`var')
}
```

In the code below, we use the standardized items when specifying the measurement model and we specify the constructs to be interacted with an asterisk sign (*).

```
plssem (CULTURE > V1A_std V1B_std V1C_std) ///
       (CURIOSITY > V2A_std V2E_std V2F_std) ///
       (H_INTEREST > V3A V3B), ///
       structural(H_INTEREST CULTURE*CURIOSITY) ///
       boot(1000) seed(123456)
```

This specification will automatically generate the product terms between the items of the two constructs (CULTURE and CUROSITY) and attach them to the interaction term (CULTURECURIOSITY) itself. Finally, the endogenous variable (HOLIDAYING INTEREST) will be regressed onto the two constructs and the interaction term. When estimating the model, we also ask for the bootstrap standard errors of the slopes using 1000 resamples. These standard errors are then used to arrive at the required p-values for testing statistically significance of the estimates. The results from this estimation are provided in Figure 6.7.

```
Partial least squares SEM                   Number of obs         =      997
                                            Average R-squared     =      0.38195
                                            Average communality   =      0.52995
Weighting scheme: path                      Absolute GoF          =      0.44991
Tolerance: 1.00e-07                         Relative GoF          =      0.94418
Initialization: indsum                      Average redundancy    =      0.32551

Measurement model - Standardized loadings
-------------------------------------------------------------------------
              |  Reflective:   Reflective:   Reflective:   Reflective:
              |    CULTURE      CURIOSITY    H_INTEREST   CULTURECUR~Y
--------------+----------------------------------------------------------
      V1A_std |     0.757
      V1B_std |     0.768
      V1C_std |     0.776
      V2A_std |                  0.643
      V2E_std |                  0.846
      V2F_std |                  0.827
          V3A |                                0.920
          V3B |                                0.926
 V1A_stdV2A~d |                                              0.540
 V1A_stdV2E~d |                                              0.656
 V1A_stdV2F~d |                                              0.644
 V1B_stdV2A~d |                                              0.629
 V1B_stdV2E~d |                                              0.755
 V1B_stdV2F~d |                                              0.718
 V1C_stdV2A~d |                                              0.598
 V1C_stdV2E~d |                                              0.619
 V1C_stdV2F~d |                                              0.609
--------------+----------------------------------------------------------
     Cronbach |     0.659        0.665         0.827         0.823
           DG |     0.811        0.819         0.920         0.863
        rho_A |     0.667        0.685         0.828         0.829
-------------------------------------------------------------------------

Discriminant validity - Squared interfactor correlation vs. Average variance extracted (AVE)
-------------------------------------------------------------------------
              |    CULTURE      CURIOSITY     H_INTEREST   CULTURECUR~Y
--------------+----------------------------------------------------------
      CULTURE |     1.000        0.055         0.189         0.021
    CURIOSITY |     0.055        1.000         0.254         0.106
   H_INTEREST |     0.189        0.254         1.000         0.108
 CULTURECUR~Y |     0.021        0.106         0.108         1.000
--------------+----------------------------------------------------------
          AVE |     0.588        0.604         0.852         0.414
-------------------------------------------------------------------------

Structural model - Standardized path coefficients (Bootstrap)
-----------------------------
     Variable |  H_INTEREST
--------------+--------------
      CULTURE |     0.323
              |    (0.000)
    CURIOSITY |     0.376
              |    (0.000)
 CULTURECUR~Y |    -0.159
              |    (0.000)
--------------+--------------
         r2_a |     0.380
-----------------------------
 p-values in parentheses
```

FIGURE 6.7: Results of the estimation of our interactive PLS-SEM model.

When it comes to the assessment and interpretation of the model (both measurement and structural part) in Figure 6.7, we use the same approach and criteria that we went through in Chapter 4. Assuming that the measurement part is psychometrically sound, we can move on to interpreting the estimates. The most salient estimate to observe is that of the interaction term (CULTURECURIOSITY). As seen, this estimate is statistically significant indicating that there is a linear interaction effect. This confirms our hypothesis that as CURIOSITY increases the effect (i.e., slope) of CULTURE on HOLIDAYING INTEREST decreases, and it does so by 0.159 point for every unit increase in CURIOSITY as shown by the following equation

$$\xi_4 = \beta_1 \xi_1 + \beta_2 \xi_2 + \beta_3 (\xi_1 \times \xi_2) + \zeta, \tag{6.10}$$

which is estimated to be

$$\widehat{\text{HOLIDAYING INTEREST}} = 0.323\ \text{CULTURE} + 0.376\ \text{CURIOSITY} - 0.159\ \text{CULTURECURIOSITY},$$

with all path coefficients being highly statistically significant.

Since we do have zeros in the range of the two constructs as a result of standardization, the coefficients of the CULTURE and CURIOSITY represent the slopes at the mean of each other (simple effect) and not any longer the main effects. More specifically, every one unit (standard deviation) increase in CULTURE leads to an average increase of 0.323 point (standard deviation) in HOLIDAYING INTEREST at the mean value of CURIOSITY. In a similar manner, every one unit increase in CURIOSITY causes an average increase of 0.376 point in HOLIDAYING INTEREST at the mean value of CULTURE.

6.5.2 Application of the two-stage approach

In this section we first show how to perform the interaction analysis we presented in the previous section (product-indicator approach) using the two-stage approach. Secondly, we also show the application of the two-stage approach with a categorical moderator.

6.5.2.1 Two-stage as an alternative to product-indicator

The first step is to build up a main effect model resembling Figure 6.3 and estimate the latent variable scores for all of the components in the equation using `plssem`. In other words, we first obtain the latent variable scores for our constructs CULTURE, CURIOSITY and HOLIDAYING INTEREST after estimating the main effect model:

```
1  plssem (CULTURE2 > V1A V1B V1C) ///
2         (CURIOSITY2 > V2A V2E V2F) ///
3         (H_INTEREST2 > V3A V3B), ///
4         structural(H_INTEREST2 CULTURE2 CURIOSITY2)
```

The estimated latent variable scores will automatically be generated and added to the existing dataset in Stata. Each latent variable score will now represent an omnibus measure of the constructs CULTURE, CURIOSITY and HOLIDAYING INTEREST.

Next, we build up and estimate the interactive model following exactly the same procedure (bootstrapping, etc.) as we used when estimating the interactive model using the product-indicator approach in Section 6.5.1 using the following code. Notice that the estimated latent variable scores are included here as single items:

```
1  plssem (CULTURE > CULTURE2) ///
2         (CURIOSITY > CURIOSITY2) ///
3         (H_INTEREST > H_INTEREST2), ///
4         structural(H_INTEREST CULTURE*CURIOSITY)
```

Here too, as we observe from Figure 6.8, the coefficient (-0.139) on the product term (CULTURECURIOSITY) is statistically significant, confirming the existence of a linear interaction effect. Moreover, as shown in equation (6.11), the remaining results (simple effects, magnitudes of the coefficients, etc.) are quite similar to those obtained from the product-indicator estimation in Section 6.5.1. The interpretation of the results that we have here in the two-stage approach is done again in an identical manner as we do in the product-indicator approach.

$$\xi_4^{LS} = \beta_1 \xi_1^{LS} + \beta_2 \xi_2^{LS} + \beta_3 \left(\xi_1^{LS} \times \xi_2^{LS}\right) + \zeta, \tag{6.11}$$

which is estimated to be

$$\text{HOLIDAY}\widehat{\text{ING INTEREST}} = 0.325\ \text{CULTURE} + 0.381\ \text{CURIOSITY} - 0.139\ \text{CULTURECURIOSITY},$$

with all path coefficients being highly statistically significant.

Since the two-stage approach above is based on single items, it does not make any sense to interpret the results from the measurement part. Thus, we have only examined the estimates and their significance in the structural part in Figure 6.8. Having said, we should however examine the psychometric properties of the two-stage model based on the measurement part in the first stage (i.e., the main effect model) using the criteria learnt in Chapter 4.

6.5.2.2 Two-stage with a categorical moderator

We are in this section going to continue with the same example study. The only difference is that we now have a categorical/dummy variable representing our moderator CURIOSITY. This variable is named as CURIOSITY_D in the current dataset in which 0 represents incurious whereas 1 represents curious respondents. Needless to say, our exogenous variable is still CULTURE and our endogenous variable is HOLIDAYING INTEREST.

As shown in Figure 6.9, the first stage is to build up the main effects model including all the components except for the product term. Notice that while the constructs CULTURE and HOLIDAYING INTEREST are reflected by the same indicators as in the previous sections, CURIOSITY is now reflected by a binary indicator (CURIOSITY_D). Subsequently, we estimate this model to obtain the scores for the constructs (CULTURE and HOLIDAYING INTEREST) in the model:

```
Partial least squares SEM                         Number of obs         =     997
                                                  Average R-squared     =     0.37676
                                                  Average communality   =     1.00000
Weighting scheme: path                            Absolute GoF          =     .
Tolerance: 1.00e-07                               Relative GoF          =     .
Initialization: indsum                            Average redundancy    =     0.37676

Measurement model - Standardized loadings
---------------------------------------------------------------------------
              | Reflective:    Reflective:    Reflective:    Reflective:
              |    CULTURE      CURIOSITY      H_INTEREST    CULTURECUR~Y
--------------+------------------------------------------------------------
     CULTURE2 |      1.000
  CURIOSITY2  |                    1.000
  H_INTEREST2 |                                   1.000
  CULTURE2CU~2|                                                  1.000
--------------+------------------------------------------------------------
     Cronbach |      1.000         1.000          1.000          1.000
           DG |      1.000         1.000          1.000          1.000
        rho_A |      1.000         1.000          1.000          1.000
---------------------------------------------------------------------------

Discriminant validity - Squared interfactor correlation vs. Average variance extracted (AVE)
---------------------------------------------------------------------------
              |    CULTURE      CURIOSITY      H_INTEREST    CULTURECUR~Y
--------------+------------------------------------------------------------
      CULTURE |      1.000         0.055          0.189          0.019
    CURIOSITY |      0.055         1.000          0.254          0.109
   H_INTEREST |      0.189         0.254          1.000          0.096
  CULTURECUR~Y|      0.019         0.109          0.096          1.000
--------------+------------------------------------------------------------
          AVE |      1.000         1.000          1.000          1.000
---------------------------------------------------------------------------

Structural model - Standardized path coefficients
-----------------------------
     Variable |   H_INTEREST
--------------+--------------
      CULTURE |      0.325
              |    (0.000)
    CURIOSITY |      0.381
              |    (0.000)
  CULTURECUR~Y|     -0.139
              |    (0.000)
--------------+--------------
         r2_a |      0.375
-----------------------------
 p-values in parentheses
```

FIGURE 6.8: Results of the interactive model using two-stage approach.

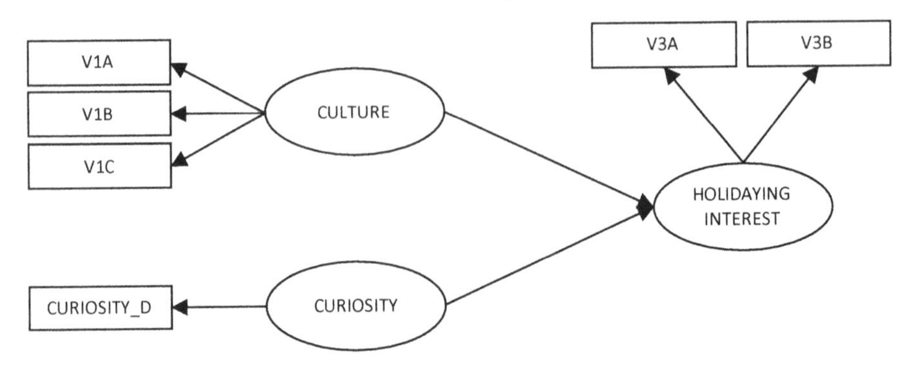

FIGURE 6.9: Stage one of the two-stage approach with a categorical moderator.

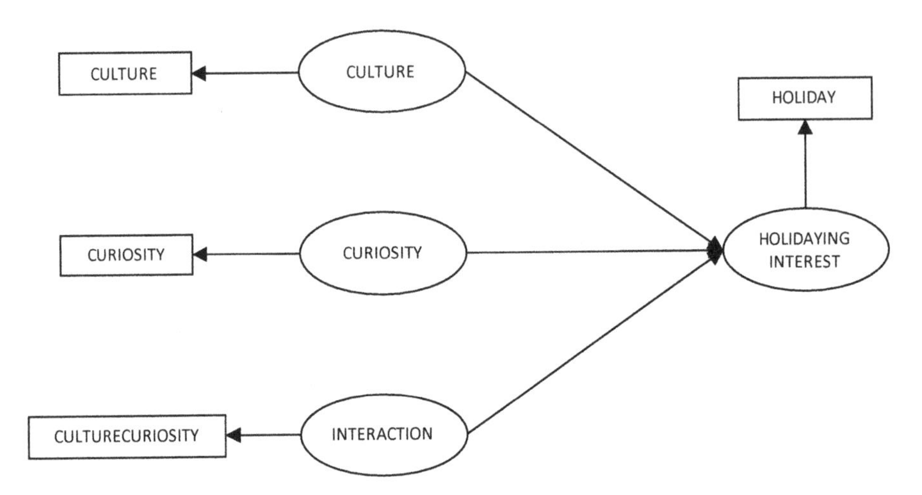

FIGURE 6.10: Stage two of the two-stage approach with a categorical moderator.

```
1  plssem (CULTURE3 > V1A V1B V1C) ///
2         (CURIOSITY3 > CURIOSITY_D) ///
3         (H_INTEREST3 > V3A V3B), ///
4         structural(H_INTEREST3 CULTURE3 CURIOSITY3)
```

In the second stage, we build up the interactive model by attaching the latent variable scores obtained in the first stage to their respective constructs as single indicators as depicted in Figure 6.10.

Finally, we estimate the interaction model with `plssem` using the code below:

```
1  plssem (CULTURE > CULTURE3) ///
2         (CURIOSITY > CURIOSITY3) ///
3         (H_INTEREST > H_INTEREST3), ///
4         structural(H_INTEREST CULTURE*CURIOSITY)
```

```
Partial least squares SEM                      Number of obs          =      998
                                               Average R-squared      =      0.27137
                                               Average communality    =      1.00000
Weighting scheme: path                         Absolute GoF           =      .
Tolerance: 1.00e-07                            Relative GoF           =      .
Initialization: indsum                         Average redundancy     =      0.27137

Measurement model - Standardized loadings
--------------------------------------------------------------------------
             | Reflective:    Reflective:    Reflective:    Reflective:
             |     CULTURE      CURIOSITY     H_INTEREST   CULTURECUR~Y
-------------+------------------------------------------------------------
    CULTURE3 |       1.000
  CURIOSITY3 |                     1.000
 H_INTEREST3 |                                    1.000
 CULTURE3CU~3 |                                                   1.000
-------------+------------------------------------------------------------
    Cronbach |       1.000         1.000          1.000           1.000
          DG |       1.000         1.000          1.000           1.000
       rho_A |       1.000         1.000          1.000           1.000
--------------------------------------------------------------------------

Discriminant validity - Squared interfactor correlation vs. Average variance extracted (AVE)
--------------------------------------------------------------------------
             |     CULTURE      CURIOSITY     H_INTEREST   CULTURECUR~Y
-------------+------------------------------------------------------------
     CULTURE |       1.000         0.006          0.188           0.000
   CURIOSITY |       0.006         1.000          0.064           0.001
  H_INTEREST |       0.188         0.064          1.000           0.029
 CULTURECUR~Y |      0.000         0.001          0.029           1.000
-------------+------------------------------------------------------------
         AVE |       1.000         1.000          1.000           1.000
--------------------------------------------------------------------------

Structural model - Standardized path coefficients
----------------------------
    Variable |  H_INTEREST
-------------+--------------
     CULTURE |      0.419
             |    (0.000)
   CURIOSITY |      0.227
             |    (0.000)
 CULTURECUR~Y |    -0.187
             |    (0.000)
-------------+--------------
        r2_a |      0.269
----------------------------
 p-values in parentheses
```

FIGURE 6.11: Results of the two-stage approach with a categorical moderator.

As we see in the output (structural part) in Figure 6.11, the p-value for the coefficient on the product term is smaller than 0.001. This again indicates that there is a significant interaction effect in that CURIOSITY moderates the relationship between CULTURE and HOLIDAYING INTEREST as curious group (coded as 1) has a significantly lower coefficient (0.187 point less) than the incurious group (coded as 0). Put it more directly, this finding suggests that CULTURE has a significantly larger positive effect on HOLIDAYING INTEREST among incurious group as compared to curious group. As in the previous example, we do not use the numbers from the output in Figure 6.11 as a basis for judging the quality of the measurement part. Instead, we use the main effect model estimates to examine the goodness of the measurement model.

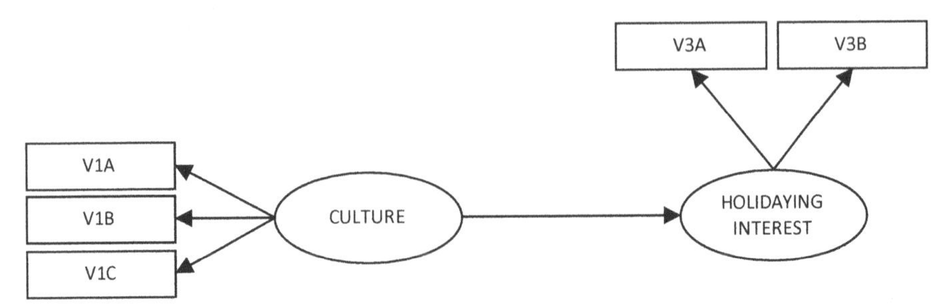

FIGURE 6.12: The main effect model of the multi-sample approach.

Furthermore, we see from equation (6.12) that the coefficient on CULTURE is 0.419, which is simply showing the slope at the value of 0 of CURIOSITY, namely incurious group. The difference between 0.419 and 0.187 (i.e., 0.232) would then provide the slope at the value 1 of CURIOSITY, which is the curious group. We can also observe that the coefficient on CURIOSITY is 0.227, which is showing the slope for those at the average of CULTURE.

$$\xi_4^{LS} = \beta_1 \xi_1^{LS} + \beta_2 x_1 + \beta_3 \left(\xi_1^{LS} \times x_1\right) + \zeta, \tag{6.12}$$

which is estimated to be

$$\text{HOLIDAY}\widehat{\text{ING INTEREST}} = 0.419\ \text{CULTURE} + 0.227\ \text{CURIOSITY} - 0.187\ \text{CULTURECURIOSITY},$$

with all path coefficients being highly statistically significant.

6.5.3 Application of the multi-sample approach

In the previous section we tested our moderation/interaction hypothesis using the two-stage approach. In this section, we are going to show how the same hypothesis can be tested using the multi-sample approach. In the multi-sample approach, we as usual build up the main effect model for the entire sample as depicted in Figure 6.12. Since CURIOSITY_D is now going to be our grouping variable, it should naturally not be included in the main effect model.

Having done so, we now estimate this main effect model for both samples (incurious and curious) simultaneously. To do so, we specify the `group` option in the `plssem` command by providing the grouping variable CURIOSITY_D. The `group` option accepts the following suboptions:

- `method()` indicates the type of test to perform for the comparisons; allowed values are `normal`, for normal-based theory, `bootstrap` and `permutation`,
- `reps(#)` specifies the number of bootstrap or permutation replications,
- `plot` indicates that the results must also be represented graphically,
- `alpha(#)` indicates the significance level to use (default to 0.05),

- `unequal` specifies whether to assume unequal variances in the parametric tests (i.e., normal-based and bootstrap),
- `groupseed(#)` provides the random number seed for reproducibility.

In this example we ask for 1000 bootstrap resamples for the parametric test to compare the path coefficients and loadings of the incurious and curious groups assuming unequal variances:

```
1  plssem (CULTURE > V1A V1B V1C) ///
2         (H_INTEREST > V3A V3B), ///
3         structural(H_INTEREST CULTURE) ///
4         group(CURIOSITY_D, unequal reps(1000) ///
5         groupseed(123456) method(bootstrap))
```

The path coefficients estimated for both groups can be observed in Figure 6.13. The slopes for the incurious and curious groups are respectively 0.492 and 0.355. The difference between these two slopes is about 0.137. The results show that the slopes for the incurious and curious groups do significantly differ from each other. This is a clear indication of an interaction effect in that CULTURE has a significantly larger effect on HOLIDAYING INTEREST among incurious group compared to curious group. As we clearly see, the results from the multi-sample approach to examining an interaction hypothesis are close to those obtained from the two-stage approach described in Section 6.5.2.2. As a general rule of thumb, we suggest to use the two-stage approach when you have a large and complex model. Otherwise, the multi-sample approach can be used instead.

6.6 Measurement Model Invariance

When we compare path coefficients of two independent samples, what we formally are doing is testing the invariance of the structural model. If our hypothesis is about assuming significant differences between the slopes, what we ideally will hope for is the existence of a variant structural model. The other type of invariance concerns the measurement model endeavouring to find out whether the indicators reflecting the different constructs mean the same thing to people in different samples. We naturally hope for an invariant/equivalent measurement model prior to testing the invariance of structural model. Otherwise, findings of differences between the slopes in two different samples cannot be unambiguously interpreted (Horn and McArdle, 1992). In other words, we cannot be sure whether the slope differences represent the real differences or occur as a result of different perceptions of the items by people.

Thus, we should ideally precede the multi-sample analysis in the earlier section with a testing of the measurement model invariance. There is unfortunately scarce treatment of measurement invariance in the partial least squares structural equation

```
Partial least squares SEM

Weighting scheme: path
Tolerance: 1.00e-07
Initialization: indsum

Multigroup comparison (CURIOSITY_D) - Parametric test
------------------------------------------------------------------------------------------
Measurement effect |   Global    Group_1    Group_2   Abs_Diff   Statistic    P-value
-------------------+----------------------------------------------------------------------
    CULTURE -> V1A |    0.756      0.754      0.673      0.081       0.854      0.393
    CULTURE -> V1B |    0.768      0.776      0.648      0.127       1.436      0.152
    CULTURE -> V1C |    0.774      0.765      0.888      0.123       2.533      0.011
 H_INTEREST -> V3A |    0.921      0.919      0.899      0.020       0.518      0.605
 H_INTEREST -> V3B |    0.926      0.927      0.887      0.041       1.497      0.135
------------------------------------------------------------------------------------------
 number of replications: 1000
 unequal variances assumed
 group labels:
   Group 1: incurious
   Group 2: curious
 group sizes:
   Group 1: 616
   Group 2: 382

Multigroup comparison (CURIOSITY_D) - Parametric test
----------------------------------------------------------------------------------------------
     Structural effect |   Global    Group_1    Group_2   Abs_Diff   Statistic    P-value
-----------------------+----------------------------------------------------------------------
 CULTURE -> H_INTEREST |    0.434      0.492      0.355      0.137       2.451      0.014
----------------------------------------------------------------------------------------------
 number of replications: 1000
 unequal variances assumed
 group labels:
   Group 1: incurious
   Group 2: curious
 group sizes:
   Group 1: 616
   Group 2: 382
```

FIGURE 6.13: The results of the multi-sample approach for the model shown in Figure 6.12.

modelling literature. Thus, one commonly employs the procedures suggested in the covariance-based structural equation modelling (CB-SEM) domain for testing measurement invariance in PLS-SEM as well. Nevertheless, not all of the invariance forms suggested in CB-SEM are directly transferrable to the PLS-SEM context.

As a pragmatic solution we still suggest at least the two basic forms of invariance used in CB-SEM also for examining measurement invariance in PLS-SEM. The most basic form of invariance to establish is the typically called *configural invariance* which may also be referred to as equal form invariance or even pattern invariance (Wang and Wang, 2012). The idea here is simply to check whether a proposed factor structure is the same in different samples/groups under study. Our suggestion for testing the configural invariance in PLS-SEM context is to apply the dimensionality criterion that we treated in Chapter 4. Dimensionality is about finding out the number of constructs that may be reflected by a set of items. If you have same dimensionality pattern regarding the constructs in both samples, we may conclude that the factor structure is invariant/equivalent in both samples.

The examination of measurement invariance is a sequential one in that we first need to establish the configural invariance prior to checking the second form of invariance, namely construct-level *metric invariance* (see for example Kline, 2016), which is also referred to as equal factor loadings or even weak measurement invariance. If factor loadings can be shown to be equivalent across samples/groups, then measures across groups can be considered to be on the same scale (Wang and Wang, 2012).

In Stata, to check for the dimensionality of the constructs (CULTURE and HOLIDAYING INTEREST) of the measurement model, you simply need to run a factor analysis with the extraction method of principal component on each of the construct's items for each sample and examine the eigenvalues. With regards to our last example, we would use the following code (output not reported):

```
factor V1A V1B V1C if CURIOSITY_D == 0, pcf
factor V1A V1B V1C if CURIOSITY_D == 1, pcf
factor V3A V3B if CURIOSITY_D == 0, pcf
factor V3A V3B if CURIOSITY_D == 1, pcf
```

In an equal form measurement model, one should expect the same type of eigenvalue and factor relation in both samples. For instance, if a factor is associated with one eigenvalue larger than one in one sample but with two eigenvalues larger than one in the other, then this would be a sign of variant/unequal form. As for the construct-level metric invariance, using `plssem`, we compare the differences between the standardized loadings using the bootstrap test that we use when comparing slopes/coefficients. To be able to establish construct-level metric invariance, we want non-significant results, which would indicate that the loadings are close to each other[8].

6.7 Summary

A moderating/interaction effect occurs when a third variable influences the relation between an exogenous variable and an endogenous variable. There are three main approaches to examining moderating/interaction effects in PLS-SEM: product-indicator approach, two-stage approach and multi-sample approach. The product-indicator approach is commonly used when we have a continuous moderator variable while the other two approaches are mainly used when there is a categorical moderator. The two-stage approach is further the only option for examining interactions effects with formative measures. Finally, we note that an equal measurement model should be established before comparing the slopes of different samples/groups.

[8] We have incidentally checked the equal form and loadings conditions for our example study. Both CULTURE and HOLIDAYING INTEREST are each associated with one eigenvalue larger than 1 in both incurious and curious samples. Further, nearly half of the loadings are significantly different between the two samples. This further suggests that equal loading assumption can partially be supported.

Appendix: R Commands

In this appendix we discuss how to fit PLS-SEM models that include a moderating/interaction effect using the `cSEM` package in `R`[9]. In this appendix, we are going to reuse the same examples discussed in the chapter highlighting what `cSEM` can additionally provide compared to Stata's `plssem`. These examples analyse some data that have been collected to study the determinants of the propensity to travel abroad for holiday. The data are available in the `ch6_CultureCuriosity.dta` file, which we import in `R` using the `haven` package:

```
if (!require(haven, quietly = TRUE)) {
  install.packages("haven")
}

curios_data <- as.data.frame(read_stata(file =
  file.path(path_data, "ch6_CultureCuriosity.dta")))
```

Application of the product-indicator approach

To use the product-indicator approach with the `cSEM` package we need to explicitly create the indicators for the interactions in the model. We can do that in `R` following different approaches, but a quick one uses the `model.matrix()` function. `model.matrix()` creates a matrix containing all the relevant columns given the dataset and the formula for the model we want to fit[10]. In the first example, whose path diagram is shown in Figure 6.2, this is accomplished by the following code chunk, which first standardizes the indicators and removes the missing values:

```
curios_data_std <- scale(curios_data[, 1:8])
colnames(curios_data_std) <- paste0(colnames(curios_data_std), "_std")
curios_data <- cbind(curios_data, curios_data_std)
curios_data <- na.omit(curios_data)

curios_data_pi <- model.matrix(
  object = ~ (V1A_std + V1B_std + V1C_std + V2A_std + V2E_std + V2F_std)^2
    + V3A + V3B - 1, data = curios_data)
```

The "`^2`" syntax means to create all the interactions between the terms within parentheses, while the final "`-1`" indicates to avoid adding to the matrix a column of ones, which typically is used to estimate the intercept in a regression context. Then, we can fit the model using the `csem()` function (output not reported):

[9]The `plspm` package also includes a function, called `plspm.groups()`, that performs multi-group analysis, but it is limited to the comparison of two groups only. In case you are interested in understanding how it works, we suggest to look at Chapter 6 of Sanchez (2013).

[10]To effectively use the `model.matrix()` function, you should know something about `formula` objects. You can find more details by typing `?formula`.

```
curios_mod_pi_1 <- "
  # structural model
  H_INTEREST ~ CULTURE + CURIOSITY + CULTURE_CURIOSITY
  # measurement model
  CULTURE =~ V1A_std + V1B_std + V1C_std
  CURIOSITY =~ V2A_std + V2E_std + V2F_std
  H_INTEREST =~ V3A + V3B
"
curios_mod_pi_2 <- "
  CULTURE_CURIOSITY =~ V1A_std:V2A_std + V1A_std:V2E_std +
    V1A_std:V2F_std + V1B_std:V2A_std + V1B_std:V2E_std +
    V1B_std:V2F_std + V1C_std:V2A_std + V1C_std:V2E_std +
    V1C_std:V2F_std
"
curios_mod_pi <- paste0(curios_mod_pi_1, gsub("\n", "", curios_mod_pi_2))
library(cSEM)
curios_res_pi <- csem(.data = curios_data_pi,
  .model = curios_mod_pi, .PLS_weight_scheme_inner = "path",
  .disattenuate = FALSE, .tolerance = 1e-07,
  .resample_method = "bootstrap", .R = 1000, .seed = 101)
# summarize(curios_res_pi)
```

Note that in the code above, we separated the specification of the model in two parts, `curios_mod_pi_1` and `curios_mod_pi_2`, to avoid showing the long formula for `CULTURE_CURIOSITY` in a single line.

Application of the two-stage approach

The two-stage approach can be used with `cSEM` following the same two steps we described in the chapter using Stata, that is:

Step 1. fit a main effect model (i.e., without interactions),

Step 2. using the predicted construct scores from the previous step, fit the interactive model using the scores as the unique indicators.

These steps are easily implemented in `R` as shown in the following code (output not reported):

```
## first stage
curios_mod_2s_step1 <- "
  # measurement model
  CULTURE2 =~ V1A + V1B + V1C
  CURIOSITY2 =~ V2A + V2E + V2F
  H_INTEREST2 =~ V3A + V3B
  # structural model
  H_INTEREST2 ~ CULTURE2 + CURIOSITY2
"

curios_res_2s_step1 <- csem(.data = curios_data,
  .model = curios_mod_2s_step1, .PLS_weight_scheme_inner = "path",
```

```
12     .disattenuate = FALSE, .tolerance = 1e-07,
13     .resample_method = "bootstrap", .R = 1000, .seed = 101)
14   # summarize(curios_res_2s_step1)

15   ## second stage
16   curios_data_2s <- as.data.frame(
17     getConstructScores(curios_res_2s_step1)$Construct_scores)
18   curios_data_2s <- model.matrix(
19     object = ~ (CULTURE2 + CURIOSITY2)^2 + H_INTEREST2 - 1,
20     data = curios_data_2s)

21   curios_mod_2s_step2 <- "
22     # measurement model
23     CULTURE =~ CULTURE2
24     CURIOSITY =~ CURIOSITY2
25     H_INTEREST =~ H_INTEREST2
26     CULTURE_CURIOSITY =~ CULTURE2:CURIOSITY2

27     # structural model
28     H_INTEREST ~ CULTURE + CURIOSITY + CULTURE_CURIOSITY
29   "

30   curios_res_2s_step2 <- csem(.data = curios_data_2s,
31     .model = curios_mod_2s_step2, .PLS_weight_scheme_inner = "path",
32     .disattenuate = FALSE, .tolerance = 1e-07,
33     .resample_method = "bootstrap", .R = 1000, .seed = 101)
34   # summarize(curios_res_2s_step2)
```

The same identical steps apply also to the case of a categorical moderator (see Section 6.5.2.2) so we do not show them here.

Before moving to the last example, we highlight that the `cSEM` package has built-in capabilities to deal with non-linear terms such as polynomials and interactions, which implements the approach described in Dijkstra and Schermelleh-Engel (2014). In addition, after the estimation of a model that includes non-linear terms, the package also provides the function `doNonlinearEffectsAnalysis()`, which allows to estimate the expected value of the dependent variable conditional on the values of an independent variables and a moderator variable, keeping all other variables at their mean levels. We now show how to perform the same analysis described above but directly using these functionalities. To do that, the model specification must use the "`.`" operator to include the non-linear terms. Our example involves the interaction between `CULTURE` and `CURIOSITY` so that the required code is

```
1   ## direct use of csem() with an interaction
2   curios_mod_int <- "
3     # measurement model
4     CULTURE =~ V1A + V1B + V1C
5     CURIOSITY =~ V2A + V2E + V2F
6     H_INTEREST =~ V3A + V3B

7     # structural model
```

```
8     H_INTEREST ~ CULTURE + CURIOSITY + CULTURE.CURIOSITY
9   "
```

Then, the model can be fitted as usual with the `csem()` function:

```
1 curios_res_int <- csem(.data = curios_data,
2   .model = curios_mod_int, .PLS_weight_scheme_inner = "path",
3   .disattenuate = FALSE, .tolerance = 1e-07,
4   .resample_method = "bootstrap", .R = 1000, .seed = 101)
5 summarize(curios_res_int)
```

```
________________________________________________________________________________
----------------------------------- Overview -----------------------------------

General information:
------------------------
Estimation status                  = Ok
Number of observations             = 997
Weight estimator                   = PLS-PM
Inner weighting scheme             = "path"
Type of indicator correlation      = Pearson
Path model estimator               = OLS
Second-order approach              = NA
Type of path model                 = Nonlinear
Disattenuated                      = No

Resample information:
---------------------
Resample method                    = "bootstrap"
Number of resamples                = 1000
Number of admissible results       = 1000
Approach to handle inadmissibles   = "drop"
Sign change option                 = "none"
Random seed                        = 101

Construct details:
------------------
Name          Modeled as      Order         Mode

CULTURE       Common factor   First order   "modeA"
CURIOSITY     Common factor   First order   "modeA"
H_INTEREST    Common factor   First order   "modeA"

----------------------------------- Estimates ----------------------------------

Estimated path coefficients:
============================
                                                                         CI_percentile
  Path                                 Estimate  Std. error   t-stat.   p-value         95%
  H_INTEREST ~ CULTURE                   0.3251      0.0269   12.0832    0.0000 [ 0.2720; 0.3784 ]
  H_INTEREST ~ CURIOSITY                 0.3809      0.0281   13.5726    0.0000 [ 0.3278; 0.4345 ]
  H_INTEREST ~ CULTURE.CURIOSITY        -0.1156      0.0281   -4.1184    0.0000 [-0.1677;-0.0618 ]

Estimated loadings:
===================
                                                                CI_percentile
  Loading              Estimate  Std. error   t-stat.   p-value         95%
  CULTURE =~ V1A         0.7564      0.0257   29.4397    0.0000 [ 0.7008; 0.8009 ]
  CULTURE =~ V1B         0.7675      0.0245   31.3522    0.0000 [ 0.7153; 0.8088 ]
  CULTURE =~ V1C         0.7759      0.0229   33.8115    0.0000 [ 0.7293; 0.8182 ]
  CURIOSITY =~ V2A       0.6428      0.0352   18.2557    0.0000 [ 0.5665; 0.7064 ]
  CURIOSITY =~ V2E       0.8460      0.0153   55.4736    0.0000 [ 0.8137; 0.8737 ]
  CURIOSITY =~ V2F       0.8268      0.0217   38.0185    0.0000 [ 0.7781; 0.8643 ]
  H_INTEREST =~ V3A      0.9185      0.0104   88.4414    0.0000 [ 0.8952; 0.9360 ]
  H_INTEREST =~ V3B      0.9277      0.0076  121.9789    0.0000 [ 0.9111; 0.9410 ]

Estimated weights:
==================
                                                                CI_percentile
  Weight               Estimate  Std. error   t-stat.   p-value         95%
  CULTURE <~ V1A         0.3841      0.0298   12.8698    0.0000 [ 0.3224; 0.4380 ]
  CULTURE <~ V1B         0.3840      0.0274   14.0371    0.0000 [ 0.3299; 0.4338 ]
  CULTURE <~ V1C         0.5344      0.0324   16.5095    0.0000 [ 0.4715; 0.6013 ]
  CURIOSITY <~ V2A       0.3752      0.0371   10.1239    0.0000 [ 0.3079; 0.4545 ]
```

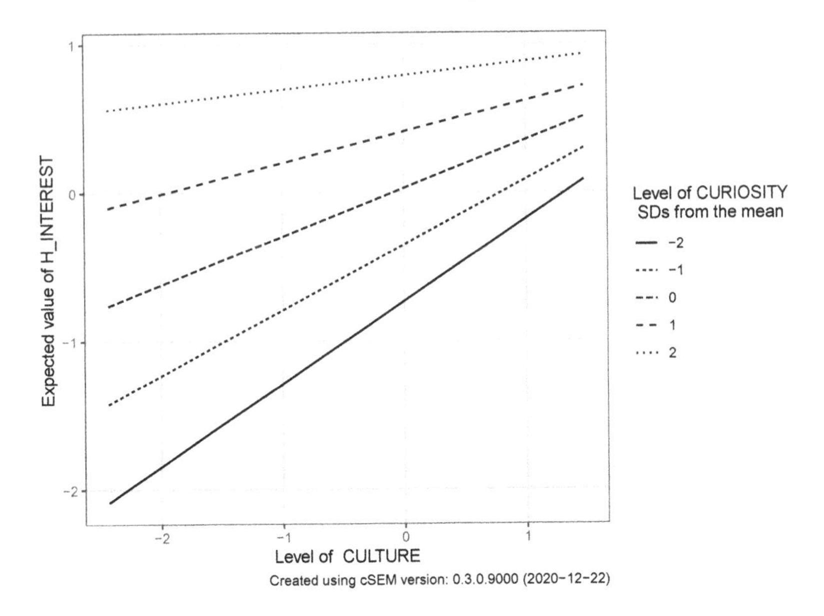

FIGURE 6.14: The diagram shows how the impact of CULTURE on H_INTEREST is affected by the values of CURIOSITY for the example discussed in Section 6.5.2.

```
  CURIOSITY <~ V2E        0.4117      0.0206   20.0069    0.0000 [ 0.3726; 0.4521 ]
  CURIOSITY <~ V2F        0.4965      0.0260   19.0985    0.0000 [ 0.4465; 0.5481 ]
  H_INTEREST <~ V3A       0.5260      0.0132   39.7687    0.0000 [ 0.5000; 0.5520 ]
  H_INTEREST <~ V3B       0.5571      0.0153   36.4587    0.0000 [ 0.5299; 0.5888 ]

Estimated construct correlations:
=================================
                                                                    CI_percentile
  Correlation              Estimate  Std. error   t-stat.   p-value        95%
  CULTURE ~~ CURIOSITY       0.2356      0.0334    7.0523    0.0000 [ 0.1728; 0.2990 ]
________________________________________________________________________________
```

Note that all the estimates match with those provided by Stata apart from the estimated coefficient of the interaction term. This is due to a different approach implemented within the csem() function.

After the model with non-linear terms has been estimated, we can assess how the effect of CULTURE on H_INTEREST is affected by the values of the CURIOSITY moderator using the doNonlinearEffectsAnalysis() function and plot the corresponding results (see Figure 6.14):

```
1  neffects <- doNonlinearEffectsAnalysis(curios_res_int,
2    .dependent = "H_INTEREST",
3    .moderator = "CURIOSITY",
4    .independent = "CULTURE")
5  # neffects
6  plot(neffects, .plot_type = "simpleeffects")
```

As we already commented in the chapter, the diagram shows that as CURIOSITY

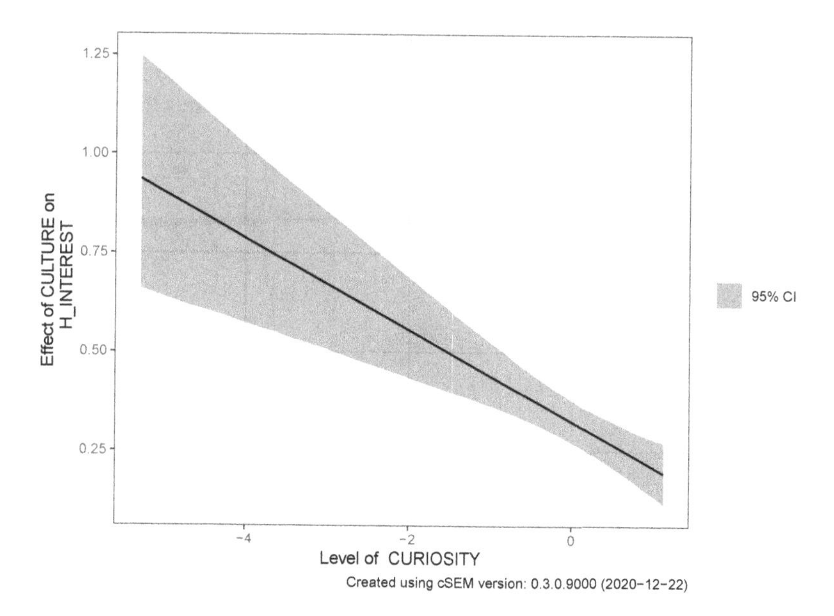

FIGURE 6.15: The diagram shows the floodlight analysis (Spiller et al., 2013) for the example introduced in Section 6.5.2.

increases the effect of `CULTURE` on `H_INTEREST` decreases. Note that, since the constructs are standardized, the values of the moderator equal the deviation from its mean (i.e., zero) measured in standard deviations. This conclusion is further confirmed by the so called *floodlight analysis* (Spiller et al., 2013), that can be obtained by executing the command `plot(neffects, .plot_type = "floodlight")`, whose output is shown in Figure 6.15.

Application of the multi-sample approach

We conclude this appendix showing how to perform a multi-sample (also called multi-group) analysis with the `cSEM` package. To do that, we need to provide the `csem()` function with the further argument `.id`, which indicates the name of the column containing the groups[11]. The `csem()` function will then split the data by groups and run the estimation for each group separately. If the number of groups is large, the computational burden can critically increase, especially if resampling is requested. To speed up the computation in these cases, it is possible to set the `.eval_plan` argument to `"multiprocess"` thus enabling parallel computing.

Once the model has been fitted separately for each group, we can compare the estimates across groups with the `testMGD()` function, which allows to test the differences using different methods (for more details on the available approach, see

[11]Alternatively, we may provide a list of datasets corresponding to the subsets of the sample data that refer to each group.

the function help page). Inference on the parameter differences is performed using both bootstrap and permutation tests. After reloading the data, the following code replicates the example we presented in Section 6.5.3 and compares the results using the approach described in Chin and Dibbern (2010):

```
1  curios_data <- as.data.frame(read_stata(file =
2    file.path(path_data, "ch6_CultureCuriosity.dta")))
3  curios_data <- na.omit(curios_data)

4  curios_mod_mga <- "
5    # measurement model
6    CULTURE =~ V1A + V1B + V1C
7    H_INTEREST =~ V3A + V3B

8    # structural model
9    H_INTEREST ~ CULTURE
10 "

11 curios_res_mga <- csem(.data = curios_data,
12   .id = "CURIOSITY_D", .model = curios_mod_mga,
13   .PLS_weight_scheme_inner = "path",
14   .disattenuate = FALSE, .tolerance = 1e-07,
15   .resample_method = "bootstrap", .R = 1000, .seed = 101)
16 summarize(curios_res_mga)

17 curios_mga <- testMGD(curios_res_mga, .R_bootstrap = 1000,
18   .R_permutation = 1000, .seed = 1406)
19 print(curios_mga, .approach_mgd = "Chin")
```

```
$`0`
________________________________________________________________________________
----------------------------------- Overview -----------------------------------

General information:
------------------------
Estimation status                  = Ok
Number of observations             = 616
Weight estimator                   = PLS-PM
Inner weighting scheme             = "path"
Type of indicator correlation      = Pearson
Path model estimator               = OLS
Second-order approach              = NA
Type of path model                 = Linear
Disattenuated                      = No

Resample information:
---------------------
Resample method                    = "bootstrap"
Number of resamples                = 1000
Number of admissible results       = 1000
Approach to handle inadmissibles   = "drop"
Sign change option                 = "none"
Random seed                        = 101

Construct details:
------------------
Name         Modeled as      Order          Mode

CULTURE      Common factor   First order    "modeA"
```

```
H_INTEREST  Common factor  First order   "modeA"

----------------------------------- Estimates ----------------------------------

Estimated path coefficients:
============================
                                                                  CI_percentile
  Path                    Estimate  Std. error   t-stat.   p-value        95%
  H_INTEREST ~ CULTURE      0.4920      0.0308   15.9585    0.0000 [ 0.4365; 0.5539 ]

Estimated loadings:
===================
                                                               CI_percentile
  Loading              Estimate  Std. error   t-stat.   p-value        95%
  CULTURE =~ V1A         0.7539      0.0309   24.3913    0.0000 [ 0.6881; 0.8025 ]
  CULTURE =~ V1B         0.7759      0.0269   28.8055    0.0000 [ 0.7199; 0.8226 ]
  CULTURE =~ V1C         0.7652      0.0288   26.6031    0.0000 [ 0.7033; 0.8164 ]
  H_INTEREST =~ V3A      0.9188      0.0123   74.7279    0.0000 [ 0.8911; 0.9385 ]
  H_INTEREST =~ V3B      0.9273      0.0099   93.8775    0.0000 [ 0.9066; 0.9447 ]

Estimated weights:
==================
                                                               CI_percentile
  Weight               Estimate  Std. error   t-stat.   p-value        95%
  CULTURE <~ V1A         0.3651      0.0325   11.2293    0.0000 [ 0.2970; 0.4264 ]
  CULTURE <~ V1B         0.4092      0.0309   13.2376    0.0000 [ 0.3504; 0.4691 ]
  CULTURE <~ V1C         0.5322      0.0368   14.4748    0.0000 [ 0.4627; 0.6053 ]
  H_INTEREST <~ V3A      0.5272      0.0223   23.6043    0.0000 [ 0.4857; 0.5713 ]
  H_INTEREST <~ V3B      0.5560      0.0234   23.7645    0.0000 [ 0.5120; 0.6036 ]

------------------------------------ Effects -----------------------------------

Estimated total effects:
========================
                                                                  CI_percentile
  Total effect            Estimate  Std. error   t-stat.   p-value        95%
  H_INTEREST ~ CULTURE      0.4920      0.0308   15.9585    0.0000 [ 0.4365; 0.5539 ]
________________________________________________________________________________

$`1`
________________________________________________________________________________
----------------------------------- Overview -----------------------------------

General information:
------------------------
Estimation status                  = Ok
Number of observations             = 381
Weight estimator                   = PLS-PM
Inner weighting scheme             = "path"
Type of indicator correlation      = Pearson
Path model estimator               = OLS
Second-order approach              = NA
Type of path model                 = Linear
Disattenuated                      = No

Resample information:
---------------------
Resample method                    = "bootstrap"
Number of resamples                = 1000
Number of admissible results       = 1000
Approach to handle inadmissibles   = "drop"
Sign change option                 = "none"
Random seed                        = 101

Construct details:
------------------
Name          Modeled as     Order         Mode

CULTURE       Common factor  First order   "modeA"
H_INTEREST    Common factor  First order   "modeA"

----------------------------------- Estimates ----------------------------------
```

```
Estimated path coefficients:
============================
                                                                    CI_percentile
  Path                          Estimate  Std. error   t-stat.   p-value      95%
  H_INTEREST ~ CULTURE            0.3566      0.0466    7.6546    0.0000 [ 0.2692; 0.4475 ]

Estimated loadings:
===================
                                                                CI_percentile
  Loading                  Estimate  Std. error   t-stat.   p-value      95%
  CULTURE =~ V1A             0.6747      0.0870    7.7571    0.0000 [ 0.4550; 0.7804 ]
  CULTURE =~ V1B             0.6468      0.0830    7.7943    0.0000 [ 0.4363; 0.7682 ]
  CULTURE =~ V1C             0.8903      0.0385   23.1046    0.0000 [ 0.8224; 0.9709 ]
  H_INTEREST =~ V3A          0.8983      0.0267   33.6684    0.0000 [ 0.8337; 0.9370 ]
  H_INTEREST =~ V3B          0.8869      0.0264   33.5509    0.0000 [ 0.8268; 0.9270 ]

Estimated weights:
==================
                                                                CI_percentile
  Weight                   Estimate  Std. error   t-stat.   p-value      95%
  CULTURE <~ V1A             0.3420      0.0814    4.2015    0.0000 [ 0.1356; 0.4449 ]
  CULTURE <~ V1B             0.2189      0.0802    2.7306    0.0063 [ 0.0392; 0.3512 ]
  CULTURE <~ V1C             0.7050      0.0836    8.4376    0.0000 [ 0.5699; 0.8848 ]
  H_INTEREST <~ V3A          0.5741      0.0449   12.7894    0.0000 [ 0.4899; 0.6670 ]
  H_INTEREST <~ V3B          0.5460      0.0475   11.5056    0.0000 [ 0.4515; 0.6402 ]

----------------------------------- Effects -----------------------------------

Estimated total effects:
========================
                                                                    CI_percentile
  Total effect                  Estimate  Std. error   t-stat.   p-value      95%
  H_INTEREST ~ CULTURE            0.3566      0.0466    7.6546    0.0000 [ 0.2692; 0.4475 ]
________________________________________________________________________________
```

```
________________________________________________________________________________
----------------------------------- Overview -----------------------------------

Total permutation runs               = 1000
Admissible permutation results       = 1000
Permutation seed                     = 1406

Total bootstrap runs                 = 1000
Admissible bootstrap results:

Group        Admissibles
0                 1000
1                 1000

Bootstrap seed:

Group           Seed
0                    1406
1                    1406

Number of observations per group:

Group        No. Obs.
0               616
1               381

Overall decision (based on alpha = 5%):

             p_adjust = 'none'
Sarstedt                 reject
Chin                     reject
Keil                     reject
Nitzl                    reject
________________________________________________________________________________
```

```
-------- Test for multigroup differences based on Chin & Dibbern (2010) --------

Null hypothesis:

                    ============================================
                    H0: Parameter k is equal across two groups.
                    ============================================

Test statistic and p-value:

Multiple testing adjustment: 'none'

  Compared groups: 0_1

Parameter              Test statistic          p-value         Decision
H_INTEREST ~ CULTURE            0.1354          0.0140           reject
CULTURE =~ V1A                  0.0792          0.2040    Do not reject
CULTURE =~ V1B                  0.1291          0.0260           reject
CULTURE =~ V1C                 -0.1252          0.0060           reject
H_INTEREST =~ V3A               0.0205          0.3840    Do not reject
H_INTEREST =~ V3B               0.0404          0.0620    Do not reject
________________________________________________________________________________
```

These results confirm the findings we reported in Figure 6.13, in particular that there is an interaction of `CURIOSITY` with `CULTURE` since the slopes for the incurious and curious groups in the structural model do significantly differ from each other.

Measurement model invariance

We close the appendix with a quick mention to the `testMICOM()` function, which performs the permutation-based test for measurement invariance of composites across groups proposed by Henseler et al. (2016). This function provides a more formal approach for assessing measurement model invariance as we discussed it in Section 6.6. As an example, the code below uses the `testMICOM()` function to assess measurement model invariance for the multi-sample analysis we described above:

```
1 testMICOM(curios_res_mga, .R = 1000, .seed = 1406)
```

```
________________________________________________________________________________
-------- Test for measurement invariance based on Henseler et al (2016) --------
======================= Step 1 - Configural invariance =======================

Configural invariance is a precondition for step 2 and 3.
Do not proceed to interpret results unless
configural invariance has been established.

====================== Step 2 - Compositional invariance ======================

Null hypothesis:

        ==========================================================
        H0: Compositional measurement invariance of the constructs.
        ==========================================================
```

```
Test statistic and p-value:

  Compared groups: 0_1
                               p-value by adjustment
Construct  Test statistic        none
CULTURE          0.9753         0.0020
H_INTEREST       0.9998         0.5010

================= Step 3 - Equality of the means and variances =================

Null hypothesis:

            ======================================================
            1. H0: Difference between group means is zero
            2. H0: Log of the ratio of the group variances is zero
            ======================================================

Test statistic and critical values:

  Compared groups: 0_1

Mean
                               p-value by adjustment
Construct  Test statistic        none
CULTURE         -0.1081         0.1140
H_INTEREST      -0.1683         0.0030

Var
                               p-value by adjustment
Construct  Test statistic        none
CULTURE          0.0896         0.3100
H_INTEREST       0.2825         0.1090

Additional information:

Out of 1000 permutation runs, 1000 where admissible.
See ?verify() for what constitutes an inadmissible result.

The seed used was: 1406

Number of observations per group:

Group      No. observations
0          616
1          381
```

7

Detecting Unobserved Heterogeneity in PLS-SEM

In this chapter we illustrate the main methods developed so far in the literature to estimate unobserved heterogeneity in a PLS-SEM analysis. We start the chapter with a description of the types of heterogeneity one may encounter in a statistical analysis, namely observed and unobserved heterogeneity. Then, we focus on the latter and present some approaches for detecting it in the PLS-SEM framework. In particular, we provide a full description of two methodologies called REBUS-PLS and FIMIX-PLS. Other approaches that are less frequently used in practice will be briefly presented as well. As usual, we will illustrate the new ideas with the support of practical examples. The models included in the examples will be estimated using the `plssem` Stata command, whose results will be further inspected with the `estat unobshet` postestimation command. In the presentation we will refer to some popular statistical approaches for identifying homogenous groups of observations such as cluster analysis and latent class analysis, a review of which is available in Section 2.3.

7.1 Introduction

Heterogeneity represents to a great extent the primary focus of any statistical modelling. This general statement is typically operationalized by choosing a set of covariates (or independent variables) through which one aims at explaining the observed variability in the response (or dependent) variable. However, it is very rarely the case that we succeed in collecting data on all relevant covariates in a given problem. Thus, we usually end up the analysis with a portion of the observed response variability that is left "unexplained".

From a practical point of view, we can distinguish two types of heterogeneity, that is observed and unobserved heterogeneity. **Observed heterogeneity** is typically due to differences between groups of data that we can measure *a priori* using cultural, geographic, demographic and socio-economic variables (e.g., age, gender, country of origin, income), as well as variables that are more strictly related to the context of our analysis (e.g., product usage frequency or store loyalty in marketing; cigarettes consumption or food habits in the health sciences). On the other hand, **unobserved**

heterogeneity is related to information that we have not been able to measure a priori, implying that differences between groups of cases should be discovered indirectly using some *post hoc* analysis. Typical sources of unobserved heterogeneity are personality traits, personal lifestyle, preferences and attitudes.

The easiest way to deal with observed heterogeneity is to split the data into groups according to the values of the variables we measured (age, gender, etc.) and perform a group-specific analysis. However, more sophisticated methods are typically used in practice[1]. On the contrary, accounting for unobserved heterogeneity is more challenging. Statisticians have developed different approaches in this respect. Random effects are one example (Skrondal and Rabe-Hesketh, 2004; Bartholomew, 2013), but a similar issue arises also in other fields such as casual inference for observational studies, where a primary concern is heterogeneity originated by unmeasured confounding variables (Pearl et al., 2016; Hernán and Robins, 2021). Another method that explicitly deals with unobserved heterogeneity is finite mixture modelling (McLachlan and Peel, 2000; Frühwirth-Schnatter, 2006), in which the observed data are assumed to be generated by different subpopulations with potentially different parameter values.

No matter what the source of heterogeneity is, the crucial point in all cases is that failing to account for it may lead to a severe bias in the model estimates, thus producing potentially misleading results. As an easy example, consider the situation represented in Figure 7.1a, which reports the histogram for a sample of fictional data. The histogram shows that the data are composed by two groups, the first one on the left that is centred around the value 10, and the second one on the right that is centred around a value in between 15 and 20 instead. If we consider these data as coming from an homogeneous population, and we are interested in estimating its mean, we would obtain an estimate (sample mean) equal to 13.72. However, taking also into consideration the grouping structure, we would get 9.95 and 17.89 as the group-specific estimates. The two subpopulations that we used to generate the data are the two normal distributions shown in Figure 7.1b. In this simple example we clearly see that disregarding the grouping of the data produces a biased result.

Even in a simple context like this, the critical issue to consider regards the information we have about the grouping. If the groups refer to information we have collected/measured, we are in the case of observed heterogeneity and we can account for the group differences by performing a two-sample t-test. If the groups refer to information that we have not collected/measured (unobserved heterogeneity), we need to use one of the strategies for discovering the latent groups (cluster analysis, finite mixtures, etc.). The aim of this chapter is to address these same issues with regard to PLS-SEM.

[1] In PLS-SEM this corresponds to multi-sample analysis, that we illustrated in Chapter 6.

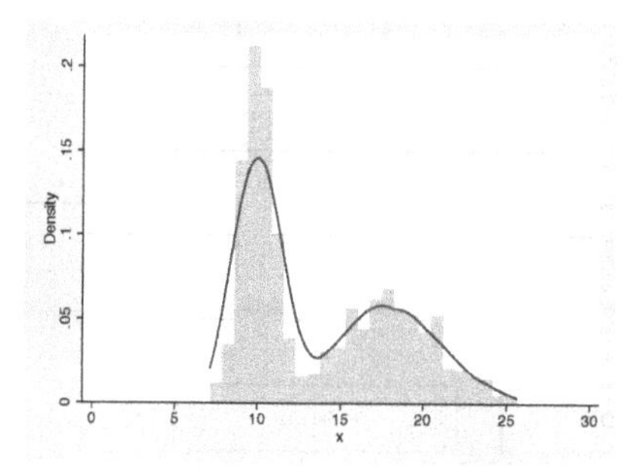

(a) Overall sample data (histogram) with a kernel density estimate overimposed (solid line).

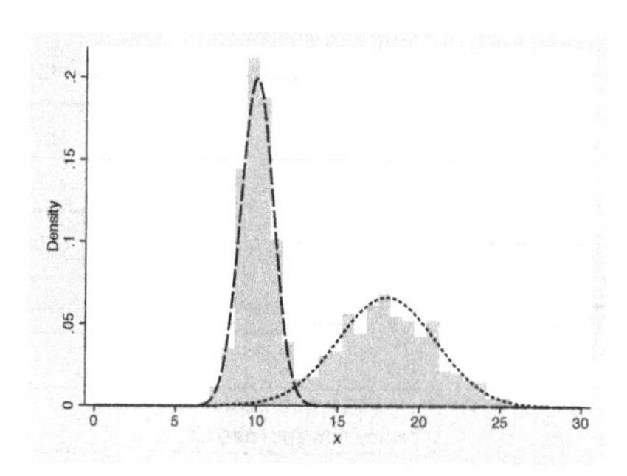

(b) Overall sample data (histogram) together with the generating populations (dashed lines).

FIGURE 7.1: An example of data originating from a mixture of two subpopulations.

7.2 Methods for the Identification and Estimation of Unobserved Heterogeneity in PLS-SEM

As we discussed above, if we don't know a priori the sources of heterogeneity, a common approach is to segment the dataset through a cluster analysis (Section 2.3.1) and run the main analysis on each of the identified segments. This practice is often used for example in CB-SEM, where separate models are estimated for the different segments obtained by an external clustering analysis. A similar strategy can be used in PLS-SEM as well. In particular, after fitting a global model using the whole sample, we may perform a cluster analysis on the indicators and/or the latent variable scores, and finally run a group-specific PLS-SEM analysis. Unfortunately this approach presents many drawbacks (Esposito Vinzi et al., 2008), and some simulation studies have shown a poor performance for the identification of group differences (Sarstedt and Ringle, 2010). Nonetheless, the crucial limitation of this procedure is that it fails to account for the measurement and structural relationships postulated by the researcher.

The aim of this section is to present the most popular methods developed so far in the literature for detecting and estimating unobserved heterogeneity in PLS-SEM. Figure 7.2 summarizes the available methods according to a taxonomy suggested in Sarstedt (2008). In the following we present the details of two of these approaches, the **response-based unit segmentation in PLS-SEM** approach (REBUS-PLS) (Trinchera, 2007; Esposito Vinzi et al., 2008) and the **finite mixture PLS** (FIMIX-PLS) approach (Hahn et al., 2002).

7.2.1 Response-based unit segmentation in PLS-SEM

Response-based unit segmentation (REBUS-PLS) is a technique that is inspired by cluster analysis in that at each iteration it assesses whether leaving an observation

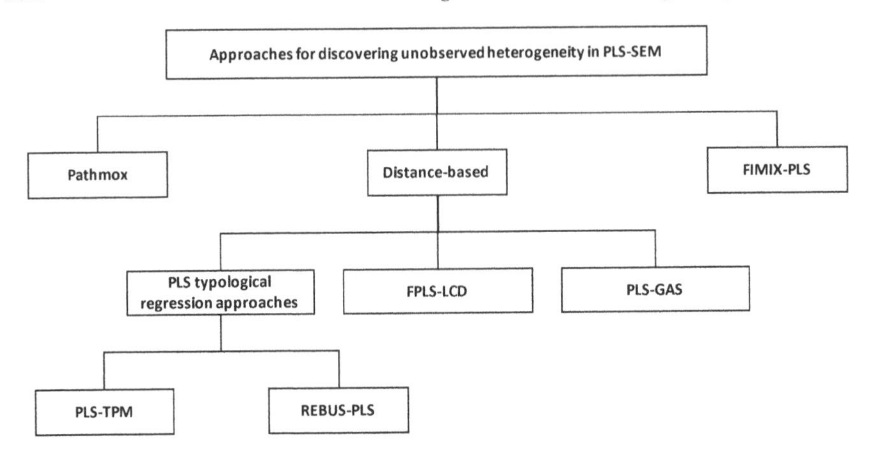

FIGURE 7.2: Methods introduced in the PLS-SEM literature for discovering the presence of unobserved heterogeneity (see Sarstedt, 2008).

in its original group or reassigning it to another one will improve the results. More specifically, the core idea of the algorithm is to calculate the so called **closeness measure** between each observation and each *local* (i.e., group-specific) *model*. The intuition behind this concept is that if latent classes do exist, observations belonging to the same class will share a similar local model. In other terms, if an observation is assigned to the "correct" (appropriate) group, the performance of the corresponding local model will be higher than that we would obtain by placing the same observation in any of the other groups.

The REBUS-PLS algorithm starts by estimating the **global model**, that is the model that uses the whole sample. Then, residuals from the global model are computed for both the measurement and the structural parts. These residuals are usually called **communality residuals** and **structural residuals** respectively. Next, using all the residuals, the number of groups K to segment the data into is chosen by running a hierarchical clustering algorithm (see Section 2.3.1.1)[2] REBUS-PLS then proceeds by estimating the **local models** corresponding to the groups identified with the cluster analysis in the previous step. This produces K group-specific PLS-SEM models each one fitted using only the observations in the corresponding segment. The group-specific parameter estimates are used to calculate the communality and the structural residuals for each observation in the sample (i.e., not only for the observations in a class) from every local model. Afterwards, using the residuals, the closeness measure for each observation from every local model is computed. This provides an assessment of how close each observation is from every group-specific model. Each unit is then assigned to the closest local model. During this reallocation, it can happen that

[2]There is no strict indication in the literature on the specific hierarchical algorithm to use. Similarly, the authors of the REBUS-PLS algorithm did not provide suggestions on how to select the number of groups from the hierarchy of solutions. The `plssem` Stata package implements the Ward's method and the number of groups can be either provided by the user or selected automatically by the software using the Caliński-Harabasz stopping rule.

some observations remain in the same clusters, while others move to another group. After the assignments of all observations, the K local models for the new allocation are estimated again and the algorithm continues until a specified stopping criterion is met. The usual suggestion (Esposito Vinzi et al., 2010, page 70) is to stop the algorithm when no more than 5% of the observations change class from one iteration to the other. The whole algorithm typically reaches convergence in few iterations (usually less than 15). The steps of the REBUS-PLS procedure are summarized in Algorithm 7.3.

When group stability is attained, the K final local models are estimated and the corresponding results are compared. In addition, Trinchera (2007) developed an index, the **group quality index** (GQI), to assess the goodness of the latent partition found by the REBUS-PLS algorithm. The GQI fundamentally represents a reformulation of the GoF index (see Chapter 4) in a multi-group context. From an intuitive point of view, the GQI can be considered as an average of the GoF indexes from the final local models. Moreover, it can be shown that the GQI for a partition with a single group that includes all the observations is mathematically equivalent to the GoF index for the global model. Therefore, to assess the goodness of the partition, it makes sense to compare the GQI with the GoF index from the global model. That is to say, if any of the local models performs better than the global one, the GQI will be larger than the global GoF index, indicating that we found some evidence of heterogeneity. In this case, it would be appropriate to perform separate PLS-SEM analyses. The technical appendix at the end of the chapter provides more details about the REBUS-PLS procedure.

On top of the GQI, the quality of the partition produced by the REBUS-PLS algorithm can be further validated using a **permutation test**. This procedure involves generating a large number M (usually at least 500) of random partitions of the data, keeping fixed the group proportions as detected by REBUS-PLS. For each random partition, the GQI is computed obtaining an empirical distribution for the index. Then, one can use this simulation-based sampling distribution of the GQI to compare the actual value of the GQI obtained in the main REBUS-PLS analysis to assess if the REBUS-PLS partition performs better than random allocations of the observations to the groups. In particular, the permutation test returns a p-value that can be used to test the null hypothesis that the actual REBUS-PLS partition is not better than a random partition. Therefore, if the p-value of the test is smaller than the significance level (e.g., 0.05), we can reject the null hypothesis concluding that there is evidence for heterogeneity in the PLS-SEM analysis. You can find a brief review of permutation tests in the technical appendix at the end of the chapter.

Finally, some authors (see for example Hair et al., 2018c, Chapter 5) suggest to further corroborate the findings by crossing the class membership indicator for the identified partition with other variables. This is usually called **ex post analysis**. This step usually involves running cross tabulations, logistic regressions or other predictive analysis (e.g., tree-based models like the CHAID algorithm or the CART approach) to check which are the covariates (so called *concomitant variables*) that can predict the class membership.

Algorithm 7.3 The REBUS-PLS algorithm.

Input: Data on indicators for n observations; specification of the measurement and structural models.

1: Estimate the global PLS-SEM model.
2: **for** $i = 1$ to n **do**
3: Compute the communality and structural residuals from the global model.
4: **end for**
5: Perform a hierarchical cluster analysis on the communality and structural residuals
6: Using a prespecified stopping rule (e.g., Caliński-Harabasz), choose the number of classes K to consider in the rest of the procedure.
7: **while** more than 5% of observations change class from one iteration to the other **do**
8: **for** $j = 1$ to K **do**
9: Estimate the local PLS-SEM model for the jth class.
10: **end for**
11: **for** $j = 1$ to K **do**
12: **for** $i = 1$ to n **do**
13: Compute the communality and structural residuals for observation ith from the jth local model.
14: Compute the closeness measure for observation ith from the jth local model.
15: **end for**
16: **end for**
17: **for** $i = 1$ to n **do**
18: Assign observation ith to the class with which it has the smallest closeness measure value.
19: **end for**
20: **end while**
21: **for** $j = 1$ to K **do**
22: Estimate the final local PLS-SEM model for the jth class using the final allocation of the observations to the classes.
23: **end for**

Output: Estimates of the K final local model parameters; class membership for all observations.

The REBUS-PLS approach has some advantages over other methods for detecting the presence of unobserved heterogeneity in PLS-SEM. First, it does not require any distributional assumption for the data, which is in line with the overall philosophy of the PLS-SEM approach. Second, in forming the partition it takes into account both the measurement and structural parts of the model. This is a nice feature because it means that the method is capable to detect heterogeneity in either one or both of the two components. Third, the approach does not require concomitant variables (e.g., socio-demographic variables) and so no external information is needed to define the partition.

On the contrary, one limitation of REBUS-PLS is that it can be applied to PLS-SEM models that include only reflective latent variables. This drawback derives from the fact that the method is based on the residuals of both the measurement and structural parts. In particular, the communality residuals (those related to the measurement model) are computed as the difference between the observed and predicted values of the manifest variables. Thus, this computation is possible only for reflective constructs. This is an unfortunate restriction especially because formative measures have attracted an increasing attention in recent years.

In Stata you can perform a REBUS-PLS analysis using the `estat unobshet` postestimation command available in the `plssem` package[3]. Currently, this command includes the following options:

- `method(`*`unobshet_method`*`)` specifies the method to use for assessing the presence of unobserved heterogeneity; currently only REBUS-PLS (`rebus`) and FIMIX-PLS (`fimix`; see Section 7.2.2) are available,
- `numclass(#)`, allows to set the number of classes to use in the REBUS-PLS algorithm; minimum is 1. If none is specified, the number of classes is automatically selected based on the Caliński-Harabasz stopping rule,
- `maxclass(#)` allows to set the maximum number of classes for the clustering stopping rule,
- `dendrogram` visualizes the dendrogram for a Ward hierarchical clustering algorithm of the residuals from the global model,
- `maxiter(#)` sets the maximum number of iterations before the REBUS-PLS algorithm stops,
- `stop(#)` allows to set the stopping rule for the REBUS-PLS algorithm with regard to the stability of the partition from one iteration to the other,
- `test` performs a permutation test for the GQI of a REBUS-PLS solution,
- `reps(#)` sets the number of replications in the permutation test on the GQI,

[3]Remind that in Stata a postestimation command works only if it is used right after the corresponding main estimation command, otherwise an error will be thrown.

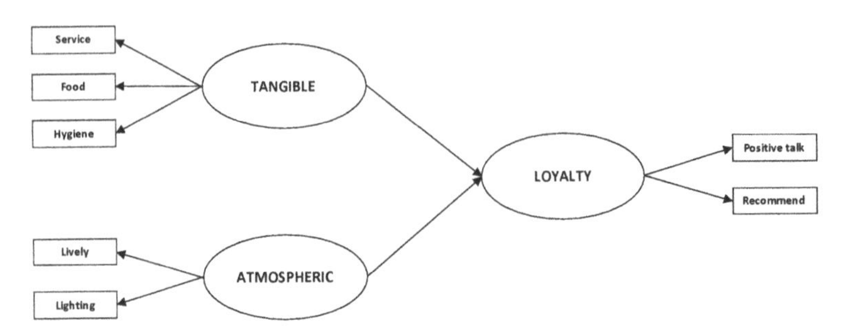

FIGURE 7.3: PLS-SEM for the REBUS-PLS example described in the text.

- `seed(#)` sets the seed for the permutation test on the GQI to allow for reproducible results,
- `plot` visualizes the GQI empirical distribution corresponding to the replications generated in the permutation test,
- `name(varname)` allows to set the name of the variable that will contain the final partition produced by the REBUS-PLS algorithm.

REBUS-PLS: A worked example using Stata

We now apply the REBUS-PLS method to real data collected in a cruise-experience setting and described in Mehmetoglu (2011). The sample contains 408 respondents who completed a questionnaire that included questions mainly about activities participated in on the cruise, the importance of motives for spending money on a cruise, the consideration of several cruise-experience attributes, as well as socio-demographic features. Figure 7.3 shows the structural model that we use in this example, where, assuming that loyalty is a key variable in the context of a cruise experience, we link both the `TANGIBLE` and `ATMOSPHERIC` attributes (exogenous) to `LOYALTY` (endogenous). We note that the model includes only reflective measures, as it is required by REBUS-PLS. The questionnaire measured the three latent variables with an ordinal scale (1 = completely disagree, ..., 7 = completely agree) by asking the respondents to indicate to what degree they agreed with the statements listed in Table 7.1. The data are available in the file `ch7_Cruise.dta`.

After loading the data, we first estimate the global model for the whole sample using the `plssem` command. Note that the dataset contains some missing values (the `misstable patterns` command allows to check that only 88% of the cases are complete). Therefore, to avoid wasting too much information, we use the `missing(mean)` option of `plssem` which imputes the missing values with the mean of the available data for each indicator before running the estimation algorithm. The corresponding code is reported below while the results are shown in Figure 7.4.

TABLE 7.1: Content of the latent and manifest variables for the REBUS-PLS example in the text.

Latent variable	Manifest variable	Content
Tangible	Service	The staff are capable of helping me with any questions that I have
	Food	The quality of the food is good
	Hygiene	The cleaning of the cabins is satisfactory
Atmospheric	Lively	The ship has an atmosphere full of warmth
	Lighting	The lighting contributes to the warm atmosphere
Loyalty	Positive talk	I shall talk positively about this cruise ship
	Recommend	I shall recommend my friends to travel with this cruise ship

```
use ch7_Cruise, clear

misstable patterns

plssem (TANGIBLE > Service Food Hygiene) ///
       (ATMOSPHERIC > Lively Lighting) ///
       (LOYALTY > Positive_talk Recommend), ///
       structural(LOYALTY TANGIBLE ATMOSPHERIC) ///
       missing(mean)
```

As we discussed in Chapter 4, the results for the global model show that convergent validity, reliability and discriminant validity are attained. The estimates of the structural model parameters also provide a significant evidence that the tangible and atmospheric attributes have nearly equal positive effects on loyalty. Finally, note that even if the structural model explains practically 50% of the loyalty variability, there is still another 50% of unexplained variation, some of which may be attributable to unobserved heterogeneity. Therefore, we now proceed to identify the presence of homogenous classes in the data, which may help the cruise businesses to improve their marketing policies.

The following code performs a REBUS-PLS analysis on the cruise data choosing automatically the number of groups through a Ward's algorithm with the Caliński-Harabasz stopping rule. In addition, a permutation test with 1000 replications is run and a graphical representation of the empirical GQI distribution is provided. The output is shown in Figures 7.5 and 7.6.

```
Partial least squares SEM                        Number of obs          =      408
                                                 Average R-squared      =      0.49259
                                                 Average communality    =      0.75189
Weighting scheme: path                           Absolute GoF           =      0.60858
Tolerance: 1.00e-07                              Relative GoF           =      0.97660
Initialization: indsum                           Average redundancy     =      0.43837

Measurement model - Standardized loadings
-----------------------------------------------------------
             |  Reflective:    Reflective:     Reflective:
             |     TANGIBLE    ATMOSPHERIC         LOYALTY
-------------+---------------------------------------------
     Service |        0.726
        Food |        0.846
     Hygiene |        0.821
      Lively |                       0.900
    Lighting |                       0.870
  Positive_t~k |                                     0.951
   Recommend |                                     0.936
-------------+---------------------------------------------
    Cronbach |        0.717          0.724           0.877
          DG |        0.841          0.878           0.942
       rho_A |        0.733          0.732           0.887
-----------------------------------------------------------

Discriminant validity - Squared interfactor correlation vs. Average variance extracted (AVE)
-----------------------------------------------------------
             |     TANGIBLE    ATMOSPHERIC         LOYALTY
-------------+---------------------------------------------
    TANGIBLE |        1.000          0.279           0.385
 ATMOSPHERIC |        0.279          1.000           0.368
     LOYALTY |        0.385          0.368           1.000
-------------+---------------------------------------------
         AVE |        0.639          0.783           0.890
-----------------------------------------------------------

Structural model - Standardized path coefficients
-----------------------------
    Variable |        LOYALTY
-------------+---------------
    TANGIBLE |          0.416
             |        (0.000)
 ATMOSPHERIC |          0.387
             |        (0.000)
-------------+---------------
        r2_a |          0.490
-----------------------------
  p-values in parentheses
```

FIGURE 7.4: Results for the estimation with `plssem` of the (global) model reported in Figure 7.3.

```
set matsize 1000
estat unobshet, test reps(1000) seed(123456) plot ///
  method(rebus)
```

Before discussing the results, we call your attention on the first line in the previous chunk. This code is necessary because `estat unobshet` internally creates some big matrices of temporary results that may be too large for the default maximum matrix size in Stata, which is 400 columns. If you forget to set it, you will get a `matsize too small` error message and the execution will stop.

The results of REBUS-PLS show that two classes have been automatically identified, with the first one that contains 259 observations (i.e., 63% of the sample) and

```
Response based unit segmentation (REBUS) solution

Weighting scheme: path
Tolerance: 1.00e-07
Initialization: indsum
Number of REBUS iterations: 7
Group Quality Index (GQI): 0.66962

REBUS classes
------------------------------------------------------------
                        |   Global     Class 1     Class 2
------------------------+-----------------------------------
           Observations |      408         259         149
             Percentage |  100.000      63.480      36.520
                    GoF |    0.609       0.673       0.664
------------------------------------------------------------

Loadings
------------------------------------------------------------
                        |   Global     Class 1     Class 2
------------------------+-----------------------------------
                Service |    0.726       0.764       0.689
                   Food |    0.846       0.764       0.906
                Hygiene |    0.821       0.778       0.881
                 Lively |    0.900       0.897       0.878
               Lighting |    0.870       0.892       0.834
          Positive_talk |    0.951       0.916       0.904
              Recommend |    0.936       0.904       0.856
------------------------------------------------------------

Path coefficients
------------------------------------------------------------
                        |   Global     Class 1     Class 2
------------------------+-----------------------------------
    TANGIBLE -> LOYALTY |    0.416       0.566       0.328
 ATMOSPHERIC -> LOYALTY |    0.387       0.350       0.546
------------------------------------------------------------

Permutation test
------------------------------------
                        |     Value
------------------------+-----------
           Replications |      1000
                P-value |     0.000
------------------------------------
```

FIGURE 7.5: Results of the REBUS-PLS algorithm for the model reported in Figure 7.3.

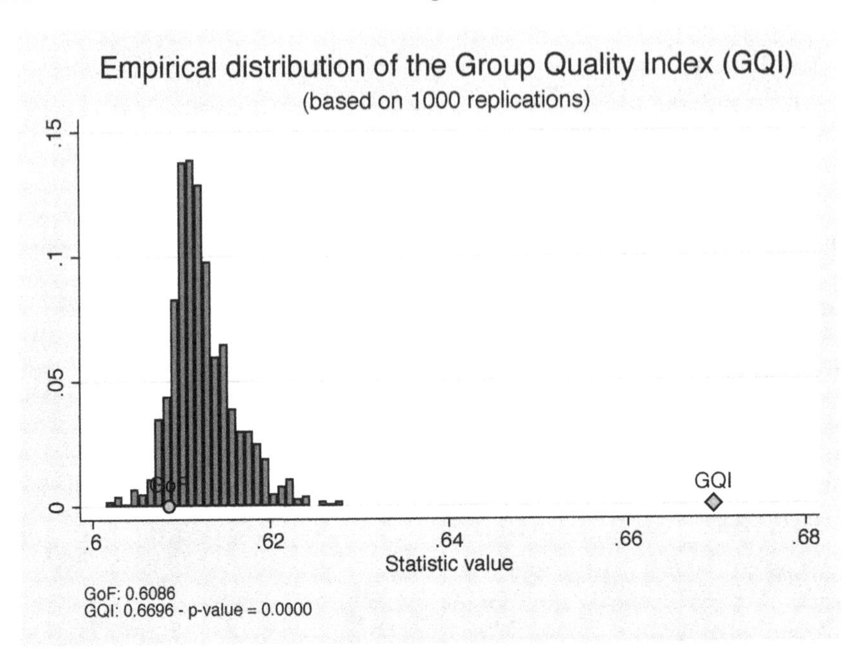

FIGURE 7.6: PLS-SEM for the REBUS-PLS example described in the text.

the second the remaining 149 (37% of the sample). The table of loadings shows that the two classes are similar for what regards the measurement model, while the table of path coefficients shows more noticeable differences. In particular, for the first class, in contrast to the atmospheric attribute, the tangible attribute has a stronger effect on loyalty. Further, this effect is larger in this class than it is for both the second class and the global model. The opposite situation holds instead for the second class, where the atmospheric attribute has a stronger effect on loyalty than in the first class.

The goodness of the partition can be assessed using the GQI together with the corresponding permutation test. By comparing the GQI (0.6696) with the GoF for the global model (0.6086) we conclude that the REBUS-PLS solution provides evidence for unexplained heterogeneity. The GoF measure for the global model is in fact equivalent to the GQI for a partition that includes a single class, so the fact that the former is larger implies that the two-class partition found by the algorithm seems to provide a better explanation of the variability in the data compared to the global model. Moreover, the permutation test for the GQI has a very small p-value (reported as 0.000) confirming the goodness of the partition.

To complete the analysis, we perform a multi-sample analysis (see Chapter 6) using the class membership generated by `estat unobshet` (by default this column is called `rebus_class`). The results, reported in Figures 7.7 and 7.8, lend support in favour of the two-class partition as the bootstrap tests for the difference in the path coefficients have p-values below the conventional 0.05 level.

```
1  plssem (TANGIBLE > Service Food Hygiene) ///
2         (ATMOSPHERIC > Lively Lighting) ///
3         (LOYALTY > Positive_talk Recommend), ///
4         structural(LOYALTY TANGIBLE ATMOSPHERIC) ///
5         group(rebus_class, method(bootstrap) ///
6              reps(1000) plot seed(123456)) ///
7         missing(mean)
```

```
Partial least squares SEM

Weighting scheme: path
Tolerance: 1.00e-07
Initialization: indsum

Multigroup comparison (rebus_class) - Bootstrap t-test
---------------------------------------------------------------------------------------------
     Measurement effect |   Global    Group_1    Group_2   Abs_Diff  Statistic      P-value
------------------------+--------------------------------------------------------------------
     TANGIBLE -> Service |   0.726      0.764      0.689      0.075      1.357        0.176
        TANGIBLE -> Food |   0.846      0.764      0.906      0.142      2.614        0.009
     TANGIBLE -> Hygiene |   0.821      0.778      0.881      0.103      1.669        0.096
    ATMOSPHERIC -> Lively |   0.900      0.897      0.878      0.019      0.006        0.995
  ATMOSPHERIC -> Lighting |   0.870      0.892      0.834      0.058      1.979        0.049
 LOYALTY -> Positive_talk |   0.951      0.916      0.904      0.012      0.611        0.542
     LOYALTY -> Recommend |   0.936      0.904      0.856      0.047      2.181        0.030
---------------------------------------------------------------------------------------------
 number of replications: 1000
 group labels:
   Group 1: 1
   Group 2: 2
 group sizes:
   Group 1: 259
   Group 2: 149

Multigroup comparison (rebus_class) - Bootstrap t-test
---------------------------------------------------------------------------------------------
      Structural effect |   Global    Group_1    Group_2   Abs_Diff  Statistic      P-value
------------------------+--------------------------------------------------------------------
    TANGIBLE -> LOYALTY |   0.416      0.566      0.328      0.238      2.140        0.033
 ATMOSPHERIC -> LOYALTY |   0.387      0.350      0.546      0.196      1.980        0.048
---------------------------------------------------------------------------------------------
 number of replications: 1000
 group labels:
   Group 1: 1
   Group 2: 2
 group sizes:
   Group 1: 259
   Group 2: 149
```

FIGURE 7.7: Results of the multi-group analysis using the partition produced by the REBUS-PLS algorithm for the model reported in Figure 7.3.

7.2.2 Finite mixture PLS (FIMIX-PLS)

Another popular approach for detecting unobserved heterogeneity in a PLS-SEM is FIMIX-PLS (Hahn et al., 2002; Ringle et al., 2010; Becker et al., 2013), which represents the extension to PLS-SEM of the finite mixture model approach introduced by Jedidi et al. (1997a,b) in the CB-SEM context. FIMIX-PLS is based on more restrictive assumptions than REBUS-PLS. Indeed, it assumes that the unobserved heterogeneity involves only the structural part of the model and not the measurement part, which is then assumed to be fixed across the latent segments. Furthermore, it assumes that the endogenous latent variables are modelled as a finite mixture of normal

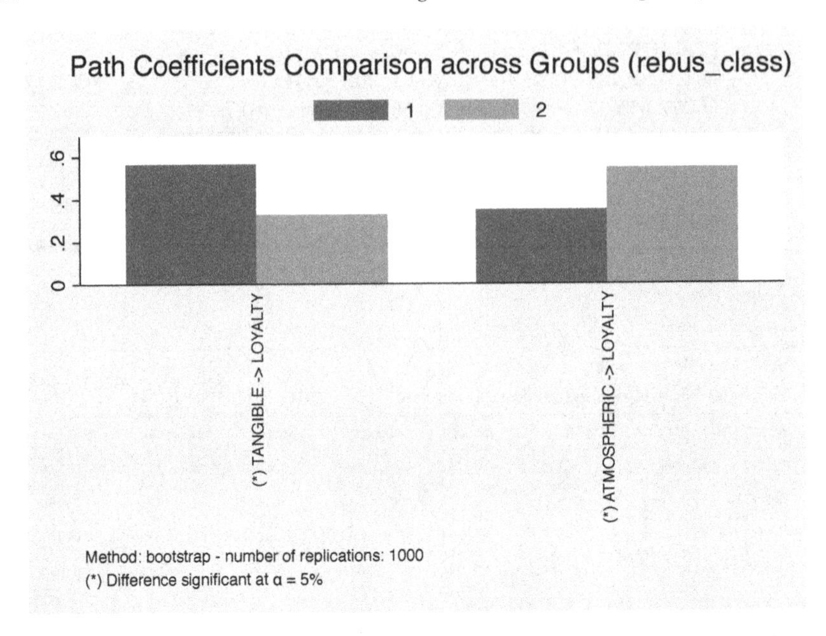

FIGURE 7.8: Results of the multi-group analysis using the partition produced by the REBUS-PLS algorithm for the model reported in Figure 7.3 (only the structural model coefficients are reported).

distributions. This is to say that *within each class* the endogenous latent variables are assumed to be normally distributed.

As for REBUS-PLS, the first step in FIMIX-PLS is to estimate the global PLS-SEM model using the entire sample. Then, based on the estimated latent scores from the global model, FIMIX-PLS uses the expectation-maximization (EM) algorithm to estimate the parameters of the mixture components[4]. Finally, the algorithm computes the class posterior probabilities, that is the probabilities for each observation to belong to the different classes[5]. It follows that an observation will be allocated to the class with the largest posterior probability. Once the classes have been identified, the final step is to estimate the segment-specific models. A multi-sample analysis may be further performed to assess the statistical significance of the differences between the local model parameters (see Section 6.4). The FIMIX-PLS steps are detailed in Algorithm 7.4.

Different from REBUS-PLS, which first performs a hierarchical cluster analysis to determine the number of classes K to use, in FIMIX-PLS the number of classes is not chosen automatically by the algorithm. A typical approach consists in running FIMIX-PLS with different values of K (usually all the values in between 1 and a given maximum K_{max}) and then comparing the results using some criteria such as

[4]More details on finite mixture models are available in Section 2.3.2.

[5]The class posterior probabilities should not be confused with the class *marginal* probabilities, which instead are represented by the estimates of the mixture weights.

Algorithm 7.4 The FIMIX-PLS method using the EM algorithm.

Input: Data on indicators for n observations; specification of the measurement and structural models; number of latent classes K; stopping criterion δ; maximum number of iterations $iter_{max}$; number of algorithm restarts R.

1: Estimate the global PLS-SEM model.
2: **for** $r = 1$ to R **do**
3: Set random starting values for the posterior mixture proportions P_{ik} with $i = 1, \dots, n$ and $k = 1, \dots, K$.
4: Set the current log-likelihood value $loglik_{curr}$ to a given number V and the log-likelihood change to Δ.
5: Perform an initial *maximization* step (M-step) to get initial parameter estimates.
6: **while** (iteration number is smaller than $iter_{max}$ **and** $\Delta \geq \delta$) **do**
7: Perform the *expectation* step (E-step) producing updated values of the posterior mixture proportions P_{ik}, for every $i = 1, \dots, n$ and $k = 1, \dots, K$.
8: Set the previous log-likelihood value $loglik_{prev}$ to the current one.
9: Perform the M-step to get updated parameter estimates.
10: Compute the updated current log-likelihood value.
11: Set $\Delta = loglik_{curr} - loglik_{prev}$.
12: **end while**
13: **end for**
14: Select as the best solution that with the largest log-likelihood value.
15: Compute the information and classification criteria.
16: **for** $i = 1$ to n **do**
17: Assign observation ith to the class with the largest posterior probability P_{ik}.
18: **end for**
19: **for** $j = 1$ to K **do**
20: Estimate the final local PLS-SEM model for the jth class using the final allocation of the observations to the classes.
21: **end for**

Output: Estimates of the K final local model parameters; class membership for all observations; likelihood-based information criterion and entropy measures.

AIC and BIC (see Section 2.3.2). Other popular indexes used in FIMIX-PLS are (Sarstedt et al., 2011):

- Modified AIC with factor 3 (AIC_3) and 4 (AIC_4), which use a penalization factor of 3 and 4 respectively instead of the usual factor 2 as in equation (2.21). These indexes, introduced by Bozdogan (1994), allow to reduce the tendency of the AIC index to overspecify the correct number of segments.
- Consistent AIC (CAIC), developed by Bozdogan (1987), which is more conservative than the AIC and prefers models with fewer classes.
- Minimum description length with factor 5 (MDL_5), developed by Liang et al. (1992). This criterion has shown weak performance in simulations because it tends to underestimate the actual number of classes.
- Entropy criterion (EN), which is a class separation index, that is a measure of how well the class-specific component densities of the mixture are separated (Ramaswamy et al., 1993). This index is bounded between 0 and 1. Values close to 1 indicates that the classes are well separated, while values of the index close to 0 indicate no class separation. Celeux and Soromenho (1996) proposed a modification of this index, the normed (or normalized) entropy criterion (NEC), but this has some drawbacks, and so it is less frequently used in practice.

Hair et al. (2016) provide the following suggestions to select the number of classes in FIMIX-PLS:

- If AIC_3 and CAIC agree in indicating the same number of segments, choose this as the best solution. Differently, consider AIC_3 (or AIC_4) jointly with BIC.
- The entropy criterion EN should be larger than 0.5.
- Check that the solution under examination produces segments with *reasonable* sample sizes. In particular, if you get some classes that contain only few observations, this may indicate that you have chosen a value of K that is too large.
- Take into considerations a priori information and also indications coming from the theory related to the problem you are analysing.

Despite the availability of so many indexes, the selection of the number of latent classes remains a difficult and generally unresolved problem, which requires knowledge of the problem and experience in the technicalities of mixture models.

A further practical aspect to take into account when running FIMIX-PLS is that it is based on the EM algorithm. As we already highlighted in Chapter 2, EM represents a general approach for fitting mixture models, but it is not free from criticisms. More specifically, even if the EM algorithm always reaches convergence, it has a tendency to get stuck in local optima, especially when the model includes a large number of parameters. One way to limit this problem is to rerun the algorithm a given number of times (in the literature it is suggested 10) and pick the best solution, that is the solution with the largest value of the log-likelihood.

Even if FIMIX-PLS represents an important advancement in the treatment of unobserved heterogeneity in PLS-SEM, it also has limitations. First, FIMIX-PLS assumes that heterogeneity only affects the structural model, while the measurement model is fixed during the iterations. In fact, the first step of the algorithm involves the estimation of the global model, whose latent scores are then used as the input of the next step. This implies that, in all local models, the outer weights are constant and equal to those obtained for the global model. As suggested by Ringle et al. (2010), a strategy to solve this problem consists in looking for external (observed) variables that are able to reproduce the classes detected by the FIMIX-PLS algorithm. Then, a multi-group analysis based on these variables is performed leading to class-specific measurement and structural models. Unfortunately, the chances to identify such variables in practice are few. A second limitation of FIMIX-PLS is that it assumes that within each class the endogenous latent variables are normally distributed. This is a strong distributional assumption, which is unlikely to hold in practice and that goes in opposite direction with respect to the distribution-free perspective at the base of PLS-SEM.

In Stata you can perform a FIMIX-PLS analysis using the `estat unobshet` postestimation command setting the `method()` option to `fimix` (the default method is `rebus`[6]). This command provides the following options:

- `numclass(#)`, allows to set the number of classes to use in the FIMIX-PLS algorithm; minimum is 1.
- `maxiter(#)` sets the maximum number of iterations before the FIMIX-PLS algorithm stops; the default is 30000 iterations.
- `stop(#)` allows to set the stopping rule for the FIMIX-PLS algorithm with regard to the log-likelihood of the underlying finite mixture model.
- `restart(#)` sets the number of times the algorithm will be executed to avoid local maxima; the default is 10 restarts.
- `seed(#)` sets the seed to allow for reproducible results.
- `name(`*`varname`*`)` allows to set the name of the variable that will contain the final partition produced by the FIMIX-PLS algorithm.
- `groups(range)` this option allows to pass a list of number of classes to use; the results of the different solutions are then compared by means of the indexes discussed above, namely AIC, AIC_3, AIC_3, BIC, CAIC, HQ, MDL_5, EN, NFI and NEC[7].

[6]Currently the FIMIX-PLS option in `plssem` is still experimental and under testing.

[7]HQ refers to the Hannan-Quinn information criterion (Hannan and Quinn, 1979), while NFI corresponds to the non-fuzzy classification criterion developed by Roubens (1978).

7.2.3 Other methods

While REBUS-PLS and FIMIX-PLS are currently the prominent methods to detect unobserved heterogeneity in the PLS-SEM framework, other approaches have been introduced in the literature for the same purpose. In this section we briefly review two of them, namely the path modelling segmentation tree algorithm, also known as the Pathmox algorithm, and the genetic algorithm segmentation for PLS-SEM method, also known as PLS-GAS.

7.2.3.1 Path modelling segmentation tree algorithm (Pathmox)

The Pathmox[8] algorithm, introduced in the literature by Sanchez and Aluja (2006) and further explored in Lamberti et al. (2016, 2017), adopts a binary segmentation approach to produce a tree structure with different path models in each of the final nodes. More specifically, the algorithm starts by fitting the global PLS-SEM model to the entire sample and then proceeds looking for the optimal split that provides path models that are as different as possible. To split the observations in groups, the Pathmox algorithm uses external concomitant (typically socio-demographic) variables. The optimal split is determined by calculating all possible binary splits obtainable from the categories of the concomitant variables, thus dividing the parent node into two segments. The generated child nodes (i.e., the corresponding local models) are then compared first with regards to the structural parts. In particular, an F-test for comparing the path coefficients of the child nodes is used. The partition showing the most significant (i.e., smaller) p-value is considered a candidate for the optimal split. The same process is then repeated for the categories of each concomitant variable, finally identifying as optimal the split corresponding to the minimum p-value among all candidates. The algorithm stops when the F-tests produce non-significant results.

The Pathmox algorithm is a sensible approach for detecting unobserved heterogeneity, but it has several limitations. First, it requires using external *observed* variables to classify the observations. On one side this implies that the partition produced by Pathmox does not involve the structural relationships directly, but it substantially comes from external information. Moreover, the ability of the algorithm to discover the heterogeneity in the data is conditioned on the availability of such external variables. If none are available in your study, you will not be able to apply the method. Second, as in FIMIX-PLS, the Pathmox algorithm only assesses the presence of heterogeneity in the structural model, disregarding what happens in the measurement part. Finally, the F-tests used for splitting presuppose that the error terms in the structural relationships are normally distributed with equal variance, assumptions that are rarely met in practice.

As for the software, currently there is only one `R` package that implements the Pathmox approach, the `pathmox` package (Sanchez and Aluja, 2013).

[8] As the authors of the method suggest, "mox" means "divide into two" in the Aztec language.

7.2.3.2 Partial least squares genetic algorithm segmentation (PLS-GAS)

The last approach we present for detecting unobserved heterogeneity in PLS-SEM is called partial least squares genetic algorithm segmentation, or simply PLS-GAS. Genetic algorithms (GAs) are a family of computational models inspired by evolution that belong to the larger class of evolutionary algorithms. These algorithms encode a potential solution to a specific problem on a simple chromosome-like data structure and apply operators such as mutation, crossover and selection to these structures so as to preserve critical information (for an introduction to GAs see Mitchell, 1996). In GAs, a population of candidate solutions (usually called individuals) to an optimization problem is evolved toward better solutions. The evolution starts from a randomly generated population of individuals, and in each generation, the *fitness* of every individual is evaluated. The fitness is usually the value of the objective function in the optimization problem being solved by the GA. Genetic algorithms are often viewed as function optimizers, although the range of problems to which they have been applied is much broader. Introduced by Ringle et al. (2013, 2014), PLS-GAS uses GAs to identify the best hidden partition of the data in a PLS-SEM analysis. PLS-GAS aims at finding the partition that minimizes the error variance of the PLS-SEM in the detected classes. Thus, its objective function corresponds to the sum of the squared residuals for both measurement and structural parts from all segments. PLS-GAS involves two stages. Following Cowgill et al. (1999), in the first stage a GA is used to identify a nearly optimal partition of the data. Since it is not guaranteed that the first stage provides the global optimum, the second stage refines the first stage solution by locally improving it. This is achieved by adopting the same idea as in REBUS-PLS, that is checking whether the partition might improve (i.e., produces a better objective function value) by reassigning each observation to a different segment based on its "distance" from each local model[9].

The main advantages of PLS-GAS are that it is a distribution-free approach and that, differently from REBUS-PLS, it can be applied even when formative constructs are included in the model[10]. However, the major drawback of the method is represented by the intense computational burden it requires.

At the time of writing, the only software implementing the PLS-GAS procedure is some GAUSS code made available by the developer of the algorithm[11] and a Python package developed by Seman (2016)[12]. Unfortunately, using them requires advanced programming skills from the side of the user.

---

[9]Note that PLS-GAS and REBUS-PLS differ in the way they measure the distance of each observations from each local model: in REBUS-PLS the closeness measure is used, while in PLS-GAS this corresponds to the algorithm's objective function, that is the sum of squared residuals.

[10]This is because formative constructs are not included in the objective function of PLS-GAS.

[11]You can download the GAUSS code for PLS-GAS from `https://www.pls-sem.net/downloads/`.

[12]The GitHub repository for the package is `https://github.com/lseman/pylspm`.

7.3 Summary

Heterogeneity in a statistical analysis is usually classified as either observed or unobserved. In the former case we can attribute the variability to quantities we have been able to measure (i.e., observed variables), while the latter refers to situations where the variability cannot be ascribed to any measured quantity. The easiest way of dealing with observed heterogeneity is to split the data into groups according to the values of the measured variables (age, gender, etc.) and perform group-specific analyses. Accounting for unobserved heterogeneity is instead more challenging, and a variety of approaches for detecting it have been developed so far (e.g., cluster analysis and finite mixture models). The presence of unobserved heterogeneity is also a major concern in PLS-SEM, because failing to account for it may lead to biased results and hence to potentially harmful decisions. Therefore, different methods for detecting it have been introduced in the last years, the most prominent ones being REBUS-PLS and FIMIX-PLS. Each of these methods has advantages and disadvantages, so that none of them has shown a clear superiority over the others. No matter which method we use, discovering the presence of unobserved heterogeneity in a PLS-SEM framework remains an arduous problem that requires a high level of technical expertise.

Appendix: R Commands

At the time of writing of this book only one of the R packages currently available for PLS-SEM provides some facilities for assessing the presence of unobserved heterogeneity, the plspm package. In particular, the package includes the rebus.pls() function for performing a response-based unit segmentation (REBUS-PLS) analysis. Our presentation here follows that available in Chapter 9 of Sanchez (2013).

The rebus.pls() performs all the steps we described in Section 7.2.1, that is estimation of the global model and iterative assignment of the observations to the K local models until convergence. The function requires a single mandatory argument, an object of class plspm as returned by the plspm() function (see page 141). In addition, the function accepts the argument stop.crit indicating the stopping criterion to apply (percentage of units changing class from one iteration to the other), and iter.max that represents the maximum number of iterations. The following code loads the data, applies mean imputation and fits the global model (output not reported):

```
1  if (!require(haven, quietly = TRUE)) {
2    install.packages("haven")
3  }
```

```
4  cruise_data <- as.data.frame(read_stata(file =
5    file.path(path_data, "ch7_Cruise.dta")))

6  # mean imputation
7  if (!require(mice, quietly = TRUE)) {
8    install.packages("mice")
9  }
10 library(mice)

11 cruise_imp <- mice(cruise_data, method = "mean",
12   print = FALSE)
13 cruise_comp <- complete(cruise_imp)

14 library(plspm)

15 TANGIBLE <- c(0, 0, 0)
16 ATMOSPHERIC <- c(0, 0, 0)
17 LOYALTY <- c(1, 1, 0)
18 cruise_path <- rbind(TANGIBLE, ATMOSPHERIC, LOYALTY)
19 colnames(cruise_path) <- rownames(cruise_path)
20 cruise_blocks <- list(c(2, 4, 5), c(1, 3), c(6, 7))
21 cruise_modes <- rep("A", 3)

22 cruise_plspm <- plspm(Data = cruise_comp,
23   path_matrix = cruise_path,
24   blocks = cruise_blocks,
25   modes = cruise_modes,
26   scheme = "path", tol = 1e-7)
27 # summary(cruise_plspm)
```

The global model is then passed to the `rebus.pls()` function. Note that `rebus.pls()` produces a dendrogram of the outer and inner residuals based on which the user must interactively provide the number of groups to use in the next step of the REBUS-PLS algorithm. For our example we decide to use two groups:

```
1  cruise_rebus <- rebus.pls(cruise_plspm, stop.crit = 0.005,
2    iter.max = 100)
3  cruise_rebus
```

```
RESPONSE-BASED UNIT SEGMENTATION (REBUS)
IN PARTIAL LEAST SQUARES PATH MODELING
----------------------------------------------

Parameters Specification
  Number of segments:    2
  Stop criterion:        0.005
  Max number of iter:    100
```

```
REBUS solution (on standardized data)
  Number of iterations:  7
  Rate of unit change:   0.004901961
  Group Quality Index:   0.6695858

REBUS Segments
                    Class.1   Class.2
number.units            259       149
proportions(%)           63        37

---------------------------------------------
$path.coef
                          Class.1   Class.2
TANGIBLE->LOYALTY          0.5658    0.3282
ATMOSPHERIC->LOYALTY       0.3500    0.5457

---------------------------------------------
$loadings
                 Class.1   Class.2
Service           0.7643    0.6895
Food              0.7644    0.9063
Hygiene           0.7782    0.8808
Lively            0.8973    0.8781
Lighting          0.8925    0.8341
Positive_talk     0.9163    0.9041
Recommend         0.9038    0.8564

---------------------------------------------
$quality
                       Class.1    Class.2
Aver.Com
  Com.TANGIBLE       0.5913396  0.6908425
  Com.ATMOSPHERIC    0.8008456  0.7333647
  Com.LOYALTY        0.8282901  0.7754492
Aver.Redu
  Red.LOYALTY        0.5218247  0.4696579
R2
  R2.LOYALTY         0.6300024  0.6056591
GoF
  GoF                0.6828628  0.6663938
```

All the results match with those reported by Stata. The object returned by the `rebus.pls()` function also includes an element called `segments`, which provide the identified group membership for each observation. This can be used to perform further analyses.

Next, the `local.models()` function can be used to extract the local models, that is the fitted models for the chosen number of groups. This function returns a list with the results of the global model as well as those of each local model. The local models can be inspected using the same methods as for `plspm` objects.

Finally, to assess the quality of the local models obtained by REBUS-PLS, we can use the `rebus.test()` function, that performs a permutation test for comparing each pair of groups. The function `rebus.pls()` requires as arguments an object of class `plspm` corresponding to the global model and the object returned by `rebus.pls()`. The number of permutations performed by `rebus.test()` in each test is fixed at 100. The output produced by `rebus.test()` is a list whose elements are the test results for the comparison of each pair of groups. The following code performs the permutation test for the example above:

```
cruise_test <- rebus.test(cruise_plspm, cruise_rebus)
cruise_test$test_1_2
```

```
$paths
                      Class.1 Class.2 diff.abs p.value sig.05
TANGIBLE->LOYALTY      0.5658  0.3282   0.2376  0.0099    yes
ATMOSPHERIC->LOYALTY   0.3500  0.5457   0.1957  0.0099    yes

$loadings
              Class.1 Class.2 diff.abs p.value sig.05
Service        0.7393  0.7393   0.0748  0.0099    yes
Food           0.8672  0.8672   0.1419  0.0099    yes
Hygiene        0.8507  0.8507   0.1026  0.0099    yes
Lively         0.8902  0.8902   0.0192  0.0099    yes
Lighting       0.8571  0.8571   0.0584  0.0099    yes
Positive_talk  0.9505  0.9505   0.0122  0.0099    yes
Recommend      0.9323  0.9323   0.0474  0.0099    yes

$gof
  Class.1 Class.2 diff.abs p.value sig.05
1  0.6133  0.6133   0.0093  0.0099    yes
```

Appendix: Technical Details

The math behind the REBUS-PLS algorithm

The core of the REBUS-PLS algorithm is represented by the calculation of the distance between each observation and the local models corresponding to the latent classes. This distance is usually referred to as the **closeness measure** (CM). In this appendix we provide the technical definition of the CM as well as that of the group quality index (GQI), which is used to assess the quality of a REBUS-PLS partition. The following presentation follows that by Trinchera (2007, Chapter 5).

Since the definition of CM used in REBUS-PLS is based on the same idea behind the goodness-of-fit (GoF) index, we first remind the definition of the latter, that we introduced in Chapter 4. The GoF index is defined as

$$\begin{aligned} GoF &= \sqrt{\overline{\text{communality}} \times \overline{R}^2} \\ &= \sqrt{\frac{\sum_{q=1}^{Q}\sum_{p=1}^{P_q} \text{cor}^2(\boldsymbol{x}_{pq}, \boldsymbol{\xi}_q)}{\sum_{q=1}^{Q} P_q} \times \frac{\sum_{j=1}^{J} R^2\left(\boldsymbol{\xi}_j, \{\boldsymbol{\xi}_q\text{'s explaining } \boldsymbol{\xi}_j\}\right)}{J}}. \end{aligned} \tag{7.1}$$

The left term inside the square root, which corresponds to the average communality, provides an assessment of the measurement model goodness, while the right term, corresponding to the average R-squared, represents an assessment of the structural model quality.

The CM used in REBUS-PLS follows the same idea as the global GoF index since it is defined using both the measurement and structural models, and it is based on the residuals for both parts of the model, that is the **measurement** (or **communality**) and **structural residuals**. Measurement residuals are computed for each manifest variable in the model[13], while structural residuals are defined for each endogenous latent variable. More specifically, the measurement residual e_{ipqk} of the ith observation for the kth local model (i.e., the model estimated using observations allocated to the kth latent class), is defined as

$$e_{ipqk} = x_{ipq} - \widehat{x}_{ipqk}, \tag{7.2}$$

where x_{ipq} is the observed value of the pth manifest variable defined by the qth latent variable for the ith observation, while $\widehat{x}_{ipqk}$ is the corresponding predicted value obtained from the kth local model. In particular, $\widehat{x}_{ipqk}$ is computed as

$$\widehat{x}_{ipqk} = \widehat{\lambda}_{pqk} \xi_{iqk}, \tag{7.3}$$

with $\widehat{\lambda}_{pqk}$ denoting the estimated class-specific loading associated with the pth manifest variable of the qth block in the kth latent class, and ξ_{iqk} being the score of the qth latent variable for the ith unit. The latent variable score ξ_{iqk} is computed using the weights $\widehat{w}_{pqk}$ estimated by performing a PLS path model on observations belonging to the kth class, that is

$$\xi_{iqk} = \sum_{p=1}^{P_q} \widehat{w}_{pqk} x_{ipq}, \tag{7.4}$$

The structural residual f_{ijk} of the ith observation for the jth endogenous latent variable in the kth local model is defined as

$$f_{ijk} = \xi_{ijk} - \widehat{\xi}_{ijk}, \tag{7.5}$$

where ξ_{ijk} is computed according to (7.4) and $\widehat{\xi}_{ijk}$ is instead obtained as

$$\widehat{\xi}_{ijk} = \sum_{\{\ell : \xi_\ell \text{ explains } \xi j\}} \widehat{\beta}_{\ell jk} \xi_{ijk}, \tag{7.6}$$

with $\widehat{\beta}_{\ell jk}$ representing the estimated path coefficient from the kth local model for the ℓth exogenous latent variable linked to the jth endogenous latent variable.

Given the measurement and structural residuals (7.2) and (7.5), the closeness measure CM_{ik} of the ith unit to the kth local model is defined as

$$CM_{ik} = \sqrt{\frac{\sum_{q=1}^{Q} \sum_{p=1}^{P_q} \left[e_{ipqk}^2 / \text{com}(\boldsymbol{x}_{pq}, \boldsymbol{\xi}_{qk}) \right]}{\frac{\sum_{i=1}^{n} \sum_{q=1}^{Q} \sum_{p=1}^{P_q} \left[e_{ipqk}^2 / \text{com}(\boldsymbol{x}_{pq}, \boldsymbol{\xi}_{qk}) \right]}{(n-2)}} \times \frac{\sum_{j=1}^{J} \left[f_{ijk}^2 / R^2 \left(\boldsymbol{\xi}_j, \{ \boldsymbol{\xi}_\ell\text{'s explaining } \boldsymbol{\xi}_j \} \right) \right]}{\frac{\sum_{i=1}^{n} \sum_{j=1}^{J} \left[f_{ijk}^2 / R^2 \left(\boldsymbol{\xi}_j, \{ \boldsymbol{\xi}_\ell\text{'s explaining } \boldsymbol{\xi}_j \} \right) \right]}{(n-2)}}}, \tag{7.7}$$

[13]Remember that REBUS-PLS can be used only for models based on reflective constructs.

where $\text{com}(\boldsymbol{x}_{pq})$ denotes the communality for the pth manifest variable in the qth block from the kth local model (for the definition of communalities, see Section 4.2).

Comparing expressions (7.1) and (7.7), we note the similarity between the GoF measure and CM, since they both take into account the quality of the measurement and structural parts. A feature worth of notice for the CM is that all residuals are computed for each observation with respect to each local model, regardless of the membership of the observation. Therefore, an observation will be allocated to the kth latent class by the REBUS-PLS algorithm if its residuals with respect the kth local model are smaller than those corresponding to the other $(K-1)$ local models.

The GQI can be used to assess whether the K local models identified by the REBUS-PLS algorithm perform better than the global model. In a sense it represents a generalization of the GoF index and it is based on the model's residuals in a similar way as the CM.

To understand the definition of the GQI it is useful to start recalling that the R-squared index for a linear regression model can be computed as one minus the ratio between the residual sum of squares (RSS) and the total sum of squares (TSS), that is

$$R^2 = 1 - \frac{RSS}{TSS} = 1 - \frac{\sum_{i=1}^{n}(Y_i - \widehat{Y}_i)^2}{\sum_{i=1}^{n}(Y_i - \overline{Y})^2} = 1 - \frac{\sum_{i=1}^{n} e_i^2}{\sum_{i=1}^{n}(Y_i - \overline{Y})^2}, \tag{7.8}$$

where $e_i = Y_i - \widehat{Y}_i$ is the residual for the ith observation. Using this fact, we can reformulate the GoF index (7.1) in terms of residuals as

$$GoF = \sqrt{\frac{\sum_{q=1}^{Q}\sum_{p=1}^{P_q}\left(1 - \frac{\sum_{i=1}^{n} e_{ipq}^2}{\sum_{i=1}^{n}(x_{ipq}-\overline{x}_{pq})^2}\right)}{\sum_{q=1}^{Q} P_q} \times \frac{\sum_{j=1}^{J}\left(1 - \frac{\sum_{i=1}^{n} f_{ij}^2}{\sum_{i=1}^{n}(\xi_{ij}-\overline{\xi}_j)^2}\right)}{J}}. \tag{7.9}$$

Denoting the total number of manifest variables in the model as P, that is $P = \sum_{q=1}^{Q} P_q$, the GoF can be finally written as

$$GoF = \sqrt{\frac{1}{P}\sum_{q=1}^{Q}\sum_{p=1}^{P_q}\left(1 - \frac{\sum_{i=1}^{n} e_{ipq}^2}{\sum_{i=1}^{n}(x_{ipq}-\overline{x}_{pq})^2}\right) \times \frac{1}{J}\sum_{j=1}^{J}\left(1 - \frac{\sum_{i=1}^{n} f_{ij}^2}{\sum_{i=1}^{n}(\xi_{ij}-\overline{\xi}_j)^2}\right)}. \tag{7.10}$$

Then, if we now consider K latent classes including n_k observations each, instead of just one as in (7.10), we arrive at the definition of the GQI given by

$$GQI = \sqrt{\sum_{k=1}^{K}\frac{n_k}{n}\left[\frac{1}{P}\sum_{q=1}^{Q}\sum_{p=1}^{P_q}\left(1 - \frac{\sum_{i=1}^{n_k} e_{ipqk}^2}{\sum_{i=1}^{n_k}(x_{ipq}-\overline{x}_{pqk})^2}\right)\right] \times \sum_{k=1}^{K}\frac{n_k}{n}\left[\frac{1}{J}\sum_{j=1}^{J}\left(1 - \frac{\sum_{i=1}^{n_k} f_{ijk}^2}{\sum_{i=1}^{n_k}(\xi_{ijk}-\overline{\xi}_{jk})^2}\right)\right]}. \tag{7.11}$$

Note that when $K = 1$, that is when we consider a single group including all the observations, the GQI reduces to the GoF measure. This allows to compare the GQI with the GoF to assess the quality of the REBUS-PLS partition. In particular, simulation studies have shown (see Trinchera, 2007, Section 6.1.3) that a GQI which is 25% larger than the global GoF provides a reasonable threshold to use in practice for preferring the solution detected by the REBUS-PLS over the global model.

Permutation tests

In statistics it is common to use the sample data to compute a statistic of interest S which is then used for testing a certain null hypothesis H_0. Typically, values of S that are too extreme from the point of view of H_0 are considered as providing evidence against the null hypothesis, in favour of the alternative hypothesis. In this classical framework, the assessment of whether the value of S must be considered extreme or not is performed using the so called **null distribution** of S, that is the probability distribution of S under the assumption that H_0 is true. Unfortunately, in most practical situations the null distribution of S is not available analytically, because it is unknown or it is too difficult to be derived. In these cases **resampling methods** turn out to be very useful, because they allow to approximate the null distribution of a statistic of interest by repeatedly drawing subsets of the data from the original sample. Common resampling methods are bootstrapping, jackknifing and permutation tests (Good, 2006) and they differ in the way observations are drawn from the original sample. Here we briefly focus on permutation tests[14].

Permutation tests have been introduced by Fisher in the early 1930s as a general method of testing hypotheses based on permuting the data in ways that do not change the distribution under the null hypothesis (i.e., ways that are consistent with the null hypothesis of interest). Permutation tests do not require any standard parametric assumptions such as normality of the data.

Given that the general theory of permutation tests is far beyond the scope of this book, we only illustrate how these tests work when one is interested in checking whether two populations have the same mean value[15]. Clearly, a parametric t-test could be used here, but if the samples are small, the normality assumption may not be easy to defend. Suppose we have two samples $x_1, x_2, \ldots, x_m$ and $y_1, y_2, \ldots, y_n$ with corresponding sample means $\bar{x}$ and $\bar{y}$. For example, these could be measurements of a given variable for a sample of males and females. Suppose also that to test the null hypothesis $H_0 : \mu_X = \mu_Y$ we decide to use the test statistic

$$t = \frac{\bar{x} - \bar{y}}{\sqrt{SE_x^2 + SE_y^2}}$$

where SE_x and SE_y are the standard errors of the sample mean for the two samples.

Note that there is nothing special about this particular statistic, so that we could use any other statistics that measure a departure from the null hypothesis. We denote with t^* the statistic value corresponding to the observed data. Now the crucial point: if it is true that the population distributions are equal (i.e., they have the same mean, as well as everything else), it should not make any difference from which one of the two each observation has been taken. This is like saying that the $(m+n)$ observations come from a single population and the way they were divided into the two groups (e.g., males and females) was fundamentally random, with any other division being

[14] The bootstrap is briefly presented in Section 2.1.

[15] In this example we assume that the two populations have distributions that may differ only with respect to the mean, while all other features (spread, shape, etc.) are the same.

equally likely. Then, to test the null hypothesis of no mean difference, we compute the statistic value corresponding to *every possible* way to split the data into two samples of sizes m and n. That is, we *permute* the combined sample $(x_1,\dots,x_m,y_1,\dots,y_n)$ in all possible ways. These permutations are clearly consistent with the null hypothesis that there are no group differences. Therefore, the distribution of the test statistic values obtained in this way provides the (exact) null distribution. From an operational point of view, enumerating all the possible permutations is unfeasible unless the sample is small enough, and thus a sampling approach is taken. In our example this means that we randomly assign m observations of the combined sample to the first group and the remaining n to the second one, repeating the procedure many times (usually at least 1000). Finally, a **permutation p-value** is computed as the fraction of test statistic values that disagree with the null hypothesis. The way the permutation p-value is computed depends on the specific alternative hypothesis we choose. In particular, for the example on the mean difference:

- if H_0 is tested against $H_1 : \mu_X > \mu_Y$, the permutation p-value corresponds to the fraction of statistic values that are greater than or equal to the observed test statistic t^*,
- if H_0 is tested against $H_1 : \mu_X < \mu_Y$, the permutation p-value corresponds to the fraction of statistic values that are lower than or equal to the observed test statistic t^*,
- if H_0 is tested against the two-sided $H_1 : \mu_X \neq \mu_Y$, the permutation p-value is computed as 2 times the smaller between the one-sided p-values above (rounding the result to 1, if necessary).

Before concluding, we stress that the choice of what to permute depends on the specific problem you are studying. For example, if the observations in the example above are assumed to be paired instead of independent, then the permutations should be taken within the pairs, independently from pair to pair.

Permutation tests provide a general approach for hypothesis testing, but they cannot be used in all situations. In particular, it must be that certain invariance properties under the null hypothesis are satisfied (see for example Boos and Stefanski, 2013, Chapter 12).

Part III

Conclusions

8

How to Write Up a PLS-SEM Study

In this chapter we mention some of the most common academic publication types and how they optimally should be structured for publication. Further, by referring to an actual PLS-SEM journal article, we show how the method and results sections of a PLS-SEM paper should be detailed and reported. In doing so, we provide templates for figures and tables to represent respectively the research model of the study and numerical findings from its estimation. The suggested templates and guidelines can be deemed as the minimum requirements for reporting academic PLS-SEM work.

8.1 Publication Types and Structure

Structural equation modelling including PLS-SEM is becoming one of the most common statistical techniques used in academic publications from many different fields such as psychology, marketing, management and education. Academic publications come in a variety of forms and structures. One of these is represented by students' theses/dissertations. The emergence of readily available and intuitive software has made it possible for students at different levels (bachelor and master) to employ PLS-SEM to test relatively advanced models in their work. Secondly and more typically, we have got researchers (doctoral students, professors, research fellows, etc.) publishing papers/articles based on PLS-SEM in international outlets such as journals and conference proceedings. Finally, research reports which is generally of an applied nature written for business or policy makers may also include results based on PLS-SEM estimations.

Regardless of publication types, any study based on PLS-SEM would generally adopt the following macro structure despite some unsubstantial deviations:

1. Introduction
2. Theory
3. Method
4. Results
5. Discussion and conclusion

The *introduction* section, by referring to previous scholarly work, provides a rationale and justification for the concerned research. In so doing, the research question of the study is clearly stated. The *theory* section reviews the literature (existing theoretical and empirical studies) pertaining to the research question. This literature should be presented and synthesized in order to derive and justify the study's hypotheses making up the research model to test. The *method* section presents the data collection, research model and item measures used as well as the data analytic procedures (including PLS-SEM) employed in the study. The *results* section includes an assessment of the measurement model and subsequently an examination of the structural model testing the study's hypotheses. In the final section, *discussion and conclusion*, the results of the study are elaborated and discussed in order to generate some new theoretical insights/contributions as well as practical implications for the concerned stakeholders, but also to highlight its limitations.

8.2 Example of PLS-SEM Publication

In this section we are going to present and discuss an actual study using PLS-SEM published in a scientific journal. This review will allow the reader to see how in particular PLS-SEM as a statistical technique is justified and used and its results are presented. That is, our focus will be on the method and result sections of this publication leaving out the bits on introduction, theory and conclusions. The study that we will go through is about examining the relationship between personality (traits) and experiential consumption (activity preferences), more specifically the effect of the former on the latter. Incidentally, we suggest the reader to keep the original version of this study (Mehmetoglu, 2012) readily available as we will frequently refer to it in our review below.

In the method section of the article, the author first explains the details of the data collection stating mainly that the necessary data were collected through telephone interviews in April-May 2011 in Norway and that 1000 out of 7465 contacted individuals agreed to participate in the survey, yielding a modest response rate of nearly 13 percent.

Next, in the method section the author diagrammatically shows the overall research model (see the extract shown in Figure 8.1) that he has derived from the literature reflecting also the individual hypotheses. As SEM papers tend to be more complex, we strongly recommend that authors/researchers provide their readers with such a diagrammatic representation so that the target audience can easily see the different relationships between the study variables before reading the results of the PLS-SEM estimation.

Prior to moving on to the results section, one final task to do is justify the use of PLS-SEM in the method section. As you would remember, we provided some guidelines as to when and what version of PLS-SEM is most optimal to use for a given study at the outset in Chapter 1. Relatedly, in our example article the author explains

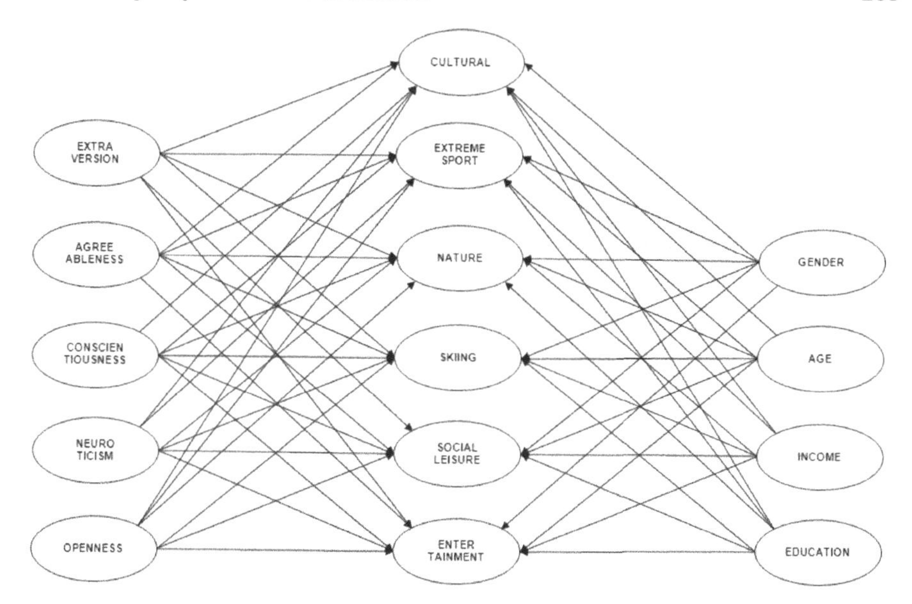

FIGURE 8.1: Research model of the PLS-SEM publication discussed in this chapter.

that PLS-SEM is chosen to estimate the study model because of non-convergence solutions encountered using CB-SEM. Although the author has used the traditional PLS-SEM, consistent PLS-SEM could have been used as it takes measurement error into account. The reason is that consistent PLS-SEM was not available at the time of publication.

In the results section of the article, the author adopts the commonly known two-stage approach used in standard covariance-based SEM to present the numerical findings (Chin, 2010): in the first stage, the focus is on the reliability and validity of the item measures used (measurement part) whereas the emphasis is put on the results from the estimations of the path coefficients (structural part) in the second stage. The first stage, where you make sure that your measures are representing the constructs of interest, is a prerequisite for interpreting the path coefficients in the second stage (ibid).

In the following extracts from our example article, we can observe how the author has adopted the two-stage approach to examine and present the results from the PLS-SEM estimation of the research model depicted in Figure 8.1. As regards the first stage, in line with the author's presentation, we propose an examination of *composite reliabilities*, *average variances extracted* (AVE), *item loadings' size* and *discriminant validity* as a minimum standard requirement for assessing a measurement model's adequacy in a PLS-SEM study (Figure 8.2).

In addition to the explanation in the extract shown in Figure 8.2, the author presents the necessary numbers in a table called measurement model (see the extract in Figure 8.3). This is a standard table that is generally expected to be included

4. Results

4.1. Reliability and validity

The research model included five latent variables representing the five personality traits and six latent variables expressing the experiential-activity preferences. Since the measurement model contained only reflective manifest variables, one should assess the measurement model on the basis of composite reliabilities, average variances extracted (AVE), item loadings' size, and discriminant validity (see for instance Liang, Saraf, Hu, & Xue, 2007). Incidentally, as far as composite reliability is concerned, Chin (1998) suggests that Dillon-Goldstein's rho (D.G. rho) is a better measurement of reliability because the D.G. rho, in contrast to Cronbach's alpha, does not assume tau equivalence.

As shown in Table 1, all of the standardised loadings were close to or above the suggested threshold of 0.7, AVE values exceeded the recommended level of 0.5, and all the D.G. rho values were above the suggested figure of 0.7 as well. These findings were indicative of reliability and convergent validity. Further, all of the average variance extracted values were larger than the squared correlations among the latent variables in the model, and thus demonstrated discriminant validity (see Table 2). As the measurement model exhibited evidence of reliability and validity, an assessment of the structural part (i.e. hypothesis testing) of the model could follow (Henseler, Ringle, & Sinkovics, 2009).

FIGURE 8.2: The results section of the PLS-SEM example publication discussed in this chapter (see Mehmetoglu, 2012).

in any PLS-SEM study. The author provides also a table showing the results of the discriminant validity analysis (see the extract in Figure 8.4).

When it comes to the second stage, the author provides quantitative evidence for testing the study's hypotheses. That is, *sign*, *size* and *significance* of the relevant path coefficients are examined. Based on this examination, the author concludes whether the suggested hypotheses are supported or not as illustrated in the extract shown in Figure 8.5.

As with the measurement model, the author provides the numerical findings related to the structural part of the PLS-SEM estimation (see the extract in Figure 8.6) in table format too. As an alternative to this table format presentation, one could show the results of the structural part (path coefficients and p-values) on the conceptual model itself (see Figure 8.1) presented in the method section.

Table 1. The measurement model

Latent variable / Manifest variables	Loadings	D.G. Rho	AVE
Extraversion		0.818	0.586
I am talkative	0.905		
I tend to be quiet (reversed)	0.610		
I am outgoing and sociable	0.752		
Agreeableness		0.775	0.526
I am helpful and unselfish with others	0.727		
I have a forgiving nature	0.623		
I am generally trusting	0.814		
Conscientiousness		0.787	0.554
I do a thorough job	0.760		
I am a reliable worker	0.747		
I make plans and follow through with them	0.726		
Neuroticism		0.815	0.587
I can be tense	0.727		
I worry a lot	0.862		
I get nervous easily	0.701		
Openness		0.846	0.637
I am original and come up with new ideas	0.876		
I am inventive	0.744		
I like to reflect and play with ideas	0.768		
Cultural		0.854	0.662
Opera	0.740		
Art exhibitions	0.906		
Museums/galleries	0.784		
Extreme sport		0.948	0.860
Base-jumping	0.925		
Paragliding	0.938		
Parachuting	0.920		
Nature		0.814	0.589
Fishing	0.803		
Hunting	0.751		
Mountain tour	0.747		
Skiing		0.875	0.696
Slalom/alpine skiing	0.857		
Skiing vacation	0.900		
Cross-country skiing	0.736		
Social leisure		0.822	0.599
Stand-up shows	0.808		
Dining out	0.701		
Clubbing	0.809		
Entertainment		0.870	0.676
Amusement parks	0.903		
Theme parks	0.855		
Circus	0.693		

FIGURE 8.3: The results section (cont.) of the PLS-SEM example publication discussed in this chapter (see Mehmetoglu, 2012).

Table 2. Discriminant validity (Squared correlations < AVE)

	EXTRA-VERSION	AGREE-ABLENESS	CONSIEN-TIOUSNESS	NEURO-TICISM	OPENNESS	CULTURAL	EXTREME SPORTS	NATURE	SKIING	SOCIAL LEISURE	ENTER-TAINMENT
EXTRAVERSION	1										
AGREEABLENESS	0.085	1									
CONSIENTIOUSNESS	0.086	0.172	1								
NEUROTICISM	0.006	0.004	0.000	1							
OPENNESS	0.079	0.091	0.062	0.002	1						
GENDER	0.027	0.007	0.023	0.025	0.011						
AGE	0.029	0.017	0.001	0.002	0.056						
INCOME	0.015	0.002	0.009	0.000	0.007						
EDUCATION	0.005	0.001	0.029	0.006	0.014						
CULTURAL	0.003	0.002	0.010	0.012	0.042	1					
EXTREME SPORTS	0.007	0.005	0.004	0.001	0.038	0.000	1				
NATURE	0.002	0.027	0.025	0.003	0.034	0.000	0.039	1			
SKIING	0.027	0.029	0.020	0.001	0.027	0.001	0.085	0.141	1		
SOCIAL LEISURE	0.059	0.017	0.005	0.000	0.049	0.000	0.089	0.028	0.144	1	
ENTERTAINMENT	0.005	0.008	0.000	0.010	0.029	0.007	0.116	0.027	0.073	0.241	1
AVE	0.586	0.526	0.554	0.587	0.637	0.662	0.860	0.589	0.696	0.599	0.676

FIGURE 8.4: The results section (cont.) of the PLS-SEM example publication discussed in this chapter (see Mehmetoglu, 2012).

4.2. Hypothesis testing

The net effects of personality traits on experiential-activity preferences are shown in table 3. Openness has a significant effect on a preference for cultural and extreme-sports activities, a finding which supports hypothesis 1. The findings show additionally that open individuals are likely to have a preference for traditional nature activities and entertainment as well. Further, neuroticism is significantly associated with a preference for non-risky activities such as cultural activities and entertainment. However, neuroticism does not have the expected effect on a preference for social-leisure activities. Thus, hypothesis 2 is partially supported. As expected, the more extravert a person is, the more likely that person is to have a preference for social-leisure activities. Hypothesis 3 is thus supported. Moreover, conscientiousness has a significant negative effect on a preference for unconventional activities such as extreme sports, a finding which provides support for hypothesis 4. Incidentally, it seems that conscientious people are likely to prefer traditional nature activities as well. Finally, the findings show that agreeableness is significantly related to a preference for traditional nature activities that require cooperation (e.g., hunting), a result that supports hypothesis 5. Furthermore, people high in agreeableness do seem to have a preference for skiing activities too.

FIGURE 8.5: The results section (cont.) of the PLS-SEM example publication discussed in this chapter (see Mehmetoglu, 2012).

Table 3. The structural model (with standardised coefficients and standard errors in parentheses)

Endogenous	Cultural activities		Extreme sport activities		Traditional nature activities		Skiing activities		Social-leisure activities		Entertainment activities	
Exogenous	β	SE	β	SE	β	SE	β	SE	β	SE	β	SE
Extraversion	-0.002	0.031	0.039	0.032	-0.017	0.033	0.056	0.032	0.145***	0.029	0.002	0.032
Agreeableness	-0.011	0.031	0.036	0.033	0.102**	0.034	0.092**	0.033	0.014	0.030	0.030	0.033
Conscientiousness	-0.027	0.031	-0.100**	0.033	0.135***	0.034	0.045	0.033	-0.015	0.030	-0.035	0.033
Neuroticism	0.100***	0.028	0.040	0.029	-0.029	0.031	-0.021	0.030	0.001	0.027	0.084**	0.030
Openness	0.271***	0.031	0.095**	0.032	0.080*	0.034	0.008	0.033	0.039	0.029	0.077*	0.032
Gender (Male)	0.189***	0.029	-0.140***	0.030	-0.256***	0.031	-0.058	0.031	-0.055*	0.027	-0.050	0.030
Age	0.310***	0.030	-0.364***	0.031	-0.076*	0.032	-0.274	0.031	-0.521***	0.028	-0.383***	0.031
Household income	0.020	0.029	-0.006	0.030	-0.005	0.031	0.104	0.030	0.003	0.027	-0.008	0.030
Educational level	0.261***	0.028	-0.003	0.030	0.014	0.031	0.179***	0.030	0.103***	0.027	-0.047	0.030
R^2	0.26		0.20		0.13		0.17		0.34		0.19	

* p < .05; ** p < .01; *** p < .001

FIGURE 8.6: The results section (cont.) of the PLS-SEM example publication discussed in this chapter (see Mehmetoglu, 2012).

8.3 Summary

In this chapter we have shown how one can structure and write up a scientific publication from a research including a model estimated using PLS-SEM. What we have presented is a minimum requirement for what to be included in an academic PLS-SEM publication. Nevertheless, additional information or different ways of presenting the results may readily be incorporated into this template. Furthermore, depending on the publication format (journal paper, dissertation, etc.) the method and results sections can be adjusted and/or extended. Finally, we encourage the reader to search for PLS-SEM publications in relevant outlets to see some more examples of reporting styles.

Part IV

Appendices

A

Basic Statistics Prerequisites

In this appendix we present some basic prerequisites for understanding the partial least squares approach to structural equation modelling. More specifically, after reviewing the definitions of covariance and linear correlation, we briefly recall the main elements of the linear regression model, which represents the fundamental building block of PLS-SEM. During the presentation we will show a number of examples using both real and simulated data. These examples will be illustrated using Stata, but in the final section we also briefly discuss how to perform the same analyses using `R`.

A.1 Covariance and Correlation

The simplest, but also most useful, exploratory analyses for assessing the linear association between two numerical variables are the covariance and correlation analyses. Covariance analysis should not be confused with "analysis of covariance" (ANCOVA), which is a variant of the analysis of variance (ANOVA) method that also takes into account the effects of other continuous variables that are not of primary interest (so called *control variables* or *covariates*).

The *sample covariance* for a pair of observed variables X and Y is defined as the average of the cross product deviations of the variables from their sample means, that is

$$\mathrm{Cov}(X,Y) = \frac{1}{n-1}\sum_{i=1}^{n}(x_i - \bar{x})(y_i - \bar{y}), \tag{A.1}$$

where n is the sample size, x_i and y_i the observed values of the two variables for the ith observation in the sample and $\bar{x}$ and $\bar{y}$ denote the corresponding sample means[1].

Practically, the covariance provides an assessment of the co-movement of the two variables. In particular:

- If the covariance is positive, it indicates that the two variables move in the *same direction* (on average); this means that when one of the two variables increases (decreases), the other one increases (decreases) as well.

[1] We recall that the unintuitive denominator $(n-1)$ has the role to make the sample covariance an unbiased estimator of the unknown population covariance.

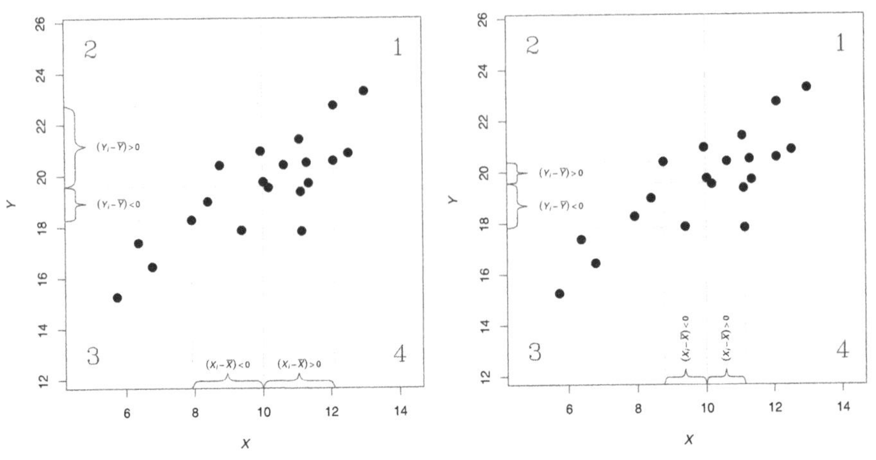

(a) Highlighted observations are from areas 1 and 3, where the deviation products are positive.

(b) Highlighted observations are from areas 2 and 4, where the deviation products are negative.

FIGURE A.1: Graphical interpretation of the sample covariance between two observed variables X and Y (simulated data).

- If the covariance is negative, this is an indication that the two variables move in *opposite direction* (on average); this instead means that when one of the two variables increases (decreases), the other one tends to decrease (increase).

The diagrams in Figure A.1 make this interpretation more explicit. These pictures show that the sample means of X and Y allow to split the graphs in four areas, that we label 1, 2, 3 and 4 respectively proceeding counterclockwise from top right. Points placed in area 1 have a positive deviation product, because both the deviations are positive, so that they contribute by increasing the covariance. The same result occurs for points in area 3, since in this region the deviations are both negative. For points placed in areas 2 and 4 we get a negative product and contribution to the covariance value instead, because in these areas one of the deviation is positive and the other one is negative. In substance, since the covariance is an average of these products, its value is positive whenever most of the points fall in areas 1 and 3, while it takes a negative value if the majority of the points fall in the other two areas[2].

The drawback of the covariance as an index of linear association between two variables is that its value depends upon the scales of the variables. For example, suppose that x_i and y_i represent the annual income and the amount spent for purchasing a new car for an individual i, which are both measured in dollars. Suppose we collect these quantities for a sample of individuals and the corresponding sample covariance is equal to 230000000. Suppose now that we rescale the variables expressing all the

[2]Clearly, the covariance considers also the magnitude of the deviations and not only how many of them are positive and negative.

values in thousands of dollars (i.e., we divide the values x_i and y_i by 1000). If we recompute the covariance we would get the value 230. Clearly, the observations didn't change, and the same is true for their association. Therefore, the magnitude of the covariance value is not useful to assess the strength of the association between two variables but only whether they are positively or negatively associated.

The information on the strength of the linear association between two variables is provided by the *sample linear correlation index*, which essentially removes from the covariance the effect of the scales. In particular, the linear correlation index for two variables X and Y is defined as

$$r_{XY} = \text{Cor}(X,Y) = \frac{\text{Cov}(X,Y)}{s_X \cdot s_Y}, \tag{A.2}$$

where s_X and s_Y denote the sample standard deviations of X and Y respectively. Intuitively, the fact that the covariance is divided by a number on the same scale makes the linear correlation index a more intuitive and useful measure for assessing the extent of the linear association between two variables. Indeed, the linear correlation index takes values in the range $[-1,1]$. In particular, the closer the index to either one of the extremes, the stronger the linear association (either negative or positive) between the variables. On the other side, the closer the index to zero, the weaker the association. We remind that a strong (weak) linear association means that we can (cannot) accurately predict the values of either one of the variables through a linear function of the other. Figure A.2 shows some examples corresponding to different situations one may encounter in practice.

We provide now a last remark on the relationship between the covariance and the linear correlation index: given two variables X and Y with sample means $\bar{x}$ and $\bar{y}$ and sample standard deviations s_X and s_Y, the linear correlation index between X and Y is equal to the covariance between their standardized versions[3], that is

$$r_{XY} = \text{Cov}\left(\frac{X-\overline{X}}{s_X}, \frac{Y-\overline{Y}}{s_Y}\right).$$

This property provides a particular case of the more general rule according to which the linear correlation index does not change if the variables are rescaled or shifted. This rule is frequently used in the PLS-SEM estimation algorithm we present in this book.

As a practical example, we use one of the datasets shipped with Stata and contained in the file `nlsw88.dta`. These data are an extract of the 1988 National Longitudinal Survey of Young Woman who were ages 14–24 in 1968 (NLSW). To load the data we run the command

[3] We remind that standardizing a variable X means subtracting its mean and dividing by the standard deviation, that is, $\widetilde{X} = (X - \bar{x})/s_X$. As a consequence, the standardized version of the variable, $\widetilde{X}$, has mean equal to zero and standard deviation equal to one. More technically, the standardization is a linear transformation of a variable that allows to center it around zero and to scale it to have unit standard deviation.

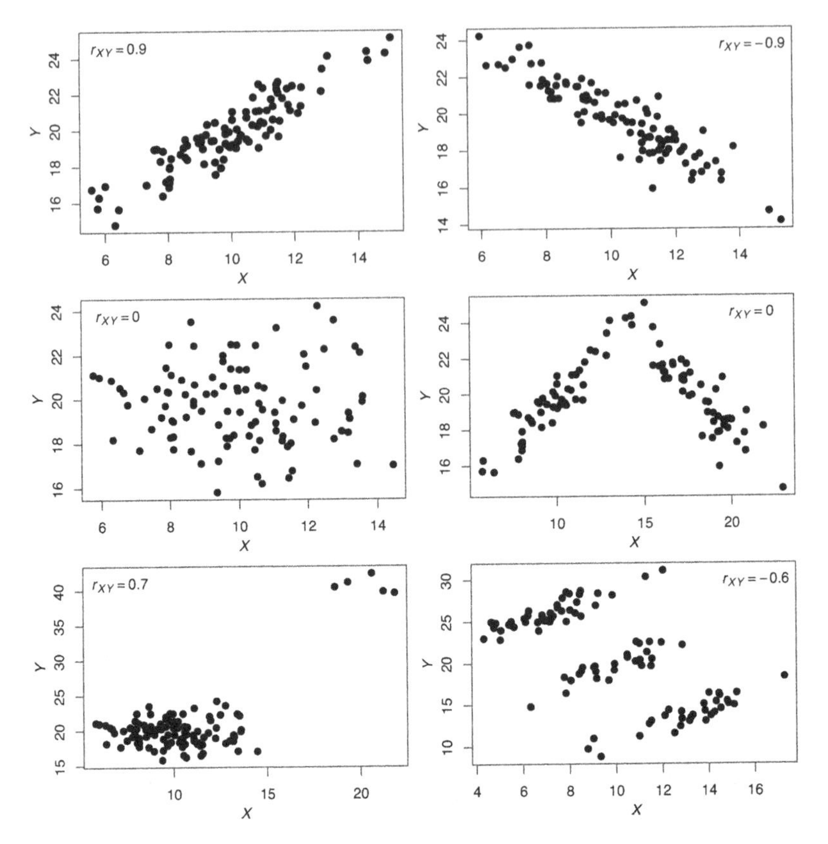

FIGURE A.2: Some examples of linear association between two variables corresponding to different extent of association (simulated data).

```
1 sysuse nlsw88, clear
```

Here, we focus on the following variables:

- `age`, that provides the respondents age in years,
- `wage`, which gives the hourly wage,
- `hours`, that contains the usual hours worked during weeks worked,
- `ttl_exp`, which provides the total work experience.

First, the following code computes some summary statistics (Figure A.3) and visualize the data with a scatter plot matrix (Figure A.4):

```
   stats |       age     hours   ttl_exp      wage
---------+----------------------------------------
N        |      2246      2242      2246      2246
mean     |  39.15316  37.21811  12.53498  7.766949
p50      |        39        40    13.125   6.27227
p25      |        36        35  9.211538  4.259257
p75      |        42        40  15.98077  9.597424
sd       |  3.060002  10.50914  4.610208  5.755523
min      |        34         1  .1153846  1.004952
max      |        46        80  28.88461  40.74659
--------------------------------------------------
```

FIGURE A.3: Summary statistics for some variables included in the `nlsw88` dataset in Stata.

```
1  tabstat age hours ttl_exp wage, ///
2          statistics(count mean median p25 p75 sd min max)
3  graph matrix age hours ttl_exp wage, half msize(small)
```

The results show that there are only four observations missing in the `hours` column and the `wage` distribution appears to be somewhat right skewed since the mean is larger than the median (reported as `p50`). Moreover, there is a fairly pronounced variability among the observed wages because the standard deviation is large compared to the mean. The scatter plots show a moderate to weak positive association among the different pairs of variables. The graphs in the last row of the picture, those reporting `wage` on the vertical axis, clearly confirm the right skewness of the wage distribution, which is mainly due to the presence of some extreme values in the upper part of the plots. A common approach in statistics to down-weight the influence of outliers and reduce the distribution asymmetry is to transform the variables with a logarithm. So, we compute a new variable, `logwage`, defined as the natural logarithm of `wage`, and we produce again the scatter plot matrix using this new variable (Figure A.5). This operation confirms the positive association among the variables:

```
1  generate logwage = log(wage)
2  label variable logwage "logarithm of hourly wage"
3  graph matrix age hours ttl_exp logwage, half ///
4          msize(small)
```

Finally, we compute the linear correlation indexes for the same variables (we directly use `logwage`). In Stata there are two commands to compute correlations, `correlate` and `pwcorr`. The difference between them is that the former provides the correlation (as well as the covariance) matrix for a set of variables while the latter computes the pairwise correlations. Practically this means that `correlate` disregards the rows containing at least one missing value for any of the variables in the set (so called *casewise* or *listwise deletion*), while `pwcorr` proceeds removing

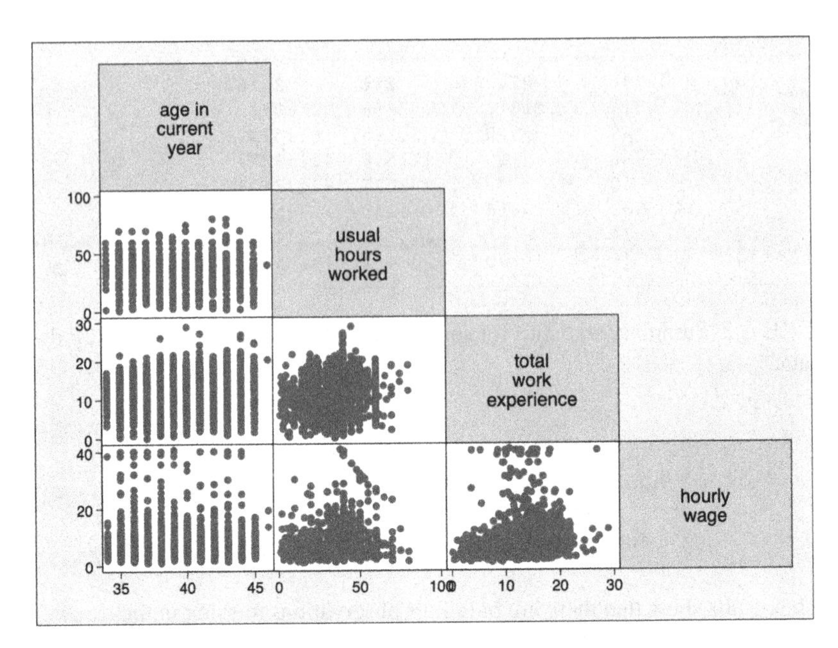

FIGURE A.4: Scatter plot matrix for some variables included in the `nlsw88` dataset in Stata.

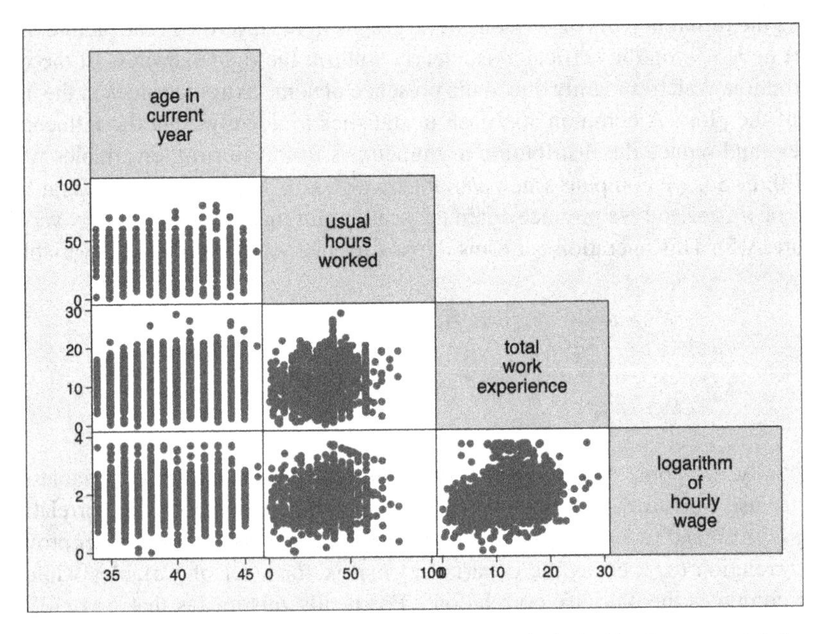

FIGURE A.5: Scatter plot matrix for some variables included in the `nlsw88` dataset in Stata.

```
. correlate age hours ttl_exp logwage
(obs=2,242)

             |      age    hours  ttl_exp  logwage
-------------+------------------------------------
         age |   1.0000
       hours |  -0.0279   1.0000
     ttl_exp |   0.1235   0.2295   1.0000
     logwage |  -0.0214   0.2049   0.3848   1.0000


. pwcorr age hours ttl_exp logwage, obs sig

             |      age    hours  ttl_exp  logwage
-------------+------------------------------------
         age |   1.0000
             |
             |     2246
             |
       hours |  -0.0279   1.0000
             |   0.1874
             |     2242     2242
             |
     ttl_exp |   0.1243   0.2295   1.0000
             |   0.0000   0.0000
             |     2246     2242     2246
             |
     logwage |  -0.0223   0.2049   0.3851   1.0000
             |   0.2909   0.0000   0.0000
             |     2246     2242     2246     2246
             |
```

FIGURE A.6: Correlation matrix and pairwise correlations for some variables included in the `nlsw88` dataset in Stata.

the missing values separately for each pair of variables[4]. Despite this difference, `pwcorr` is often more appealing than `correlate` because it allows to assess the statistical significance of the correlation indexes by computing the corresponding p-values. These are shown in the output by adding the option `sig`. The following code uses both commands to compute the correlations, which are reported in Figure A.6:

```
1 correlate age hours ttl_exp logwage
2 pwcorr age hours ttl_exp logwage, obs sig
```

The correlation indexes confirm that `logwage` is positively and significantly associated with `hours` and `ttl_exp`, while it doesn't seem to be significantly correlated with `age`. Overall, the correlations among these variables are at most moderate.

[4]We warn you that listwise deletion may dramatically reduce the effective sample size used in the calculations, especially when the missing values are highly sparse in the dataset.

A.2 Linear Regression Analysis

Linear regression is by far the most popular statistical technique applied in all fields. Since it is a vast topic, here we focus only on the main concepts we need for illustrating partial least squares path modelling. Book-length presentations of this important subject can be found for example in Kutner et al. (2005); Fahrmeir et al. (2013); Weisberg (2013) or in Fox (2016), while Mehmetoglu and Jakobsen (2016) provide a more applied perspective using Stata.

A.2.1 The simple linear regression model

Given two variables X and Y, the *simple linear regression* (SLR) *model* for the ith observation is defined as

$$y_i = \mathrm{E}(Y_i|X_i = x_i) + \varepsilon_i = \beta_0 + \beta_1 x_i + \varepsilon_i, \qquad \text{(A.3)}$$

where β_0 and β_1 represent the model's intercept and slope and ε_i denotes the unobserved error term for the same observation. Y is usually called the dependent, outcome or response variable, while X is typically referred to as the independent, explanatory or predictor variable.

The first equality in (A.3) indicates that a regression model is a way to specify how the average value of Y changes according to the value of X. More technically, the notation $\mathrm{E}(Y_i|X_i = x_i)$ represents the so called *conditional expectation* of Y given the particular value x_i taken by X[5]. In other words, SLR enriches the information provided by the linear correlation index by specifying how the value of Y can be predicted once the value of X is given. The error term ε_i represents the information about y_i that can't be accounted for by x_i, and it is assumed to be a random quantity with mean equal to zero and variance equal to a constant[6] quantity σ^2. Finally, the error terms are supposed to be uncorrelated among themselves and with the predictor X[7].

Equation (A.3) is usually referred to as the *population* regression model in the sense that it specifies the relationship that we assume is linking the average value of Y to x in the entire population of interest. For this reason, the values of the model's coefficients β_0 and β_1 (as well as those of the ε_is) are not known and need to be estimated. We denote the sample estimates of β_0 and β_1 as b_0 and b_1 respectively. The classical estimation method used in regression analysis, that sometimes is identified with the model itself, is **ordinary least squares** (OLS). To define OLS we need first

[5] More precisely, the "linearity" referred to in the name of the method refers to the model's coefficients β_0 and β_1 and not to the independent variable X. For example, the specification $y_i = \beta_0 + \beta_1 \log(x_i) + \varepsilon_i$ also is a SLR model even if the logarithm of X is used on the right hand side. An example of a non-linear regression model is $y_i = \frac{\beta_1 x_i}{\beta_0 + x_i} + \varepsilon_i$.

[6] The assumption of constant error variance is called *homoskedasticity*.

[7] The assumption that the errors and the predictor are uncorrelated is called *exogeneity*. In particular, if it holds we say that X is exogenous.

to introduce the concepts of **predictions** and **residuals**. The prediction for the ith observation from the SLR (A.3) is defined as the quantity $\widehat{y}_i$ corresponding to the predictor value x_i and computed using the coefficient estimates b_0 and b_1, that is

$$\widehat{y}_i = b_0 + b_1 x_i. \tag{A.4}$$

Equation (A.4) provides the so called *sample* (or *estimated*) regression model because it corresponds to the fitted line computed using only the sample observations. Figure A.7 provides a comparison between the population and the sample models using some fictitious data. Observations (x_i, y_i) in the hypothetical population and the corresponding model are shown as grey dots and dashed line respectively, while the sample quantities are shown using black dots and solid line. If the sample is truly random, the estimated line will be close to the unknown population line, while if the sample is biased, the difference between them can be conspicuous.

Residuals are defined as the differences between observed and predicted values for the response variable, that is for the ith observation

$$\begin{aligned} e_i &= y_i - \widehat{y}_i \qquad \text{(A.5)} \\ &= y_i - (b_0 + b_1 x_i). \end{aligned}$$

Residuals correspond to the errors made by the model in predicting the value of the response variable for a given predictor value. They are positive or negative depending on whether the model under- or over-estimates respectively the observed response value. Geometrically, residuals correspond to the distances between the points vertical coordinates and the fitted line. Figure A.8 shows a graphical representation of both predictions and residuals for a generic SLR model.

OLS is an estimation method that computes the model's coefficients by minimizing the **residual sum of squares** (RSS)[8]. More technically, RSS is defined as

$$\text{RSS} = \sum_{i=1}^{n} e_i^2 = \sum_{i=1}^{n} (y_i - \widehat{y}_i)^2, \tag{A.6}$$

which clearly depends upon b_0 and b_1 because these are embedded inside the predicted values $\widehat{y}_i$. With a little algebra it can be shown that the OLS coefficient estimates correspond to

$$\begin{aligned} b_0 &= \bar{y} - b_1 \bar{x} \qquad \text{(A.7)} \\ b_1 &= \frac{\text{Cov}(X,Y)}{s_X^2} \\ &= r_{XY} \frac{s_Y}{s_X}. \qquad \text{(A.8)} \end{aligned}$$

Equation (A.8) clearly shows the direct connection between the linear correlation index r_{XY} and the slope estimate b_1. In particular, b_1 is nothing else than the correlation between X and Y that also takes into account the scale of the variables. This implies

[8] RSS is also known as sum of squared errors of prediction (SSE).

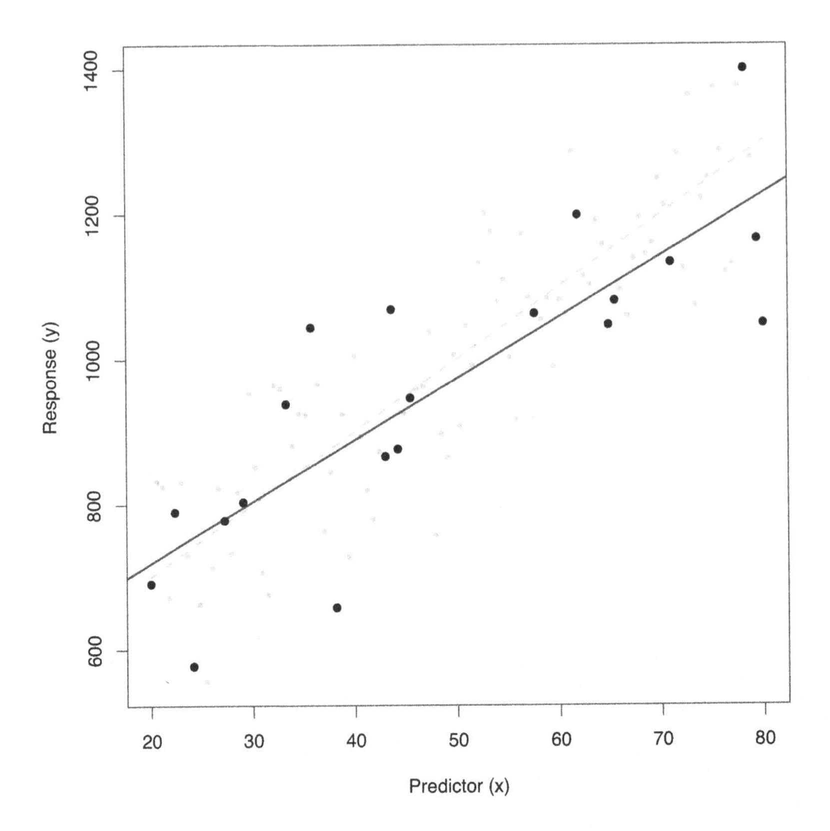

FIGURE A.7: Comparison between the population and sample regression models using simulated data. Grey dots represent the hypothetical observations in the population, the dashed grey line indicates the corresponding population model, black dots provide a random sample drawn from the population and the black solid line provides the corresponding sample regression model.

that when we perform linear regression between variables that have been standardized, the slope estimate b_1 will correspond to the correlation index r_{XY}, while the intercept estimate b_0 will be equal to zero. Moreover, both equation (A.7) and (A.8) also highlight that exchanging the role of X and Y produces different results for b_0 and b_1[9]. In this sense we can say that linear regression is not symmetric, while the correlation index is.

[9]Note that this is true unless the two variables are standardized, in which case exchanging their roles does not produce any change in the results. This is a confirmation that data alone are not able to tell us anything about the causal direction between the variables, but they are able only to provide us with an assessment of their "association". To infer any causal implications we need to postulate a casual model. For a clear and instructive discussion on the differences between association and causation we suggest to read the first chapter of Pearl and Mackenzie (2018).

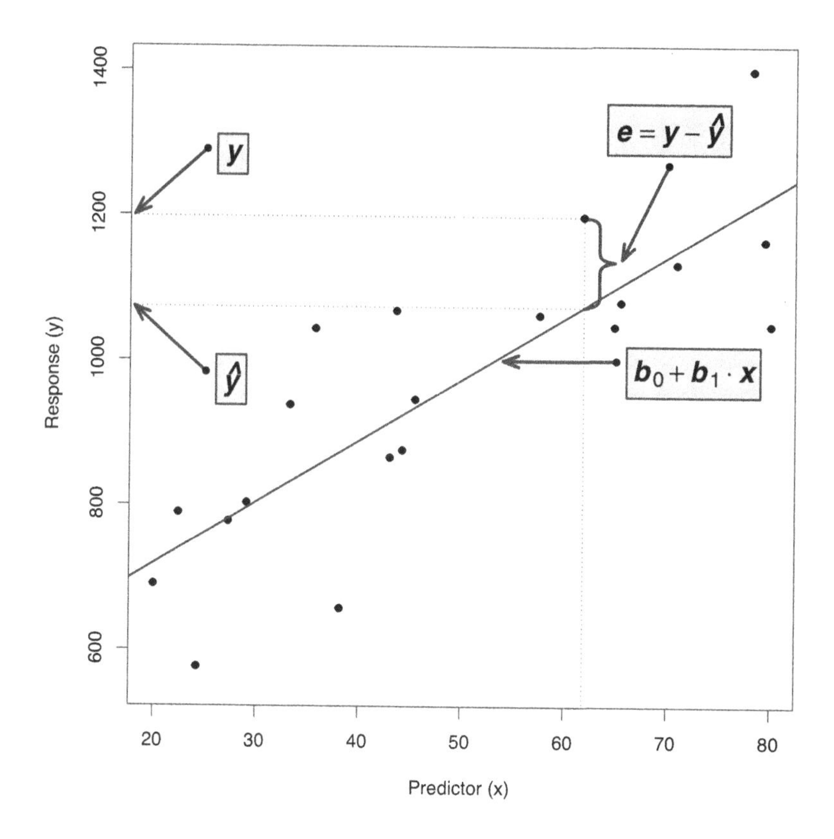

FIGURE A.8: Graphical representation of key quantities for a simple linear regression model.

Practically, estimates (A.7) and (A.8) can be interpreted as follows:

- the intercept estimate, b_0, provides the estimated mean value of the response Y when the predictor X is set to zero,
- the slope estimate, b_1, provides the change in the response estimated mean assuming the predictor value increases by one unit.

A.2.2 Goodness-of-fit

A popular way to assess the predictive ability of a linear regression model, or its goodness-of-fit, is through the **R-squared** index. To define it we need to introduce the so called deviance decomposition. The deviance of a linear regression model, also known as **total sum of squares** (TSS), corresponds to the sum of squared deviations of the response values from their mean, that is

$$\text{TSS} = \sum_{i=1}^{n} (y_i - \bar{y})^2. \tag{A.9}$$

It can be shown that the deviance of any linear regression model can be decomposed as follows

$$\text{TSS} = \text{ESS} + \text{RSS}, \tag{A.10}$$

where ESS denotes the **explained sum of squares**[10] defined as

$$\text{ESS} = \sum_{i=1}^{n} (\widehat{y}_i - \bar{y})^2, \tag{A.11}$$

and RSS is given in (A.6). Even if we do not give a formal proof of the decomposition in (A.10), an intuitive graphical justification is provided in Figure A.9.

The deviance decomposition (A.10) can be interpreted saying that only a portion of the total observed variability in the response (TSS) can be accounted for by the linear regression model (ESS), that is by the specific predictor we decide to use, while the remaining portion is left unexplained (RSS). The R-squared index provides this same information in relative terms, that is

$$R^2 = \frac{\text{ESS}}{\text{TSS}} = 1 - \frac{\text{RSS}}{\text{TSS}}. \tag{A.12}$$

Hence, R-squared takes values in the range $[0,1]$. It is close to 0 every time that the predictor is not able to explain a lot of the response variability, while it is close to 1 whenever the predictor almost perfectly predicts the response.

There is no universal threshold that we can use to assess the predictive ability of a given linear regression model. The realized value of R-squared depends clearly on the choice we made on the predictor to use in the model. Moreover, different fields may require a different level for the R-squared to qualify the model as good. Finally, the assessment of the R-squared value also depends on the intentions of the analysis. In particular, if the aim of the analysis is to provide predictions for future events, clearly we would like to base the predictions on a relatively good and robust model, while if the aim is purely descriptive, then even smaller R-squared values would be deemed as acceptable.

A.2.3 The multiple linear regression model

In practice it is unlikely that a single measure is able to explain most or all of the variation observed in the response. So, to get more accurate predictions or a better explanation of the phenomenon under investigation, the standard approach is to enrich the regression model with additional predictors. This approach gives rise to what is usually known as the *multiple linear regression* (MLR) *model*, which is defined as

$$\begin{aligned} y_i &= \text{E}(Y_i | X_{1i} = x_{1i}, X_{2i} = x_{2i}, \ldots, X_{pi} = x_{pi}) + \varepsilon_i \\ &= \beta_0 + \beta_1 x_{1i} + \beta_2 x_{2i} + \cdots + \beta_p x_{pi} + \varepsilon_i, \end{aligned} \tag{A.13}$$

[10]ESS is alternatively known as sum of squares due to regression (SSR).

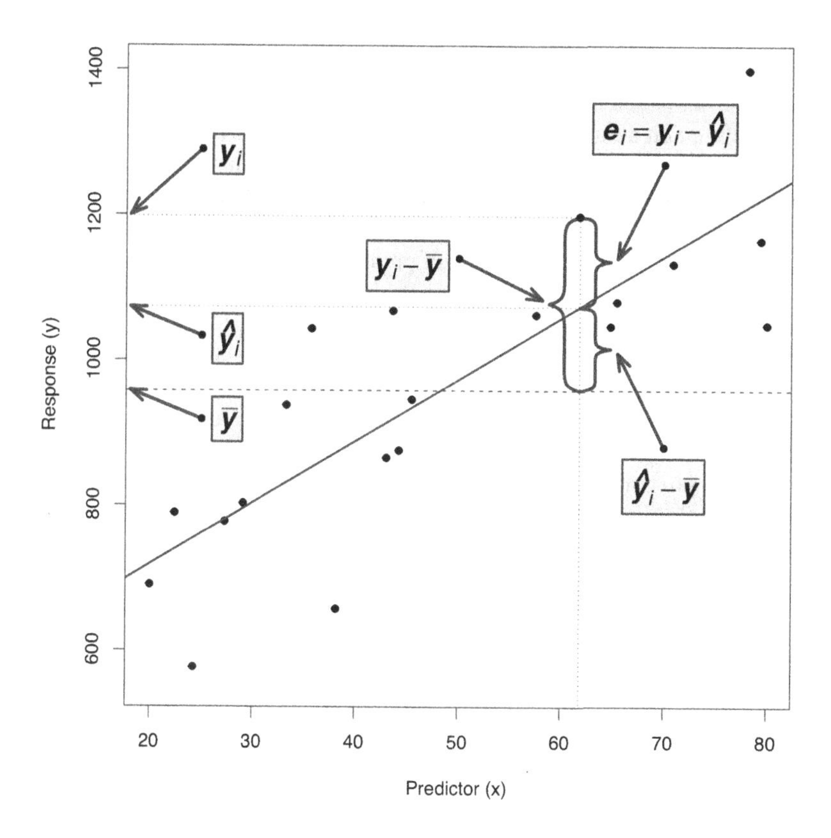

FIGURE A.9: Graphical representation of the deviance decomposition for a linear regression model.

where β_0 denotes the model's intercept, $\beta_1, \ldots, \beta_p$ correspond to the *partial* slopes with respect to each one of the p predictors $X_1, X_2, \ldots, X_p$ and ε_i still indicates the error term for the ith observation. All the concepts we introduced for SLR applies exactly in the same way in MLR. Similarly, estimation is still performed using OLS. Predictions are now computed as

$$\hat{y}_i = b_0 + b_1 x_{1i} + b_2 x_{2i} + \cdots + b_p x_{pi}. \tag{A.14}$$

However, the coefficient estimates $b_0, b_1, \ldots, b_p$ have a slightly different interpretation:

- the intercept estimate b_0 provides the estimated mean value of the response Y when the predictors $X_1, \ldots, X_p$ are jointly set to zero,
- the generic partial slope estimate b_j, with $j = 1, \ldots, p$, provides the change in the response estimated mean assuming the value of the X_j predictor increases by one unit, assuming the other predictors stay fixed.

The last part of the interpretation for the partial slope estimates is necessary to guarantee that b_j actually measures the effect of X_j on the average value of Y and it is not due to the combined effect of more than one predictor. Clearly, this statement is based on the implicit assumption that the predictors are uncorrelated of one another, which rarely applies in practice (see the discussion in Section A.2.6).

An important difference between MLR and SLR consists in how to compare the goodness-of-fit of different regression models (i.e., regression models for the same response variable but with a different set of predictors). Unfortunately, the R-squared index does not provide a fair comparison in these cases and therefore it shouldn't be used. The reason is that the R-squared numerator, ESS, is a quantity that does increase if new predictors are added to the model. This means that if we use the R-squared to select the best model, we would end up choosing the model with the largest number of predictors. A large model is not necessarily a good model, because some predictors may explain just a negligible portion of the response variability at the cost of complicating the model excessively. To correct for this drawback of R-squared, it is common to introduce in the assessment a penalization for the number of predictors included in the model. The most popular measure that goes in this direction is the **adjusted R-squared** index, which is defined as

$$R_a^2 = 1 - \frac{\frac{\text{RSS}}{n-p-1}}{\frac{\text{TSS}}{n-1}} = 1 - \frac{\text{RSS}}{\text{TSS}} \cdot \frac{n-1}{n-p-1} = 1-(1-R^2)\frac{n-1}{n-p-1}, \tag{A.15}$$

with p denoting the number of predictors in the model. Typically, the adjusted R-squared is smaller than the ordinary R-squared, even if the two differ usually by a small amount. So, the larger the adjusted R-squared, the better the model in the sense that it provides a better compromise between goodness-of-fit and complexity. We conclude this section by highlighting that in some cases the adjusted R-squared may become negative. This is typically the case when the sample size n is small. For example, assuming $n = 20$ and the R-squared index is 0.1, the adjusted R-squared is positive only when we use a single predictor model, while it becomes negative for a number of predictors equal to 2 or larger. Nevertheless, even when it is negative, we can still use it for comparing different models.

A.2.4 Inference for the linear regression model

As we already said, there is a difference between the model we postulate in the population and its sample version. The difference is due to the fact that the latter is computed using only a small subset of units from the population. Therefore, it is critically important to assess the reliability of the sample estimates as a guess of the unknown population quantities of interest. The classical approach to inference for linear regression is based on the assumption that the error terms ε_i are normally distributed. The normality assumption simplifies the calculations and allows to derive closed-form expressions for both confidence intervals and p-values. However, in few cases in practice the errors (and thus the response) can be assumed to be distributed as a normal. As a consequence, a more general approach, called the *bootstrap*, should be used. In this section we review the main results for the normal-based theory, while we already provided an introduction to the bootstrap approach in Section 2.1.

A.2.4.1 Normal-based inference

The normal-based inferential approach for the linear regression model requires the additional assumption that the error terms are normally distributed in the population with mean zero and fixed variance σ^2. More formally, we must assume that

$$\varepsilon_i \sim N(0, \sigma^2), \tag{A.16}$$

for $i = 1, \ldots, n$. As we already said, the error variance σ^2 is assumed to be a fixed unknown constant. The assumption of normally distributed errors translates in a similar assumption for the response variable, that is (A.16) is equivalent to

$$Y_i \sim N(\beta_0 + \beta_1 x_{1i} + \cdots + \beta_p x_{pi}, \sigma^2).$$

A graphical representation of this assumption in the single predictor setting is shown in Figure A.10, where the vertical axis reports the conditional probability of observing the response for three hypothetical x values on the X axis.

It is responsibility of the analyst to check that the normality assumption is satisfied, at least approximately, by the sample data. Many tools have been developed to verify this assumption, but the most popular are for sure the *normal probability plot* (also known as the normal quantile plot) and the *normality tests*. The former is a scatter plot where the ordered observed values of the response are plotted against the corresponding values of the normal distribution. If the response empirical distribution is close to a normal, then we should see the dots in the graph placed along a straight line. On the contrary, systematic departures from a straight line indicate departures from normality. For what regards the normality tests, the Shapiro-Wilk test is generally the preferred one and it is the default one implemented in most software packages indeed, but many others exist[11]. In these tests the null hypothesis corresponds to the normality of the population. As usual, we reject the null hypothesis when the p-value is smaller than the chosen significance level α[12].

Thanks to the normal assumption, it is possible to show that the $(1-\alpha)$ confidence interval for the generic β_j coefficient, with $j = 0, 1, \ldots, p$, is given by

$$\left(b_j - t_{n-p-1,\frac{\alpha}{2}} \cdot SE_j; b_j + t_{n-p-1,\frac{\alpha}{2}} \cdot SE_j\right), \tag{A.17}$$

where SE_j denotes the standard error of b_j[13] and $t_{n-p-1,\alpha/2}$ indicates the critical value of a t-distributed random variable with $(n-p-1)$ degrees of freedom and a probability of $\alpha/2$ to its right[14].

[11] For a list of popular (univariate) normality tests see for example `https://en.wikipedia.org/wiki/Normality_test`.

[12] In Stata you can use the `qnorm` command to produce the normal probability plot, while `swilk` provides the results of the Shapiro-Wilk normality test.

[13] We remind that the term *standard error* corresponds to the estimated standard deviation of the sampling distribution of a statistic, which therefore measures its variability under repeated sampling.

[14] When the sample size n is large enough, say at least 50, and $(1-\alpha)$ is 0.95, we may approximate the t critical value with the value 1.96.

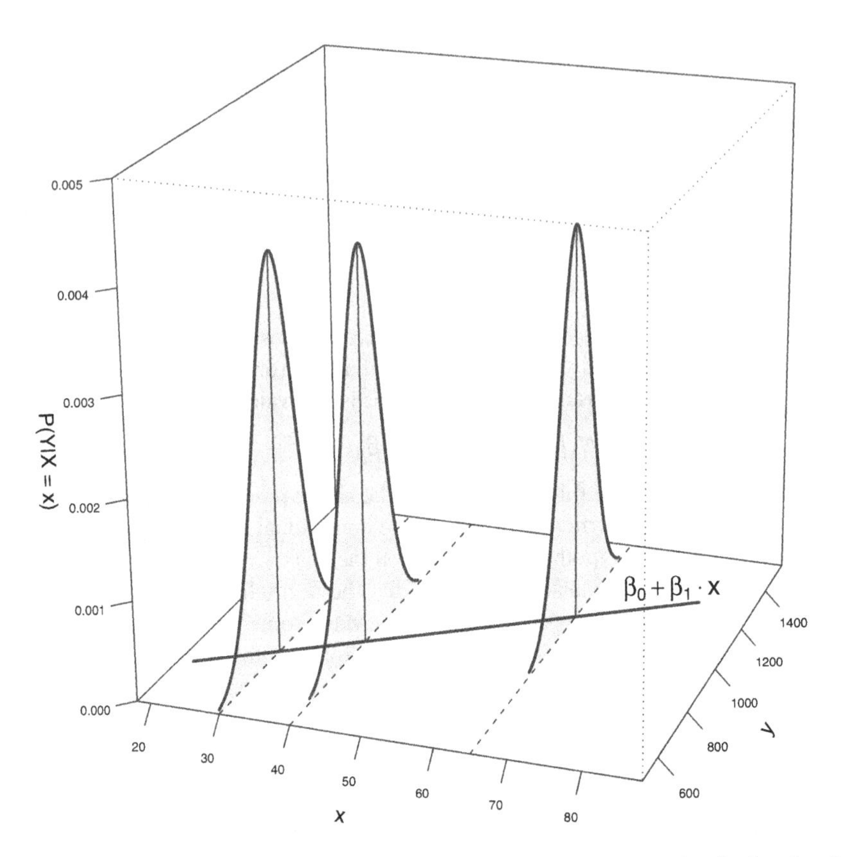

FIGURE A.10: Graphical representation of the normal assumption in the simple linear regression model for three hypothetical x values on the X axis. Note that the spread, that is the variance, of the normal curves is constant across the X values (homoskedasticity).

Using similar results, one may also prove that to test the hypothesis $H_0 : \beta_j = 0$ versus $H_1 : \beta_j \neq 0$ we can use the test statistic

$$t = \frac{b_j}{SE_j}, \tag{A.18}$$

which under the null hypothesis is distributed as a t distribution with $(n-p-1)$ degrees of freedom. As you already know, this result is used by software to compute p-values.

A.2.5 Categorical predictors

So far we discussed only the case of numerical predictors, but linear regression can also be used with categorical explanatory variables. In this section we briefly review how to deal with:

- a binary (or dichotomous) predictor, that is typically coded using a single *dummy variable*,
- a K-categories (or polytomous) predictor, which is coded in a regression model using a set of $(K-1)$ dummy variables,
- *interactions* between numerical and categorical predictors, that are coded as products between dummy variables and numerical predictors.

As we will see, these tools allow to greatly extend the flexibility of regression models.

A **dummy** is a special type of variables that may take only two possible values, 0 and 1, that is used to account for information coming from a binary categorical predictor. Values 0 and 1 refer to the categories of the binary predictor and the choice of which one codes each of the category is largely subjective and generally irrelevant for the analysis. For example, suppose we are fitting a linear regression between the annual salary and experience for a sample of male and female employees. Clearly, experience is a numerical information, while gender is categorical and we can incorporate it in the model using a single dummy variable D that takes value 0 for males and value 1 for females. Equivalently, we may decide to code the two categories the other way round, that is 0 for females and 1 for males. The only difference between these two choices is that the category coded as 0 plays the role of *baseline* (or *reference) category*, which implies a slightly different interpretation for the coefficients, but it is irrelevant from a numerical point of view.

Continuing with the example, let Y denote the annual salary, X the experience (in years) and D the gender with $D=0$ indicating females (baseline) and $D=1$ males. Then, the linear regression model introduced above is given by

$$y_i = \beta_0 + \beta_1 x_i + \beta_2 d_i + \varepsilon_i. \tag{A.19}$$

The crucial point in using dummy variables lies in the correct interpretation of the coefficients. In particular, from equation (A.19) we derive the sub-models for males and females, that is

$$\begin{aligned}
d_i = 0 \text{ (females)} \quad \Longrightarrow \quad y_i &= \beta_0 + \beta_1 x_i + \beta_2 \times 0 + \varepsilon_i \\
&= \beta_0 + \beta_1 x_i + \varepsilon_i \\
d_i = 1 \text{ (males)} \quad \Longrightarrow \quad y_i &= \beta_0 + \beta_1 x_i + \beta_2 \times 1 + \varepsilon_i \\
&= (\beta_0 + \beta_2) + \beta_1 x_i + \varepsilon_i
\end{aligned}$$

Therefore, the dummy variable coefficient β_2 corresponds to the difference in the average annual salary between males and females for any given value of experience. Geometrically, β_2 represents the difference in the intercepts between the two sub-models for males and females. In other words, including a dummy for modelling

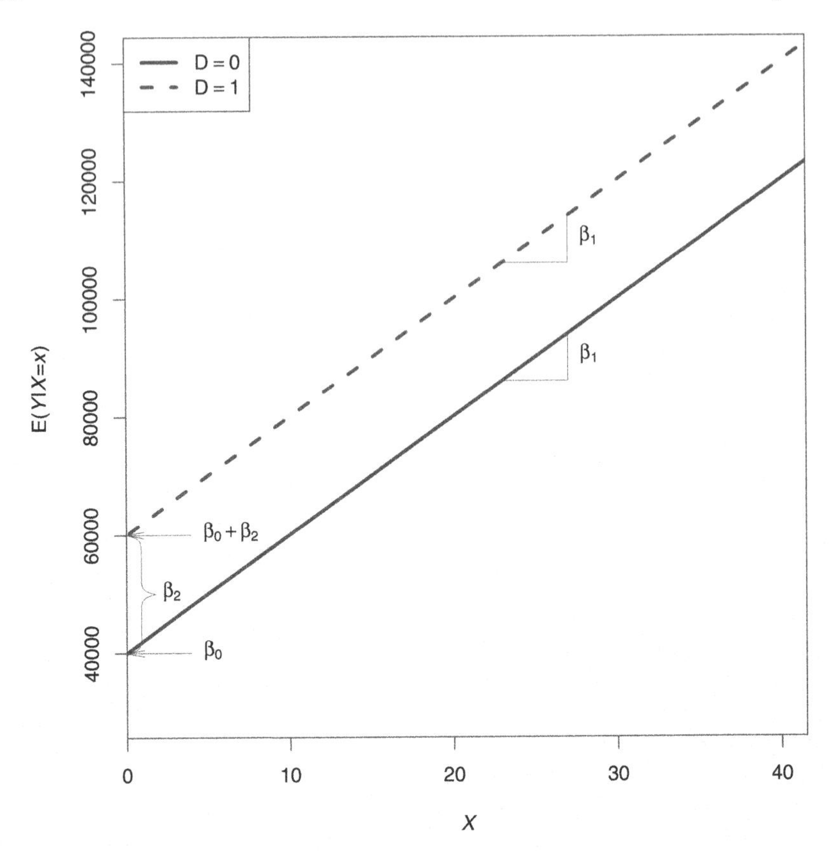

FIGURE A.11: Graphical representation for an hypothetical linear regression model that includes a numerical predictor X and a binary predictor D. The coefficient β_0 provides the intercept for the baseline category, while the dummy coefficient β_2 represents the difference in the intercepts between the two categories. In this example β_2 is positive.

the gender of the employees we have been able to fit two separate lines for males and females, which, for the moment, may only differ for their intercepts (i.e., they are parallel). A graphical representation for an hypothetical model is shown in Figure A.11.

As we said, since the baseline category (i.e., $D = 0$) here is female, β_2 provides the *male vs. female* difference. If instead we chose male as the reference, the value of β_2 would be the same but with a reversed sign, therefore we must interpret it as the *female vs. male* difference.

The next step is to show how to use dummy variables to deal with a K-category predictor, with K larger than two (otherwise we get back to the binary situation). The solution is an extension of the binary case:

1. define K dummy variables, each one taking value 1 in correspondence of one of the categories of the predictor and 0 otherwise,
2. choose the baseline category and include in the model only the remaining $(K-1)$ dummies; as for the binary case, it is indifferent which category we use as baseline; software usually use the first or the last category.

Suppose that we modify the previous example and instead of gender we decide to include in the model the age class W of the employees, which takes $K=3$ different values, that is

$$W_i = \begin{cases} \text{from 20 to 34 years} \\ \text{from 35 to 50 years} \\ \text{from 51 to 70 years} \end{cases}$$

Then, we create three dummy variables, D_1, D_2 and D_3, whose values are defined as follows:

$$d_{1i} = \begin{cases} 1 & \text{if } w_i = \text{“from 20 to 34 years”} \\ 0 & \text{otherwise} \end{cases}$$

$$d_{2i} = \begin{cases} 1 & \text{if } w_i = \text{“from 35 to 50 years”} \\ 0 & \text{otherwise} \end{cases}$$

$$d_{3i} = \begin{cases} 1 & \text{if } w_i = \text{“from 51 to 70 years”} \\ 0 & \text{otherwise.} \end{cases}$$

The model to fit now is

$$y_i = \beta_0 + \beta_1 x_i + \beta_2 d_{2i} + \beta_3 d_{3i} + \varepsilon_i, \tag{A.20}$$

where we decided to use the first category (“from 20 to 34 years”) as the baseline, which in fact is not in the model.

Interpretation of the dummy variable coefficients proceeds as in the case of a binary predictor, but keeping in mind that they all refer to the same baseline category:

$$\begin{aligned}
w_i = \text{“from 20 to 34 years”} \quad \Rightarrow \quad y_i &= \beta_0 + \beta_1 x_i + \beta_2 \times 0 + \beta_3 \times 0 + \varepsilon_i \\
&= \beta_0 + \beta_1 x_i + \varepsilon_i \\
w_i = \text{“from 35 to 50 years”} \quad \Rightarrow \quad y_i &= \beta_0 + \beta_1 x_i + \beta_2 \times 1 + \beta_3 \times 0 + \varepsilon_i \\
&= (\beta_0 + \beta_2) + \beta_1 x_i + \varepsilon_i \\
w_i = \text{“from 51 to 70 years”} \quad \Rightarrow \quad y_i &= \beta_0 + \beta_1 x_i + \beta_2 \times 0 + \beta_3 \times 1 + \varepsilon_i \\
&= (\beta_0 + \beta_3) + \beta_1 x_i + \varepsilon_i
\end{aligned}$$

So, β_2 represents the difference in the average annual salary between employees in the second age class and those in the first (baseline). Similarly, β_3 represents the difference in the average annual salary between employees in the third age class and those in the first (baseline). Geometrically, this is equivalent to saying that we are fitting three separate lines for the three groups of employees that differ only in terms of the intercepts (i.e., the lines are parallel). A graphical representation for an hypothetical model is shown in Figure A.12.

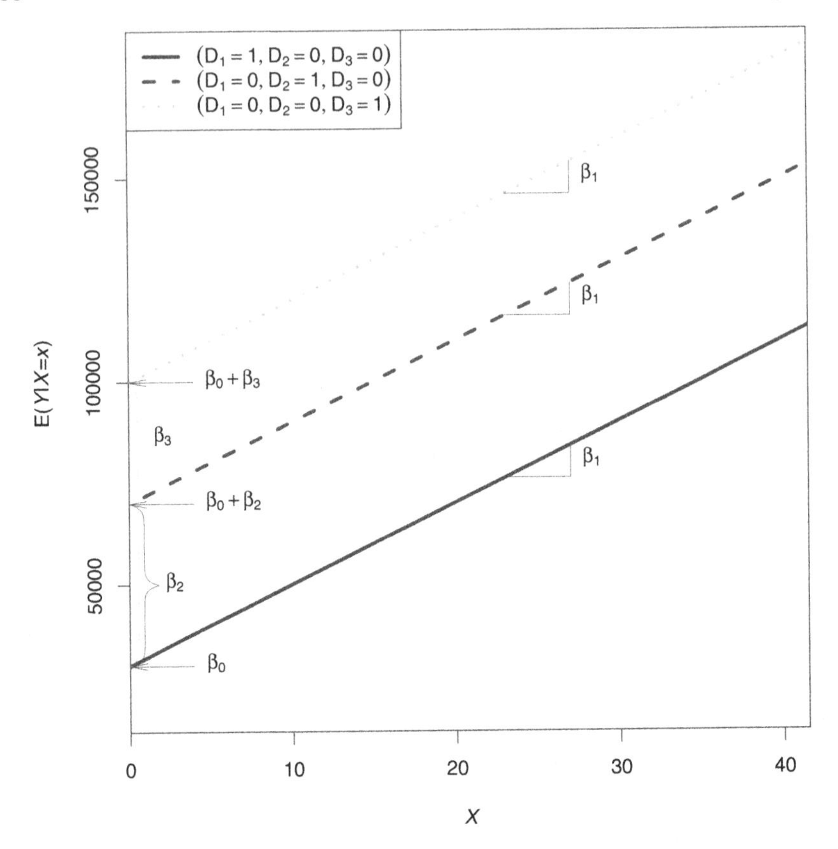

FIGURE A.12: Graphical representation for an hypothetical linear regression model that includes a numerical predictor X and a 3-category predictor W that enters the model through two dummy variables D_2 and D_3. The coefficient β_0 provides the intercept for the baseline category, while the dummy coefficients β_2 and β_3 represent the differences in the intercepts between the other categories and the baseline. In this example β_2 and β_3 are both positive and $\beta_2 < \beta_3$.

Finally, we can further extend the flexibility of a regression model by allowing also the slopes to differ across values of a predictor. This can be achieved by including in the model **interactions** between two predictors. Interactions are defined as the product of other two predictors (usually already included in the model[15]) which aim at accounting for the possibility that the effect on the response of one predictor may depend on the value of another one. Even if the two predictors that define an interaction can be of any type, we focus here on the most common case of interactions between a numerical predictor and a dummy variable. Furthermore, for simplicity,

[15]This is the so called *principle of marginality* introduced by Nelder (1977), which states that a statistical model should always include all the main effects and lower order interactions whenever a higher order interaction is included. A violation of this principle would make the model not broadly applicable.

we consider only the case of a binary categorical predictor, with that of polytomous predictors being an immediate extension.

Going back to the example involving salary (Y) versus experience (X) and gender (D), we now include in the model the interaction between the latter two variables thus getting the following specification

$$y_i = \beta_0 + \beta_1 x_i + \beta_2 d_i + \beta_3 (x_i \times d_i) + \varepsilon_i. \tag{A.21}$$

This model accounts for the possibility that the effect of X on Y may be different depending on the value of D. In fact, the sub-models corresponding to the values od D are

$$\begin{array}{llcl} d_i = 0 \text{ (females)} \quad \Longrightarrow & y_i & = & \beta_0 + \beta_1 x_i + \beta_2 \times 0 + \beta_3 (x_i \times 0) + \varepsilon_i \\ & & = & \beta_0 + \beta_1 x_i + \varepsilon_i \\ d_i = 1 \text{ (males)} \quad \Longrightarrow & y_i & = & \beta_0 + \beta_1 x_i + \beta_2 \times 1 + \beta_3 (x_i \times 1) + \varepsilon_i \\ & & = & (\beta_0 + \beta_2) + (\beta_1 + \beta_3) x_i + \varepsilon_i \end{array}$$

It follows that the interaction coefficient β_3 represents the difference in the slopes between males and females, that is the differential effect of experience on the annual salary for males versus females. Figure A.13 shows an hypothetical situation where both β_2 and β_3 are positive.

A.2.6 Multicollinearity

We already mentioned that the linear regression model is based on a set of assumptions that mainly regard the error terms. However, a further issue that may arise regards the possibility that some predictors are strongly linear related, a situation known as **multicollinearity**, or simply as collinearity. The main consequences of multicollinearity are:

1. the coefficient estimates become unstable, which means that a small change in the data may have a huge impact on the regression results. Moreover, in presence of multicollinearity it is more difficult to separate the individual effect of each predictor on the response variable, because when one predictor moves also others will do, thus making difficult to understand what is the source of the change in Y. Therefore, in these cases we can only afford measuring the aggregate effect of the predictors affected by multicollinearity. More technically, in presence of multicollinearity different combinations of the regression coefficients provide similar RSS values which are almost as optimal as the least-squares solutions

2. coefficient standard errors are biased upwards thus producing wider confidence intervals and larger p-values

The main tool used to detect multicollinearity is the **variance inflation factor** (VIF), an index computed for each predictor in the model that tells us how directly

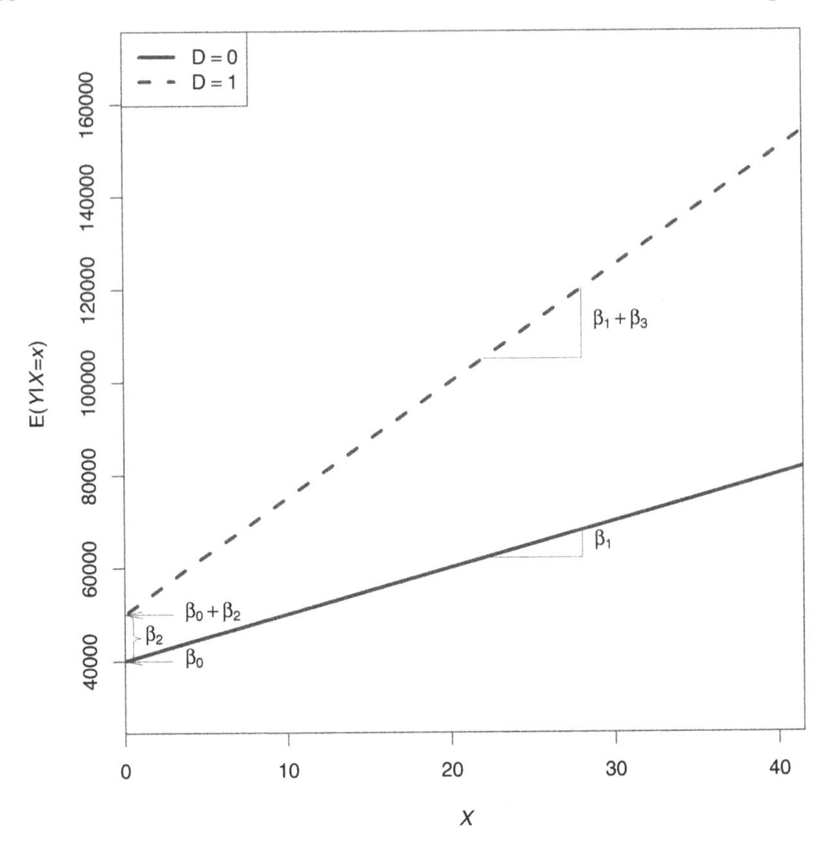

FIGURE A.13: Graphical representation for an hypothetical linear regression model that includes a numerical predictor X, a binary predictor D and their interaction. Coefficients β_0 and β_1 provide the intercept and slope for the baseline category, while coefficients β_2 and β_3 represent the differences in the intercepts and slopes between the two categories respectively. In this example β_2 and β_3 are both positive.

how much is the impact of multicollinearity on each coefficient standard error[16]. Its square root is interpreted indeed as the number of times the standard error is "inflated" due to the multicollinearity. Therefore, a VIF equal to 1 means that the predictor is not affected by multicollinearity, while a large VIF value indicates a critical situation. Formally, the VIF for the jth predictor in a model is defined as

$$VIF_j = \frac{1}{1 - R_j^2}, \tag{A.22}$$

where R_j^2 denotes the R-squared index for the regression of X_j on the other predictors.

[16] Some software also produce the *tolerance* index, which is defined as the reciprocal of the variance inflation factor.

Even if there are no theoretical guidelines on how to assess the VIF value, a popular approach consists in considering as harmless a VIF that is smaller than 5, while a value in between 5 and 10 provides a serious warning signal and a value above 10 indicates a critical situation in which multicollinearity is unduly influencing the least squares estimates.

Once we realize that our regression analysis is affected by multicollinearity, we should take counteractions to try limiting its consequences. Unfortunately, there is no general quick cure for it and in some cases the remedy may even be more harmful than the problem itself. Since multicollinearity is an issue related to the data and not to the model, one possibility is to collect more observations with the hope that the new data at least attenuate the severity of multicollinearity. Clearly, this solution is not feasible because of the costs of sampling. Other common remedies are:

- Select the variables to keep in the model, either manually or automatically, using one of the variable selection methods available (stepwise regression, best-subset regression using the adjusted R-squared or an information criterion such as the BIC)[17].
- Use a dimensionality reduction technique such as *principal component analysis* (PCA), which allow to condense most of the information available in the correlated predictors in a new set of uncorrelated components; we describe PCA in Section 3.2.
- Use a shrinkage method like ridge regression or the LASSO, which allow for a small amount of bias in the coefficient estimates in exchange to a reduction in the coefficient standard errors (see Efron and Hastie, 2016, Chapter 16).

Stata allows to compute the VIFs for a given regression problem with the postestimation command `estat vif`.

A.2.7 Example

In this section we provide a practical example of linear regression whose aim is to present the Stata commands to compute the various measures discussed above. We use again the NLSW data available in the `nlsw88.dta` Stata file and we fit different models to predict the hourly wage.

Stata's main command for linear regression is `regress`, which requires to specify first the response variable followed by the list of predictors. We start fitting the a simple model for `wage` against `grade`, the current grade completed, whose results are reported in Figure A.14:

```
1 sysuse nlsw88, clear
2 regress wage grade
```

As expected, the grade completed is positively associated with the hourly wage.

[17]For a thorough presentation of these methods we suggest to see (Hastie et al., 2008, Chapter 3).

```
      Source |       SS           df       MS      Number of obs   =     2,244
-------------+----------------------------------   F(1, 2242)      =    265.57
       Model |  7874.79847         1  7874.79847   Prob > F        =    0.0000
    Residual |   66479.532     2,242  29.6518876   R-squared       =    0.1059
-------------+----------------------------------   Adj R-squared   =    0.1055
       Total |  74354.3305     2,243  33.1495009   Root MSE        =    5.4454

------------------------------------------------------------------------------
        wage |      Coef.   Std. Err.      t    P>|t|     [95% Conf. Interval]
-------------+----------------------------------------------------------------
       grade |   .7431729   .0456033    16.30   0.000     .6537438     .832602
       _cons |  -1.965886   .6083143    -3.23   0.001    -3.158804   -.7729677
------------------------------------------------------------------------------
```

FIGURE A.14: Estimated linear regression model of the `wage` (hourly wage) variable versus `grade` (current grade completed) for the NLSW data.

The R-squared shows that `grade` explains around 11% of the observed variability in the hourly wages. Moreover, the `grade` coefficient estimate is highly reliable since its p-value is very small (shown ad `0.000`). To confirm the statistical goodness of the model we also check that the assumptions of the linear regression are satisfied. Let's remind quickly how to verify them:

- linearity can be easily checked using a scatter plot because we are using a single predictor; the scatter plot (not reported here) confirms that there is no strong evidence against the linearity assumption.
- error independence is taken for granted here because we have a cross-sectional dataset involving different individuals; the error independence assumption is more likely to be violated with time series or longitudinal data.
- constant error variance can be verified with the popular *Breusch-Pagan test*. The null hypothesis is that the error variance is constant (homoskedasticity), so the hope is to get a large p-value to avoid rejecting the null hypothesis. In Stata we can perform the test with the postestimation command `estat hettest`. In our example the p-value turns out to be very small, which provides evidence against the constant error variance assumption.
- normality of the errors can be checked with the normal probability plot of the residuals and the Shapiro-Wilk test. In the example both these tools confirm a marked deviation from normality[18].

While normality in this example is not a big issue because the sample is large enough to guarantee that the coefficient sampling distributions are close to a normal[19], the non-constant error variance is a problem since, even if the OLS estimates are still consistent (i.e., they converge to the true values as n becomes large), their standard

[18]To store the residuals in the dataset as a new column labelled `resid` use the command `predict resid, residuals`.

[19]This also implies that in this example the advantage of using the bootstrap for inference is limited, as we will show later in the section.

```
      Source |       SS           df       MS      Number of obs   =     2,244
-------------+----------------------------------   F(1, 2242)      =    469.04
       Model |  128.194579         1  128.194579   Prob > F        =    0.0000
    Residual |  612.768221     2,242  .273313212   R-squared       =    0.1730
-------------+----------------------------------   Adj R-squared   =    0.1726
       Total |  740.962799     2,243  .330344538   Root MSE        =    .52279

------------------------------------------------------------------------------
     logwage |      Coef.   Std. Err.      t    P>|t|     [95% Conf. Interval]
-------------+----------------------------------------------------------------
       grade |   .0948211   .0043782    21.66   0.000     .0862352    .1034069
       _cons |   .6267517   .0584026    10.73   0.000     .5122229    .7412805
------------------------------------------------------------------------------
```

FIGURE A.15: Estimated linear regression model of `logwage` (logarithm of hourly wage) variable versus `grade` (current grade completed) for the NLSW data.

errors are no longer valid. So, we decide to transform the response using a logarithm and create the new variable `logwage` (see Figure A.15):

```
1 generate logwage = log(wage)
2 regress logwage grade
3 estimates store model1
```

By performing the same checks we see that the situation has improved significantly, even if the p-value of the Breusch-Pagan test is still smaller than the conventional $\alpha = 0.05$. For now we stop trying other remedies because it may be the case that adding other predictors will contribute to lessen the heteroskedasticity issue[20].

We now extend the model by adding the `smsa` variable as a new predictor. This is a binary variable taking values 1 or 0 depending on whether each individual in the sample lives in a standard metropolitan statistical area (SMSA) or not[21]. Stata has some operators that allow to deal with dummy variables and interactions in a general way[22]. In particular:

- the operator `i.` in front of a variable name generates the necessary dummy variables to include in the model,
- the operator `c.` in front of a variable name tells Stata to treat the variable as numeric,
- the operator `#` includes in the model the interactions between values of two predictors,
- the operator `##` includes in the model both the interactions and the main effects.

[20] Another popular approach to deal with heteroskedasticity is to use robust estimates of the standard errors, also called *Huber-White heteroskedasticity-robust standard errors*. We can get them in Stata adding the `vce(robust)` option to `regress`. For more details see Wooldridge (2016), Chapter 8.

[21] `https://en.wikipedia.org/wiki/Metropolitan_statistical_area`.

[22] For more details type `help fvvarlist`.

The following code first fits the model that includes only smsa, stores its results, and then fits a second model including both smsa and the interaction with grade (see Figure A.16 for the results):

```
regress logwage grade i.smsa
estimates store model2
regress logwage c.grade##smsa
```

The results show that the data do not provide evidence for different slopes for the two groups of individual that live or live not in a SMSA because the p-values of the interaction in the second model is large (0.203). However, the first model provides support for different intercepts. This conclusion is also confirmed by the scatter plot shown in Figure A.17 that reports the fitted lines form the first model. So, according to these estimates, a woman living in a SMSA in 1988 would have earned about $100(e^{0.2381} - 1)\% \approx 27\%$ more per hour than a woman with the same grade but not living in a SMSA.

We now add some other predictors, in particular age, hours, ttl_exp, tenure, race, married, union, south, never_married and occupation. The results are reported in Figure A.18[23].

```
regress logwage grade i.smsa age hours ttl_exp tenure ///
  i.race i.married i.union i.south i.never_married ///
  i.occupation ///
  if occupation != 9 & occupation != 10 & occupation != 12
estimates store model3
estat hettest   // (output not reported)
estat vif       // (output not reported)
```

The model has substantially improved, with the adjusted R-squared that has more than doubled. Most coefficients are significant and their signs go in the direction that one would expect. However, note that the number of observation has decreased from about 2244 to approximately 1856. This is due mainly to the union variable that contains a lot of missing values. We also highlight that for this model heteroskedasticity is no longer an issue (the p-value is 0.3524). Similarly, multicollinearity is not critical because all the VIFs (not reported) are below the conventional threshold 5.

We conclude the example by comparing the results of the last model, which uses normal-based theory, with the corresponding bootstrap results using 2000 replications (Figure A.19):

```
quietly regress logwage grade i.smsa age hours ttl_exp ///
  tenure i.race i.married i.union i.south ///
  i.never_married i.occupation ///
  if occupation != 9 & ///
```

[23]We exclude some of the occupation categories because too rare and may cause numerical instabilities in the calculations.

```
. regress logwage grade i.smsa

      Source |       SS           df       MS      Number of obs   =     2,244
-------------+----------------------------------   F(2, 2241)      =    294.37
       Model |  154.162551         2  77.0812755   Prob > F        =    0.0000
    Residual |  586.800248     2,241    .2618475   R-squared       =    0.2081
-------------+----------------------------------   Adj R-squared   =    0.2074
       Total |  740.962799     2,243  .330344538   Root MSE        =    .51171

------------------------------------------------------------------------------
     logwage |      Coef.   Std. Err.      t    P>|t|     [95% Conf. Interval]
-------------+----------------------------------------------------------------
       grade |   .0886297   .0043303    20.47   0.000     .0801378    .0971215
             |
        smsa |
       SMSA  |   .2381447   .0239137     9.96   0.000     .1912494    .2850399
       _cons |   .5401754   .0578218     9.34   0.000     .4267856    .6535652
------------------------------------------------------------------------------

. regress logwage c.grade##smsa

      Source |       SS           df       MS      Number of obs   =     2,244
-------------+----------------------------------   F(3, 2240)      =    196.85
       Model |  154.587906         3  51.5293022   Prob > F        =    0.0000
    Residual |  586.374893     2,240  .261774506   R-squared       =    0.2086
-------------+----------------------------------   Adj R-squared   =    0.2076
       Total |  740.962799     2,243  .330344538   Root MSE        =    .51164

------------------------------------------------------------------------------
     logwage |      Coef.   Std. Err.      t    P>|t|     [95% Conf. Interval]
-------------+----------------------------------------------------------------
       grade |   .0801407   .0079432    10.09   0.000     .0645639    .0957176
             |
        smsa |
       SMSA  |   .0838435   .1233867     0.68   0.497    -.1581206    .3258077
             |
smsa#c.grade |
       SMSA  |   .0120772   .0094744     1.27   0.203    -.0065024    .0306568
             |
       _cons |   .6466321   .1015729     6.37   0.000     .4474452     .845819
------------------------------------------------------------------------------
```

FIGURE A.16: Estimated linear regression model of `logwage` (logarithm of hourly wage) variable versus `grade` (current grade completed) and `smsa` (standard metropolitan statistical area) for the NLSW data, with and without interaction.

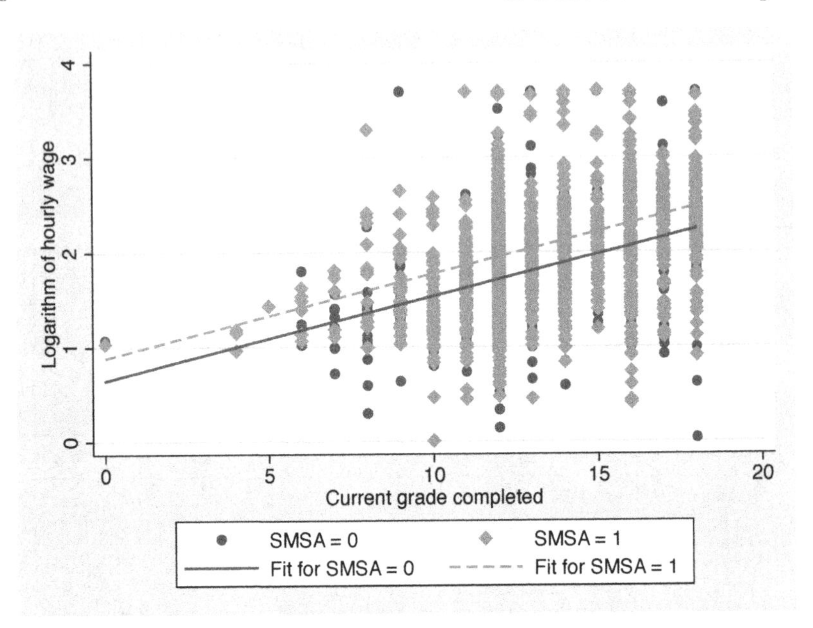

FIGURE A.17: Scatter plot and fitted lines for linear regression model of `logwage` (logarithm of hourly wage) variable versus `grade` (current grade completed) and `smsa` (standard metropolitan statistical area) for the NLSW data.

```
5         occupation != 10 & ///
6         occupation != 12, ///
7       vce(bootstrap, reps(2000) seed(101) ///
8          saving(appA_nlws_reg, replace))
9    estimates store model3b
10   estimates table model*, b(%9.4f) se(%9.4f) stats(N r2_a)
```

Note that the standard errors for the last two models (`model3` and `model3b`) are very similar. This is a confirmation that in this case the bootstrap is not strictly necessary, because thanks to the large sample size the sampling distributions of the coefficients are close to a normal. For the same reason, the bootstrap confidence intervals (not reported here, but you can get them by executing the command `estat bootstrap, all`) are also similar. We partially prove our statement by reporting the bootstrap distribution of the `tenure` coefficient and the bivariate bootstrap distributions of the coefficients related to the numerical predictors (plus the constant) included in the model (see Figures A.20 and A.21).

```
1  use appA_nlws_reg, clear
2  histogram _b_tenure, frequency normal ///
3    normopts(lwidth(thick))
4  graph matrix _b_*, half
```

```
      Source |       SS           df       MS      Number of obs   =     1,847
-------------+----------------------------------   F(21, 1825)     =     76.63
       Model |  228.017158        21  10.8579599   Prob > F        =    0.0000
    Residual |   258.57359     1,825  .141684159   R-squared       =    0.4686
-------------+----------------------------------   Adj R-squared   =    0.4625
       Total |  486.590749     1,846  .263591955   Root MSE        =    .37641

-------------------------------------------------------------------------------------
            logwage |      Coef.   Std. Err.      t    P>|t|     [95% Conf. Interval]
--------------------+----------------------------------------------------------------
              grade |   .0580982   .0045687    12.72   0.000     .0491377    .0670587
                    |
               smsa |
               SMSA |   .1980121   .0202338     9.79   0.000     .1583283    .2376959
                age |  -.0048872   .0029435    -1.66   0.097    -.0106602    .0008858
              hours |   .0009899   .0009592     1.03   0.302    -.0008913    .0028711
            ttl_exp |   .0268968   .0024905    10.80   0.000     .0220123    .0317813
             tenure |   .0112598   .0019487     5.78   0.000     .0074379    .0150817
                    |
               race |
              black |  -.0697331   .0223199    -3.12   0.002    -.1135084   -.0259578
              other |   .0295896   .0779276     0.38   0.704     -.123247    .1824263
                    |
            married |
            married |  -.0170925   .0217239    -0.79   0.431    -.0596989    .0255138
                    |
              union |
              union |   .1604566   .0218353     7.35   0.000     .1176317    .2032815
            1.south |  -.0851979   .0193013    -4.41   0.000    -.1230529    -.047343
    1.never_married |  -.0547122   .0327602    -1.67   0.095    -.1189636    .0095393
                    |
         occupation |
     Managers/admin |   .0513316   .0350093     1.47   0.143     -.017331    .1199942
              Sales |  -.1986713   .0288094    -6.90   0.000    -.2551742   -.1421684
 Clerical/unskilled |   -.174097   .0501827    -3.47   0.001    -.2725186   -.0756753
          Craftsmen |  -.1325747   .0594448    -2.23   0.026    -.2491617   -.0159876
         Operatives |  -.2405205    .039101    -6.15   0.000    -.3172079   -.1638331
          Transport |  -.4753187   .0932906    -5.10   0.000    -.6582862   -.2923511
           Laborers |  -.3441774   .0386245    -8.91   0.000    -.4199303   -.2684245
            Service |  -.0148908   .1125175    -0.13   0.895    -.2355673    .2057858
              Other |  -.2686168    .038958    -6.90   0.000    -.3450238   -.1922098
                    |
              _cons |   .9232746   .1439897     6.41   0.000     .6408727    1.205676
-------------------------------------------------------------------------------------
```

FIGURE A.18: Estimated linear regression model of `logwage` versus `grade`, `smsa`, `age`, `hours`, `ttl_exp`, `tenure`, `race`, `married`, `union`, `south`, `never_married` and `occupation` for the NLSW data.

Variable	model1	model2	model3	model3b
grade	0.0948	0.0886	0.0581	0.0581
	0.0044	0.0043	0.0046	0.0052
smsa				
SMSA		0.2381	0.1980	0.1980
		0.0239	0.0202	0.0197
age			-0.0049	-0.0049
			0.0029	0.0030
hours			0.0010	0.0010
			0.0010	0.0013
ttl_exp			0.0269	0.0269
			0.0025	0.0027
tenure			0.0113	0.0113
			0.0019	0.0021
race				
black			-0.0697	-0.0697
			0.0223	0.0208
other			0.0296	0.0296
			0.0779	0.0883
married				
married			-0.0171	-0.0171
			0.0217	0.0213
union				
union			0.1605	0.1605
			0.0218	0.0224
south				
1			-0.0852	-0.0852
			0.0193	0.0201
never_marr~d				
1			-0.0547	-0.0547
			0.0328	0.0357
occupation				
Managers/~n			0.0513	0.0513
			0.0350	0.0368
Sales			-0.1987	-0.1987
			0.0288	0.0282
Clerical/~d			-0.1741	-0.1741
			0.0502	0.0687
Craftsmen			-0.1326	-0.1326
			0.0594	0.0575
Operatives			-0.2405	-0.2405
			0.0391	0.0385
Transport			-0.4753	-0.4753
			0.0933	0.1015
Laborers			-0.3442	-0.3442
			0.0386	0.0416
Service			-0.0149	-0.0149
			0.1125	0.1118
Other			-0.2686	-0.2686
			0.0390	0.0374
_cons	0.6268	0.5402	0.9233	0.9233
	0.0584	0.0578	0.1440	0.1508
N	2244	2244	1847	1847
r2_a	0.1726	0.2074	0.4625	0.4625

legend: b/se

FIGURE A.19: Comparison of the models fitted in the example with the NLSW data.

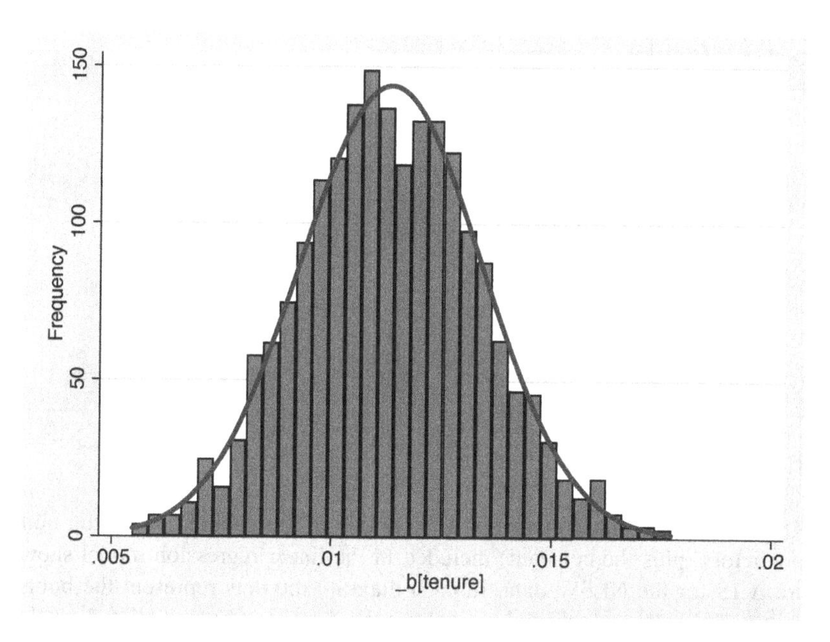

FIGURE A.20: Bootstrap distribution of the `tenure` coefficient for the linear regression model shown in Figure A.18 for the NLSW data. The solid black line indicates the fitted normal distribution with mean and standard deviation equal to those of the bootstrap replicates.

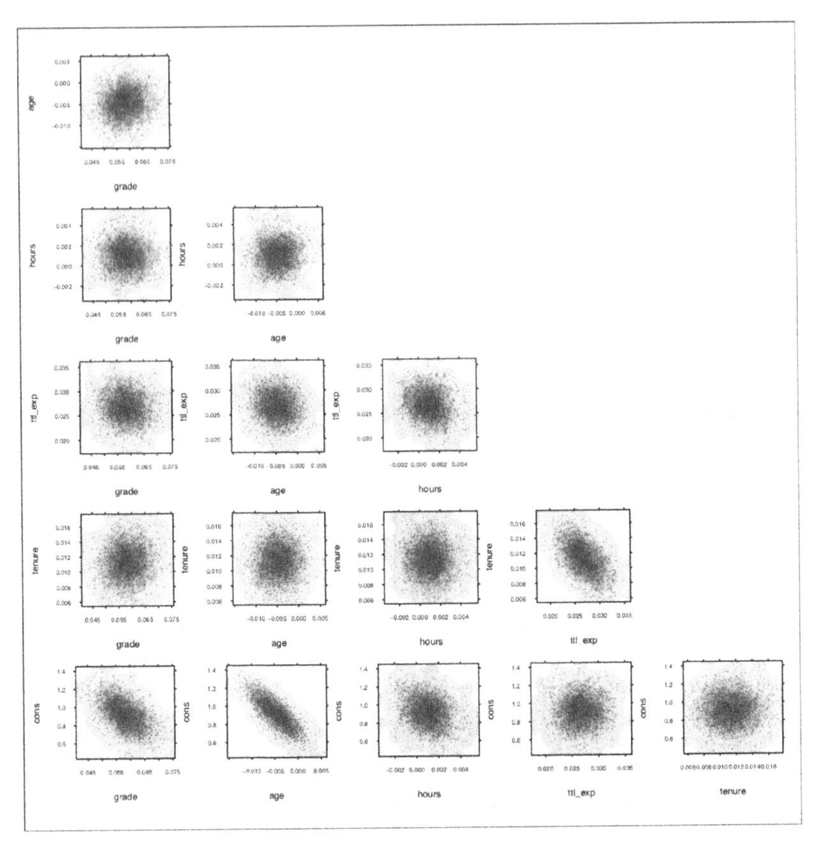

FIGURE A.21: Bivariate bootstrap distributions of the coefficients of the numerical predictors (plus the constant) included in the linear regression model shown in Figure A.18 for the NLSW data. In each diagram the dots represent the bootstrap replicates, while the grey shaded regions are the corresponding bivariate kernel density estimate. In Stata you can get it with the user-written `kdens2` command.

A.3 Summary

In this chapter we reviewed the basic statistical concepts that are at the core of the methods we present in the book. In particular, we provided a concise presentation of linear correlation and a more detailed discussion of linear regression, one of the most successful statistical model applied in practically all fields.

Appendix: R Commands

In this final section we briefly present some of the R commands for performing the analyses discussed in the appendix.

Covariance and correlation

R provides the `cov()` and `cor()` functions respectively for computing covariances and correlations between pairs of variables. They both require a data frame or a matrix object containing only numerical variables. The additional `use` argument specifies the method to use for computing covariances and correlations in the presence of missing values. Allowed values are:

- `"everything"`, which propagates NA values (the R flag for missing data) through the computation, that is a resulting value will be NA whenever one of its contributing observations is NA,
- `"all.obs"`, produces an error in case of any missing observation,
- `"complete.obs"`, which handles missing values by listwise deletion (see Section 3.6); this choice corresponds to Stata's `correlate` command,
- `"na.or.complete"` is the same as `"complete.obs"` unless there are no complete cases, in which case it gives NA,
- `"pairwise.complete.obs"`, which uses pairwise deletion (see Section 3.6); this choice corresponds to Stata's `pwcorr` command.

Through the `method` argument, the `cor()` function also allows to specify whether to compute the standard Pearson's linear correlation index (the default) or other non-parametric versions, namely the Spearman's or Kendall's indexes.

The basic graphical command to produce a scatterplot matrix in R is `pairs()`, which is simple but not very flexible. If you need more customization options, you need to use other packages like `lattice` (Sarkar, 2008) or `GGally` (Schloerke et al., 2020).

The linear independence test is provided by the `cor.test()` function.

The following code, whose output is not reported, replicates the same results as in Section A.1:

```
1  if (!require(haven, quietly = TRUE)) {
2    install.packages("haven")
3  }
4  library(haven)
5  if (!require(dplyr, quietly = TRUE)) {
6    install.packages("dplyr")
7  }
```

```
8  library(dplyr)

9  nlsw88 <- read_dta(file = file.path(path_data,
10   "nlsw88.dta"))
11 vars <- c("age", "hours", "ttl_exp", "wage")
12 pairs(nlsw88[, vars], pch = 20, col = gray(.4))

13 nlsw88 <- mutate(nlsw88, logwage = log(wage))
14 vars <- c("age", "hours", "ttl_exp", "logwage")
15 pairs(nlsw88[, vars], pch = 20, col = gray(.4))

16 cor(nlsw88[, vars], use = "complete.obs")
17 cor(nlsw88[, vars], use = "pairwise.complete.obs")

18 cor.test(x = nlsw88$logwage, y = nlsw88$age)
```

Linear regression analysis

The main function for fitting linear models in R is `lm()`[24]. This function requires a `formula` object specifying the model to fit and a `data.frame` object containing the variables involved in the model. The function returns an object of class `lm`, which can be printed, summarized or visualized using methods for generic functions such as `summary()`, `residuals()`, or `predict()` (for the full list of methods available for inspecting `lm` objects, run the command `methods(class = "lm")` and the corresponding documentation). Of these methods, `summary()` is by far the most important since it prints in a neat way the results of the fitted model.

The specification of `formula` objects may appear difficult to grasp at the beginning since it requires a special syntax. For our presentation here it suffices to mention two of the available operators (for more details see `?formula`), namely the "tilde" operator `~`, which joins the left and right hand sides of the model equation, and the "star" operator `*`, which instead is needed to include interactions between two or more predictors. To include categorical predictors in a linear regression in R you must encode the corresponding variable in the data frame as a `factor` variable. You can transform a variable's type to `factor` using the `factor()` function.

The following code shows how to fit in R the last model presented in Section A.2.7, where we also use the `subset` argument is used for selecting the data to include in the analysis:

```
1 nlsw88$smsa <- factor(nlsw88$smsa)
2 nlsw88$race <- factor(nlsw88$race)
3 nlsw88$married <- factor(nlsw88$married)
4 nlsw88$union <- factor(nlsw88$union)
5 nlsw88$south <- factor(nlsw88$south)
```

[24]For a complete discussion of linear regression in R we suggest to see Chapters 3 and 4 of Kleiber and Zeileis (2008).

```
nlsw88$never_married <- factor(nlsw88$never_married)
nlsw88$occupation <- factor(nlsw88$occupation)

nlsw88_lm <- lm(formula = logwage ~ grade + smsa + age +
  hours + ttl_exp + tenure + race + married + union +
  south + never_married + occupation, data = nlsw88,
  subset = (occupation != 9 & occupation != 10 &
              occupation != 12))
summary(nlsw88_lm)
```

We conclude this brief presentation of linear regression with R mentioning how to produce some diagnostics. More specifically, you can check if any of the predictors are affected by multicollinearity using the `vif()` function from the `car` package (Fox and Weisberg, 2019). To check if the model suffers from heteroskedasticity, you can instead use the `bptest()` function from the `lmtest` package (Zeileis and Hothorn, 2002).

Bibliography

Acock, A. C. (2013). *Discovering Structural Equation Modeling Using Stata.* Stata Press, revised edition.

Agarwal, R. and Karahanna, E. (2000). Time flies when you're having fun: Cognitive absorption and beliefs about information technology usage. *MIS Quarterly*, 24(4):665–694.

Aguirre-Urreta, M. I. and Rönkkö, M. (2018). Statistical inference with PLSc using bootstrap confidence intervals. *MIS Quarterly*, 42(3):1001–1020.

Alwin, D. F. and Hauser, R. M. (1975). The decomposition of effects in path analysis. *American Sociological Review*, 40:37–47.

Amato, S., Esposito Vinzi, V., and Tenenhaus, M. (2005). A global goodness-of-fit index for PLS structural equation modeling. Technical report, HEC School of Management, France.

Anderson, J. C. and Gerbing, D. W. (1988). Structural equation modeling in practice: A review and recommended two-step approach. *Psychological Bulletin*, 103(3):411–423.

Bagozzi, R. P. and Yi, Y. (1994). Advanced topics in structural equation models. In Bagozzi, R. P., editor, *Advanced Methods of Marketing Research*, pages 1–51. Oxford: Basil Blackwell Ltd.

Bandalos, D. L. (2018). *Measurement Theory and Applications for the Social Sciences.* Guilford Press.

Banfield, J. D. and Raftery, A. E. (1993). Model-based Gaussian and non-Gaussian clustering. *Biometrics*, 49:803–821.

Baron, R. M. and Kenny, D. A. (1986). The moderator–mediator variable distinction in social psychological research: Conceptual, strategic, and statistical considerations. *Journal of Personality and Social Psychology*, 51:1173–1182.

Barroso, C., Carriòn, G. C., and Roldàn, J. L. (2010). Applying maximum likelihood and PLS on different sample sizes: Studies on SERVQUAL model and employee behavior model. In Esposito Vinzi, V., Chin, W. W., Henseler, J., and Wang, H., editors, *Handbook of Partial Least Squares: Concepts, Methods and Applications*, pages 427–447. Springer.

Bartholomew, D., Knott, M., and Moustaki, I. (2011). *Latent Variable Models and Factor Analysis. A Unified Approach.* Wiley, 3rd edition.

Bartholomew, D. J. (2013). *Unobserved Variables. Models and Misunderstandings.* SpringerBriefs in Statistics. Springer.

Bearden, O., Netemeyer, R. G., and Haws., K. L. (2011). *Handbook of Marketing Scales: Multi-item Measures for Marketing and Consumer Behavior Research.* SAGE, 3rd edition.

Beaujean, A. A. (2014). *Latent Variable Modeling Using* R. Routledge.

Becker, J.-M., Rai, A., and Rigdon, E. (2013). Predictive validity and formative measurement in structural equation modeling: Embracing practical relevance. *Proceedings of the International Conference on Information Systems (ICIS).*

Benaglia, T., Chauveau, D., Hunter, D. R., and Young, D. (2009). `mixtools`: An `R` package for analyzing finite mixture models. *Journal of Statistical Software*, 32(6):1–29.

Benitez, J., Henseler, J., Castillo, A., and Schuberth, F. (2020). How to perform and report an impactful analysis using partial least squares: Guidelines for confirmatory and explanatory is research. *Information & Management*, 57(2):103168.

Bleske-Rechek, A. and Fritsch, A. (2011). Student consensus on ratemyprofessors.com. *Practical Assessment, Research & Evaluation*, 16(18).

Boker, S. M., Neale, M. C., Maes, H. H., Wilde, M. J., Spiegel, M., Brick, T. R., Estabrook, R., Bates, T. C., Mehta, P., von Oertzen, T., Gore, R. J., Hunter, M. D., Hackett, D. C., Karch, J., Brandmaier, A. M., Pritikin, J. N., Zahery, M., Kirkpatrick, R. M., Wang, Y., Goodrich, B., Driver, C., Massachusetts Institute of Technology, Johnson, S. G., Association for Computing Machinery, Kraft, D., Wilhelm, S., Medland, S., Falk, C. F., Keller, M., Manjunath B G, The Regents of the University of California, Ingber, L., Shao Voon, W., Palacios, J., Yang, J., Guennebaud, G., and Niesen, J. (2020). *`OpenMx` 2.17.3 User Guide.*

Bollen, K. A. (1989). *Structural Equation with Latent Variables.* John Wiley & Sons.

Boos, D. D. and Stefanski, L. A. (2013). *Essential Statistical Inference. Theory and Methods.* Springer.

Bozdogan, H. (1987). Model selection and Akaike's information criterion (AIC): The general theory and its analytical extensions. *Psychometrika*, 52:345–370.

Bozdogan, H. (1994). Mixture-model cluster analysis using model selection criteria and a new information measure of complexity. In Bozdogan, H., editor, *Proceedings of the First US/Japan Conference on Frontiers of Statistical Modelling: An Informational Approach*, volume 2, pages 69–113. Kluwer Academic Publishers.

Brock, G., Pihur, V., Datta, S., and Datta, S. (2008). `clValid`: An `R` package for cluster validation. *Journal of Statistical Software*, 25(4):1–22.

Brown, T. A. (2015). *Confirmatory Factor Analysis for Applied Research*. Guilford Press, 2nd edition.

Caliński, R. B. and Harabasz, J. (1974). A dendrite method for cluster analysis. *Communications in Statistics*, 3:1–27.

Cameron, A. C. and Trivedi, P. K. (2009). *Microeconometrics Using Stata*. Stata Press.

Canty, A. and Ripley, B. D. (2019). *`boot`: Bootstrap `R` (S-Plus) Functions*. `R` package version 1.3-24.

Celeux, G. and Soromenho, G. (1996). An entropy criterion for assessing the number of clusters in a mixture model. *Journal of Classification*, 13:195–212.

Charrad, M., Ghazzali, N., Boiteau, V., and Niknafs, A. (2014). `NbClust`: An `R` package for determining the relevant number of clusters in a data set. *Journal of Statistical Software*, 61(6):1–36.

Chin, W. (1998a). Issues and opinion on structural equation modeling. *MIS Quarterly*, 22(1):vii–xvi.

Chin, W. W. (1998b). The partial least squares approach for structural equation modeling. In Marcoulides, G. A., editor, *Modern Methods for Business Research*, pages 295–336. Lawrence Erlbaum Associates.

Chin, W. W. (2010). How to write up and report PLS analyses. In Esposito Vinzi, V., Chin, W. W., Henseler, J., and Wang, H., editors, *Handbook of Partial Least Squares: Concepts, Methods and Applications*, pages 655–690. Springer.

Chin, W. W. and Dibbern, J. (2010). An introduction to a permutation based procedure for multi-group PLS analysis: Results of tests of differences on simulated data and a cross cultural analysis of the sourcing of information system services between germany and the USA. In Esposito Vinzi, V., Chin, W. W., Henseler, J., and Wang, H., editors, *Handbook of Partial Least Squares: Concepts, Methods and Applications*, pages 171–193. Springer.

Chin, W. W. and Gopal, A. (1995). Adoption intention in GSS: Relative importance of beliefs. *Data Base Advances*, 26(2/3):42–64.

Chin, W. W., Marcolin, B. L., and Newsted, P. R. (2003). A partial least squares latent variable modeling approach for measuring interaction effects: Results from a Monte Carlo simulation study and an electronic-mail emotion/adoption study. *Information Systems Research*, 14:189–217.

Chin, W. W. and Newsted, P. R. (1999). Structural equation modeling analysis with small samples using partial least squares. In Hoyle, R., editor, *Statistical Strategies for Small Sample Research*, pages 307–341. SAGE.

Cihara, L. M. and Hesterberg, T. C. (2019). *Mathematical Statistics with Resampling and R*. Wiley, 2nd edition.

Cohen, J. (1988). *Statistical Power Analysis for Behavioral Sciences*. Erlbaum Associates, Hillside, NJ, 2nd edition.

Cohen, J. (1992). A power primer. *Psychological Bulletin*, 112:155–159.

Cowgill, M. C., Harvey, R. J., and Watson, L. T. (1999). A genetic algorithm approach to cluster analysis. *Computers & Mathematics with Applications*, 37(7):99–108.

Davison, A. C. and Hinkley, D. V. (1997). *Bootstrap Methods and Their Applications*. Cambridge University Press.

Deng, L., Yang, M., and Marcoulides, K. M. (2018). Structural equation modeling with many variables: A systematic review of issues and developments. *Frontiers in Psychology*, 9(580).

Dijkstra, T. and Schermelleh-Engel, K. (2014). Consistent partial least squares for nonlinear structural equation models. *Psychometrika*, 79(4):585–604.

Dijkstra, T. K. (1983). Some comments on maximum likelihood and partial least squares methods. *Journal of Econometrics*, 22(1/2):67–90.

Dijkstra, T. K. and Henseler, J. (2015a). Consistent and asymptotically normal PLS estimators for linear structural equations. *Computational Statistics & Data Analysis*, 81:10–23.

Dijkstra, T. K. and Henseler, J. (2015b). Consistent partial least squares path modeling. *MIS Quarterly*, 39(2):297–316.

Dimitriadou, E., Dolnicar, S., and Weingessel, A. (2002). An examination of indexes for determining the number of clusters in binary data sets. *Psychometrika*, 67:137–159.

DiStefano, C., Zhu, M., and Mîndrilă, D. (2009). Understanding and using factor scores: Considerations for the applied researcher. *Practical Assessment, Research & Evaluation*, 14(20):1–11.

Dolnicar, S., Gr un, B., and Leisch, F. (2018). *Market Segmentation Analysis*. Springer.

Duda, R. O., Hart, P. E., and Stork, D. G. (2001). *Pattern Classification*. Wiley, 2nd edition.

Efron, B. and Hastie, T. (2016). *Computer Age Statistical Inference*. Cambridge University Press.

Esposito Vinzi, V. and Russolillo, G. (2013). Partial least squares algorithms and methods. *WIREs Computational Statistics*, 5:1–19.

Esposito Vinzi, V., Trinchera, L., and Amato, S. (2010). PLS path modeling: From foundations to recent developments and open issues for model assessment and improvement. In Esposito Vinzi, V., Chin, W. W., Henseler, J., and Wang, H., editors, *Handbook of Partial Least Squares: Concepts, Methods and Applications*, pages 47–82. Springer.

Esposito Vinzi, V., Trinchera, L., Squillacciotti, S., and Tenenhaus, M. (2008). REBUS-PLS: A response-based procedure for detecting unit segments in PLS path modeling. *Applied Stochastic Models in Business and Industry*, 24:439–458.

Estabrook, R. and Neale, M. (2013). A comparison of factor score estimation methods in the presence of missing data: Reliability and an application to nicotine dependence. *Multivariate Behavioral Research*, 48:1–27.

Everitt, B. S. and Hothorn, T. (2011). *An Introduction to Applied Multivariate Analysis with* R. Springer.

Everitt, B. S., Landau, S., Leese, M., and Stahl, D. (2011). *Cluster Analysis*. Wiley, 5th edition.

Fahrmeir, L., Kneib, T., Lang, S., and Marx, B. (2013). *Regression. Models, Methods and Applications*. Springer.

Falk, R. F. and Miller, N. B. (1992). *A Primer for Soft Modeling*. The University of Akron Press.

Finch, W. H. and French, B. F. (2015). *Latent Variable Modeling with* R. Routledge.

Fornell, C. and Larcker, D. (1981). Evaluating structural equation models with unobservable variables and measurement errors. *Journal of Marketing Research*, 18:39–50.

Fox, J. (2016). *Applied Regression Analysis and Generalized Linear Models*. SAGE, 3rd edition.

Fox, J., Nie, Z., and Byrnes, J. (2017). `sem`*: Structural Equation Models*. R package version 3.1-9.

Fox, J. and Weisberg, S. (2019). *An* R *Companion to Applied Regression*. SAGE, Thousand Oaks CA, 3rd edition.

Fraley, C. and Raftery, A. E. (2002). Model-based clustering, discriminant analysis, and density estimation. *Journal of the American Statistical Association*, 97:611–631.

Frühwirth-Schnatter, S. (2006). *Finite Mixture and Markov Switching Models*. Springer.

Gana, K. and Broc, G. (2019). *Structural Equation Modeling with* `lavaan`. Wiley.

Gefen, D., Rigdon, E. E., and Straub, D. (2011). An update and extension to sem guidelines for administrative and social science research. *MIS Quarterly*, 35(2):iii–xiv.

Gerbing, D. W. and Anderson, J. C. (1987). Improper solutions in the analysis of covariance structures: Their interpretability and a comparison of alternate respecifications. *Psychometrika*, 52(1):99–111.

Good, P. I. (2006). *Resampling Methods. A Practical Guide to Data Analysis.* Springer, 3rd edition.

Goodhue, D. L., Lewis, M., and Thompson, R. (2012). Does PLS have advantages for small sample size or non-normal data? *MIS Quarterly*, 36(3):981–1001.

Gordon, A. D. (1999). *Classification*. Chapman & Hall/CRC, 2nd edition.

Gould, W. W. (2018). *The Mata Book: A Book for Serious Programmers and Those Who Want to Be*. Stata Press.

Gower, J. C. and Hand, D. J. (1996). *Biplots*. Chapman & Hall/CRC.

Grace, J. B. (2006). *Structural Equation Modeling and Natural Systems*. Cambridge University Press.

Greene, W. H. (2018). *Econometric Analysis*. Pearson, 8th edition.

Grice, J. W. (2001). Computing and evaluating factor scores. *Psychological Methods*, 6:430–450.

Grolemund, G. (2014). *Hands-On Programming with* R. O'Reilly.

Haenlein, M. and Kaplan, A. M. (2004). A beginner's guide to partial least squares analysis. *Understanding Statistics*, 3(4):283–297.

Hagenaars, J. A. and McCutcheon, A. L. (2002). *Applied Latent Class Analysis*. Cambridge University Press.

Hahn, C., Johnson, M. D., Herrmann, A., and Huber, F. (2002). Capturing customer heterogeneity using a finite mixture PLS approach. *Schmalenbach Business Review*, 54:243–269.

Hair, J. F., Black, W. C., Babin, B. J., and Anderson, R. E. (2018a). *Multivariate Data Analysis*. Pearson, 8th edition.

Hair, J. F., Hult, G. T. M., Ringle, C. M., and Sarstedt, M. (2017). *A Primer on Partial Least Squares Structural Equation Modeling (PLS-SEM)*. SAGE, 2nd edition.

Hair, J. F., Risher, J. J., Sarstedt, M., and Ringle, C. M. (2018b). When to use and how to report the results of PLS-SEM. *European Business Review*, 31(1):2–24.

Hair, J. F., Sarstedt, M., Matthews, L. M., and Ringle, C. M. (2016). Identifying and treating unobserved heterogeneity with FIMIX-PLS: part i – method. *European Business Review*, 28(1):63–76.

Hair, J. F., Sarstedt, M., and Ringle, C. M. (2019). Rethinking some of the rethinking of partial least squares. *European Journal of Marketing*, 53(4):566–584.

Hair, J. F., Sarstedt, M., Ringle, C. M., and Gudergan, S. P. (2018c). *Advanced Issues in Partial Least Squares Structural Equation Modeling*. SAGE.

Hair, J. F., Sarstedt, M., Ringle, C. M., and Mena, J. A. (2012). An assessment of the use of partial least squares structural equation modeling in marketing research. *Journal of the Academy of Marketing Science*, 40(3):414–433.

Hanafi, M. (2007). PLS path modelling: computation of latent variables with the estimation mode b. *Computational Statistics*, 22(2):275–292.

Hannan, E. J. and Quinn, B. G. (1979). The determination of the order of an autoregression. *Journal of the Royal Statistical Society B*, 41:190–195.

Harlow, L. L. (2014). *The Essence of Multivariate Thinking: Basic Themes and Methods*. Routledge.

Hastie, T., Tibshirani, R., and Friedman, J. (2008). *The Elements of Statistical Learning. Data Mining, Inference, and Prediction*. Springer, 2nd edition.

Hayes, A. F. (2013). *Introduction to Mediation, Moderation, and Conditional Process Analysis: A Regression-based Approach*. Guilford Press.

Hennig, C. (2020). `fpc`*: Flexible Procedures for Clustering*. `R` package version 2.2-5.

Henningsen, A. and Hamann, J. D. (2007). `systemfit`: A package for estimating systems of simultaneous equations in `R`. *Journal of Statistical Software*, 23(4):1–40.

Henseler, J. (2010). On the convergence of the partial least squares path modeling algorithm. *Computational Statistics*, 25:107–120.

Henseler, J. (2017). Bridging design and behavioral research with variance-based structural equation modeling. *Journal of Advertising*, 46(1):178–192.

Henseler, J. and Chin, W. W. (2010). A comparison of approaches for the analysis of interaction effects between latent variables using partial least squares path modelling. *Structural Equation Modeling*, 17:82–109.

Henseler, J., Ringle, C. M., and Sarstedt, M. (2016). Testing measurement invariance of composites using partial least squares. *International Marketing Review*, 33(3):405–431.

Henseler, J. and Sarstedt, M. (2013). Goodness-of-fit indices for partial least squares path modeling. *Computational Statistics*, 28(2):565–580.

Hernán, M. and Robins, J. (2021). *Causal Inference*. Chapman & Hall/CRC. (forthcoming).

Hesterberg, T. C. (2015). What teachers should know about the bootstrap: Resampling in the undergraduate statistics curriculum. *The American Statistician*, 69(4):371–386.

Horn, J. L. and McArdle, J. J. (1992). A practical and theoretical guide to measurement invariance in aging research. *Experimental Aging Research*, 18:117–144.

Hoyle, R. H. (2011). *Structural Equation Modeling for Social and Personality Psychology*. SAGE.

Hsiao, Y.-Y., Kwok, O.-M., and Lai, M. H. C. (2018). Evaluation of two methods for modeling measurement errors when testing interaction effects with observed composite scores. *Educational and Psychological Measurement*, 78(2):181–202.

Hui, B. S. and Wold, H. O. A. (1982). Consistency and consistency at large of partial least squares estimates. In Jöreskog, K. G. and Wold, H. O. A., editors, *Systems Under Indirect Observations - Part II*. North Holland.

Husson, F., Lê, S., and Pagès, J. (2017). *Exploratory Multivariate Analysis by Example Using R*. CRC Press, 2nd edition.

Hwang, H. and Takane, Y. (2014). *Generalized Structured Component Analysis: A Component-Based Approach to Structural Equation Modeling*. Chapman & Hall/CRC.

Iacobucci, D., Saldanha, N., and Deng, X. (2007). A mediation on mediation: Evidence that structural equation models perform better than regressions. *Journal of Consumer Psychology*, 17(2):140–154.

Jedidi, K., Jagpal, S. H., and De Sarbo, W. S. (1997a). Finite-mixture structural equation models for response-based segmentation and unobserved heterogeneity. *Marketing Science*, 16:39–60.

Jedidi, K., Jagpal, S. H., and De Sarbo, W. S. (1997b). STEMM: A general finite mixture structural equation model. *Journal of Classification*, 14:23–50.

Johnson, R. and Wichern, D. (2014). *Applied Multivariate Statistical Analysis*. Pearson, 6th edition.

Jöreskog, K. G. (1969). A general approach to confirmatory maximum likelihood factor analysis. *Psychometrika*, 34:183–202.

Jöreskog, K. G., Olsson, U. H., and Wallentin, F. Y. (2016). *Multivariate Analysis with LISREL*. Springer.

Jöreskog, K. G. and Sörbom, D. (1989). *LISREL 7: A Guide to the Program and Applications*. SPSS Inc., Chicago, Ill.

Jose, P. E. (2013). *Doing Statistical Mediation and Moderation*. Guilford Press.

Kaufman, L. and Rousseeuw, P. J. (1990). *Finding Groups in Data: An Introduction to Cluster Analysis*. Wiley.

Keith, T. Z. (2016). *Multiple Regression and Beyond*. Pearson Education, Inc., 2nd edition.

Kenny, D. A. (2016). Mediation. `http://davidakenny.net/cm/mediate.htm`.

Kleiber, C. and Zeileis, A. (2008). *Applied Econometrics with* `R`. Use `R`! Springer.

Kline, R. B. (2016). *Principles and Practice of Structural Equation Modeling*. The Guilford Press, 4th edition.

Kock, N. (2019). From composites to factors: Bridging the gap between PLS and covariance-based structural equation modelling. *Information Systems Journal*, 29(3):674–706.

Kock, N. and Hadaya, P. (2018). Minimum sample size estimation in PLS-SEM: the inverse square root and gamma-exponential methods. *Information Systems Journal*, 28(1):227–261.

Kotler, P. and Keller, K. (2015). *Marketing Management*. Pearson, 15th edition.

Kramer, N. (2007). *Analysis of high-dimensional data with partial least squares and boosting*. PhD thesis, Technische Universit at Berlin, Berlin, Germany.

Kutner, M. H., Nachtsheim, C. J., Neter, J., and Li, W. (2005). *Applied Linear Statistical Models*. McGraw-Hill, 5th edition.

Lamberti, G., Aluja, T. B., and Sanchez, G. (2016). The pathmox approach for PLS path modeling segmentation. *Applied Stochastic Models in Business and Industry*, 32(4):453–468.

Lamberti, G., Banet Aluja, T., and Sanchez, G. (2017). The pathmox approach for PLS path modeling: Discovering which constructs differentiate segments. *Applied Stochastic Models in Business and Industry*, 33(6):674–689.

Leisch, F. (2004). `FlexMix`: A general framework for finite mixture models and latent class regression in `R`. *Journal of Statistical Software*, 11(8):1–18.

Leisch, F. (2006). A toolbox for K-centroids cluster analysis. *Computational Statistics and Data Analysis*, 51(2):526–544.

Levene, H. (1960). Robust tests for equality of variances. In Olkin, I., Ghurye, S. G., Hoeffding, W., Madow, W. G., and Mann, H. B., editors, *Contributions to Probability and Statistics: Essays in Honor of Harold Hotelling*, pages 278–292. Stanford University Press.

Liang, Z., Jaszak, R. J., and Coleman, R. E. (1992). Parameter estimation of nite mixtures using the em algorithm and information criteria with applications to medical image processing. *IEEE Transactions on Nuclear Science*, 39(4):1126–1133.

Linzer, D. A. and Lewis, J. B. (2011). `poLCA`: An `R` package for polytomous variable latent class analysis. *Journal of Statistical Software*, 42(10):1–29.

Litman, J. A. and Spielberger, C. D. (2003). Measuring epistemic curiosity and its diversive and specific components. *Journal of Personality Assessment*, 80(1):75–86.

Lohmöller, J. B. (1989). *Latent Variable Path Modeling with Partial Least Squares*. Physica.

Lyttkens, E. (1973). The fix-point method for estimating interdependent systems with the underlying model specification. *Journal of the Royal Statistical Society A*, 135(3):353–375.

MacKinnon, D. P. (2008). *Introduction to Statistical Mediation Analysis*. Erlbaum Associates.

Maechler, M., Rousseeuw, P., Struyf, A., Hubert, M., and Hornik, K. (2019). *`cluster`: Cluster Analysis Basics and Extensions*. `R` package version 2.1.0.

Mardia, K. V., Kent, J. T., and Bibby, J. M. (1979). *Multivariate Analysis*. Academic Press.

McCutcheon, A. L. (1987). *Latent Class Analysis*. SAGE.

McLachlan, G. and Peel, D. (2000). *Finite Mixture Models*. Wiley.

Mehmetoglu, M. (2011). Model-based post hoc segmentation (with REBUS-PLS) for capturing heterogeneous consumer behaviour. *Journal of Targeting, Measurement and Analysis for Marketing*, 19(3):165–172.

Mehmetoglu, M. (2012). Personality effects on experiential consumption. *Personality and Individual Differences*, 52(1):94–99.

Mehmetoglu, M. and Jakobsen, T. G. (2016). *Applied Statistics Using Stata. A Guide for the Social Sciences*. SAGE.

Mevik, B.-H., Wehrens, R., and Liland, K. H. (2019). *`pls`: Partial Least Squares and Principal Component Regression*. `R` package version 2.7-2.

Milligan, G. W. and Cooper, M. C. (1985). An examination of procedures for determining the number of clusters in a data set. *Psychometrika*, 50:159–179.

Mitchell, M. (1996). *An Introduction to Genetic Algorithms*. MIT Press.

Molenberghs, G., Fitzmaurice, G., Kenward, M. G., Tsiatis, A., and Verbeke, G. (2015). *Handbook of Missing Data Methodology*. Chapman & Hall/CRC.

Monecke, A. and Leisch, F. (2012). `semPLS`: Structural equation modeling using partial least squares. *Journal of Statistical Software*, 48(3):1–32.

Mulaik, S. A. (2010). *Foundations of Factor Analysis*. CRC Press, 2nd edition.

Muthén, L. K. and Muthén, B. O. (2002). How to use a Monte Carlo study to decide on sample size and determine power. *Structural Equation Modeling*, 9(4):599–620.

Nelder, J. A. (1977). A reformulation of linear models. *Journal of the Royal Statistical Society A*, 140(1):48–76.

Nunnally, J. C. and Bernstein, I. H. (1994). *Psychometric Theory*. McGraw-Hill, 3rd edition.

Paxton, P., Curran, P. J., Bollen, K. A., Kirby, J., and Chen, F. (2001). Monte Carlo experiments: Design and implementation. *Structural Equation Modeling*, 8(2):287–312.

Pearl, J., Glymour, M., and Jewell, N. P. (2016). *Causal Inference in Statistics. A Primer*. Wiley.

Pearl, J. and Mackenzie, D. (2018). *The Book of Why. The New Science of Cause and Effect*. Basic Books.

Rademaker, M. E. and Schuberth, F. (2020). *cSEM: Composite-Based Structural Equation Modeling*. Package version: 0.3.0.9000.

Ramaswamy, V., Desarbo, W. S., and Reibstein, D. J. (1993). An empirical pooling approach for estimating marketing mix elasticities with pims data. *Marketing Science*, 12:103–124.

Reinartz, W., Haenlein, M., and Henseler, J. (2009). An empirical comparison of the efficacy of covariance-based and variance-based sem. *International Journal of Research in Marketing*, 26:332–344.

Rencher, A. C. and Christensen, W. F. (2012). *Methods of Multivariate Analysis*. Wiley, 3rd edition.

Rigdon, E. E. (2013). Partial least squares path modeling. In Hancock, G. R. and Mueller, R. O., editors, *Structural Equation Modeling – A Second Course*, chapter 3. Information Age Publishing, Inc., 2nd edition.

Rindskopf, D. (1984). Structural equation models: Empirical identification, heywood cases, and related problems. *Sociological Methods & Research*, 13(1):109–119.

Ringdon, E. E. (2012). Rethinking partial least squares path modeling: In praise of simple methods. *Long Range Planning*, 45:341–358.

Ringle, C. M., Sarstedt, M., and Schlittgen, R. (2014). Genetic algorithm segmentation in partial least squares structural equation modeling. *OR Spectrum*, 36(1):251–276.

Ringle, C. M., Sarstedt, M., Schlittgen, R., and Taylor, C. R. (2013). PLS path modeling and evolutionary segmentation. *Journal of Business Research*, 66(9):1318–1324.

Ringle, C. M., Sarstedt, M., and Straub, D. W. (2012). A critical look at the use of PLS-SEM in MIS quarterly. *MIS Quarterly*, 36(1):iii–xiv.

Ringle, C. M., Wende, S., and Will, A. (2010). Finite mixture partial least squares analysis: Methodology and numerical examples. In Esposito Vinzi, V., Chin, W. W., Henseler, J., and Wang, H., editors, *Handbook of Partial Least Squares: Concepts, Methods and Applications*, pages 195–218. Springer.

Rohart, F., Gautier, B., Singh, A., and Le Cao, K.-A. (2017). `mixOmics`: An `R` package for omics feature selection and multiple data integration. *PLoS computational biology*, 13(11):e1005752.

Rönkkö, M. (2020). *`matrixpls`: Matrix-based Partial Least Squares Estimation.* `R` package version 1.0.9.

Rönkkö, M. and Evermann, J. (2013). A critical examination of common beliefs about partial least squares path modeling. *Organizational Research Methods*, 16(3):425–448.

Rosseel, Y. (2012). `lavaan`: An `R` package for structural equation modeling. *Journal of Statistical Software*, 48(2):1–36.

Rossi, P. E. (2014). *Bayesian Non- and Semi-parametric Methods and Applications.* Princeton University Press.

Roubens, M. (1978). Pattern classification problems and fuzzy sets. *Fuzzy Sets and Systems*, 1:239–253.

Sanchez, G. (2013). *PLS Path Modeling with `R`.* Trowchez Editions.

Sanchez, G. (2020). *The Saga of PLS.* Leanpub.

Sanchez, G. and Aluja, T. (2006). PATHMOX: A PLS-PM segmentation algorithm. In Esposito Vinzi, V., Lauro, C., Braverman, A., Kiers, H., and Schmiek, M. G., editors, *Proceedings of KNEMO 2006*, page 69, Tilapia, Anacapri.

Sanchez, G. and Aluja, T. (2013). `pathmox`: Pathmox approach of segmentation trees in partial least squares path modeling. `R` package version 0.2.0.

Sanchez, G., Trinchera, L., and Russolillo, G. (2017). *plspm: Tools for Partial Least Squares Path Modeling (PLS-PM)*. R package version 0.4.9.

Sarkar, D. (2008). *Lattice: Multivariate Data Visualization with R*. Springer, New York.

Sarstedt, M. (2008). A review of recent approaches for capturing heterogeneity in partial least squares path modelling. *Journal of Modelling in Management*, 3(2):140–161.

Sarstedt, M., Becker, J.-M., Ringle, C. M., and Schwaiger, M. (2011). Uncovering and treating unobserved heterogeneity with FIMIX-PLS: Which model selection criterion provides an appropriate number of segments? *Schmalenbach Business Review*, 63:34–62.

Sarstedt, M., Hair, J. F., Cheah, J.-H., Becker, J.-M., and Ringle, C. M. (2019). How to specify, estimate, and validate higher-order constructs in PLS-SEM. *Australasian Marketing Journal*, 27(3):197–211.

Sarstedt, M., Hair, J. F., Ringle, C. M., Thiele, K. O., and Gudergan, S. P. (2016). Estimation issues with PLS and CBSEM: Where the bias lies! *Journal of Business Research*, 69(10):3998–4010.

Sarstedt, M. and Ringle, C. M. (2010). Treating unobserved heterogeneity in PLS path modeling: A comparison of FIMIX-PLS with different data analysis strategies. *Journal of Applied Statistics*, 37(8):1299–1318.

Satterthwaite, F. E. (1946). An approximate distribution of estimates of variance components. *Biometrics Bulletin*, 2(6):110–114.

Schloerke, B., Cook, D., Larmarange, J., Briatte, F., Marbach, M., Thoen, E., Elberg, A., and Crowley, J. (2020). *GGally: Extension to ggplot2*. R package version 2.0.0.

Schneeweiss, H. (1993). Consistency at large in models with latent variables. In Haagen, K., Bartholomew, D. J., and Deistler, M., editors, *Statistical Modelling and Latent Variables*, pages 299–320. North-Holland.

Schumacker, R. E. and Lomax, R. G. (2016). *A Beginner's Guide to Structural Equation Modeling*. Routledge, 4th edition.

Scrucca, L., Fop, M., Murphy, T. B., and Raftery, A. E. (2016). mclust 5: Clustering, classification and density estimation using Gaussian finite mixture models. *The R Journal*, 8(1):205–233.

Seman, L. O. (2016). PyLS-PM Library. Python package version 1.0.

Sirohi, N., McLaughlin, E. W., and Wittink, D. R. (1998). A model of consumer perceptions and store loyalty intentions for a supermarket retailer. *Journal of Retailing*, 74(2):223–245.

Skrondal, A. and Rabe-Hesketh, S. (2004). *Generalized Latent Variable Modeling. Multilevel, Longitudinal, and Structural Equation Models*. Chapman & Hall/CRC.

Sobel, M. E. (1987). Direct and indirect effect in linear structural equation models. *Sociological Methods and Research*, 16:155–176.

Spiller, S. A., Fitzsimons, G. J., Lynch, J. G., and Mcclelland, G. H. (2013). Spotlights, floodlights, and the magic number zero: Simple effects tests in moderated regression. *Journal of Marketing Research*, 50(2):277–288.

Steinmetz, H. (2013). Analyzing observed composite differences across groups: Is partial measurement invariance enough? *European Journal of Research Methods for the Behavioral and Social Sciences*, 9(1):1–12.

Tenenhaus, A. and Tenenhaus, M. (2011). Regularized generalized canonical correlation analysis. *Psychometrika*, 76:257–284.

Tenenhaus, M., Amato, S., and Esposito Vinzi, V. (2004). A global goodness-of-fit index for PLS structural equation modelling. In *Proceedings of the XLII SIS Scientific Meeting*, volume Contributed Papers, pages 739–742, Padova (Italy). CLEUP.

Tenenhaus, M., Esposito Vinzi, V., Chatelin, Y.-M., and Lauro, C. (2005). PLS path modeling. *Computational Statistics & Data Analysis*, 48:159–205.

Tibshirani, R., Walther, G., and Hastie, T. (2001). Estimating the number of data clusters via the gap statistic. *Journal of the Royal Statistical Society B*, 63:411–423.

Trinchera, L. (2007). *Unobserved Heterogeneity in Structural Equation Models: A new approach to latent class detection in PLS Path Modeling*. PhD thesis, University of Naples "Federico II".

Tutz, G. and Ramzan, S. (2015). Improved methods for the imputation of missing data by nearest neighbor methods. *Computational Statistics & Data Analysis*, 90:84–99.

van Buuren, S. (2018). *Flexible Imputation of Missing Data*. Chapman & Hall/CRC, 2nd edition.

van Buuren, S. and Groothuis-Oudshoorn, K. (2011). `mice`: Multivariate imputation by chained equations in `R`. *Journal of Statistical Software*, 45(3):1–67.

Venturini, S. and Mehmetoglu, M. (2019). `plssem`: A Stata package for structural equation modeling with partial least squares. *Journal of Statistical Software*, 88(8):1–35.

Wang, J. and Wang, X. (2012). *Structural Equation Modeling. Applications Using Mplus*. Wiley.

Wedel, M. and Kamakura, W. A. (2000). *Market Segmentation. Conceptual and Methodological Foundations*. Kluwer Academic Publishers, 2nd edition.

Wehrens, R. (2011). *Chemometrics with R*. Use R! Springer.

Weisberg, S. (2013). *Applied Linear Regression*. Wiley, 4th edition.

Welch, B. L. (1947). The generalization of Student's problem when several different population variances are involved. *Biometrika*, 34(1):28–35.

Wetzels, M., Odekerken-Schröder, G., and van Oppen, C. (2009). Using PLS path modeling for assessing hierarchical construct models: Guidelines and empirical illustration. *MIS Quarterly*, 33(1):177–195.

Wickham, H. (2019). *Advanced R*. CRC Press, 2nd edition.

Wind, Y. (1978). Issues and advances in segmentation research. *Journal of Marketing Research*, 15(3):317–337.

Wold, H. O. A. (1965). A fix-point theorem with econometric background i-ii. *Arkiv für Matematik*, 6(12):209–240.

Wold, H. O. A. (1975). Path models with latent variables: The NIPALS approach. In Blalock, H. M., Aganbegian, A., Borodkin, F. M., Boudon, R., and Cappecchi, V., editors, *Quantitative Sociology*, pages 307–359. Academic Press.

Wold, H. O. A. (1982). Soft modeling: The basic design and some extensions. In Jöreskog, K. G. and Wold, H. O. A., editors, *Systems Under Indirect Observations, Part II*, pages 1–54. North-Holland.

Wooldridge, J. M. (2016). *Introductory Econometrics*. Cengage Learning, 6th edition.

Zaki, M. J. and Meira, W. (2020). *Data Mining and Machine Learning. Fundamental Concepts and Algorithms*. Cambridge University Press, 2nd edition.

Zeileis, A. and Hothorn, T. (2002). Diagnostic checking in regression relationships. *R News*, 2(3):7–10.

Zhao, X., Lynch, J. G. J., and Chen, Q. (2010). Reconsidering baron and kenny: Myths and truths about mediation analysis. *Journal of Consumer Research*, 37:197–206.

Index

Printed in the United States
by Baker & Taylor Publisher Services